MEDICINAL SEEDS AND PLANTS

MEDICINAL SEEDS AND PLANTS

K. VANANGAMUDI
Former Dean (Agriculture)
Agricultural College and Research Institute
Tamil Nadu Agricultural University, Coimbatore - 641 003.

V. ANBUKKARASI
Research Associate
Horticultural College and Research Institute
Tamil Nadu Agricultural University, Coimbatore - 641 003

M. PRABHU
Assistant Professor (Horticulture)
Horticultural College and Research Institute,
Tamil Nadu Agricultural University, Periyakulam -625 604.

Published by:

Scientific Publishers (India)
5 A, New Pali Road, P.O. Box 91
Jodhpur 342 001 (India)

Branch Office
Scientific Publishers (India)
4806/24, Ansari Road, Daryaganj
New Delhi - 110 002 (India)

E-mail: info@scientificpub.com
Website: www.scientificpub.com

Print : 2021

ISBN: 978-81-7233-875-6

Typeset by Rajesh Ojha
Printed in India

PREFACE

The World Health Organization has listed 21,000 plants that have reported medicinal uses around the world. Other estimates indicate the number to range between 35,000 and 70,000 worldwide. The last decade has witnessed a great revival of interest in the use of herbal medicines both in the developed as well as developing countries. Almost 50 per cent of the medicines we use today are derived straight from plants and 25 per cent of the prescription drugs have their genesis of tropical plants. According to World Health Organization, 80 per cent of the world population relies chiefly on plant based traditional medicines especially for their primary health care needs. Medicinal and aromatic plants have a high market potential with the world demand of herbal products growing at the rate of seven per cent per annum. The major importers of medicinal plants are the European Union, Russia and the USA, while the major exporters of medicinal plants or its value added products are China, Japan and Russia. These countries also show strong intellectual property protection. It is noteworthy that active Intellectual Property Right Protection has helped countries like China to establish supremacy in the world trade. Forty five per cent of all patents on herb or herbal based or related medicine are with China, closely followed by Japan with 28 per cent share.

In India, the evergreen forests of Western Ghats and the Himalayas are rich in herbs useful in treatment of common ailments to lethal diseases like blood cancer. Satawari, Safed Musli, Kali Musli and Ashwagandha for promising tonic, Kukronda and Adusa for respiratory trouble, Gudmar and Sadasuhagan for Diabetes, Bramhi and Bach for memory, Kalmegh for chronic fever, Sarpagandha and Arjun for heart troubles are some examples of miracle herbs. Sarpagandha, Aloe and Gloriosa once in abundance, now they have became rare plants in some parts of India. The availability of wonderful herb Safed Musli is also continuously increasing in the country. The climatic and edaphic conditions of India are suitable for commercial cultivation of more than 50 potential plants including Safed Musli, Satawari, Ashwagandha, Akarkara, Sarpagandha, Bach, Ratanjot, Aloe, Senna, Gymnema, Mucuna etc.

Considering above facts, the Medicinal Seeds and Plants has been prepared for the benefit of scientists, students and extension personnel. This book contains Global and National scenario of medicinal plants, Industrial applications of medicinal plants and detailed description of taxonomy, botany, propagation, seed biology and uses for important species.

The authors extend their sincere thanks to the Scientific Publishers (India), Jodhpur for publishing this book.

K. VANANGAMUDI
V. ANBUKKARASI
M. PRABHU

CONTENTS

Part IV- MEDICINAL CLIMBERS

Part V- PERENNIAL MEDICINAL PLANTS

PART - I

GENERAL ASPECTS

1.1. GLOBAL SCENARIO OF MEDICINAL PLANTS

Plant derived medicines are used in all civilizations and cultures and hence, plants have always played a key role in health care systems Worldwide. Many countries, especially Asia, Africa and East European countries have officially recognized the use of plant based traditional medicine in their health care delivery systems. In most developing countries, the indigenous modes of herbal treatments are a part of the culture and the dominant method of healing therapy. These remedies, with a considerable extent of effectiveness, are socially accepted, economically viable and mostly are the only available source. Plants used in traditional medicine, therefore, have a critical role in the maintenance of health all over the World.

The archaeological studies showed that the plant *Ephedra* has used in Iraq during ancient period. During 300 BC, the uses of medicinal plants were documented in Egypt, Middle East, India and China. The ancient expert Saraka wrote a book *Saraka Samritham* about traditional medicine and he described 350 species during 700 BC. The Greek medicinal expert Hippocrates (460 BC – 377 BC) is known as father of medicine. A book *Materia medica* was written by Greek doctor Dioscoretes during 100 AD and he described 600 medicinal plants along with pictures of 400 species. This book has been translated from Greek to English, Persian language and Hebrew and it is a primary reference book upto 1700 AD. In China, the kings were documented the medicinal plants during 200 BC to 100 AD. Chinese traditional compendium described 5757 drugs, which are mostly derived from herbs. The important Chinese medicinal plants are *Gingo*, Ginseng and *Ephedra*. During 500 AD – 1300 AD, the Arabians used traditional medicines and perfumes in collaboration with India and China. During 1100 AD, the plants *viz., Papaver somniferum, Cannabis sativa* were used as a pain killer and anesthesia by monastery hospitals of Scotland. During Colonial period, the rare herbs (*Alstonia, Cinchona)* were introduced in Europe from African countries to treat fever, malaria, syphilis and small pox. During 1800 AD, the active ingredients were extracted from traditional herbs in Europe. In 1970, the scientists of Oklagama University in USA studied the traditional Indian herbal plants and they found that *Digitalis purpurea* have the property of curing heart ailments.

The World Health Organization has listed 21,000 plants that have reported medicinal uses around the world. Other estimates indicate the number to range between 35,000 and 70,000 worldwide. The last decade has witnessed a great revival of interest in the use of herbal medicines both in the developed as well as developing countries. Almost 50 per cent of the medicines used today are derived straight from plants and 25 per cent of the prescription drugs have their genesis of tropical plants. According to World Health Organization, 80 per cent of the world population relies chiefly on plant based traditional medicines especially for their primary health care. Medicinal and aromatic plants have a high market potential with the world demand of herbal products growing at the rate of seven per cent per annum. In order to reduce the tremendous pressure on forests generated due to heavy demand of theses herbs in National and International drug market, the conservation of medicinal and aromatic plants has become a need of the day. The shift in demand of consumers towards herbal and natural product in the International markets has prompted a widespread research to develop medicine and health care products from medicinal plants.

The coverage of Intellectual Property Rights (IPR) has expanded to cover even resources that were previously deemed to be part of the public domain. Several declarations at National and International level emphasize the need for identification, evaluation, conservation, and sustainable utilization of plant genetic resources (PGR) particularly medicinal plants. These declarations have assumed greater significance under post-WTO era, where the TRIPS Agreement requires patent protection for at least 20 years for any invention of pharmaceutical product or process.

For exploiting the opportunities in the World market, which is growing at the rate of 10-15 per cent per annum; the country need strive towards utilizing the flexibility available under the Convention on Biological Diversity (CBD) and the Doha Declaration. In the absence of such policy framework, the illegal intellectual property protection and misuse of plant genetic resources would continue and our share in the global market would remain meager. Optimal institutional framework is necessary for carrying out species-specific and socio-economic environment specific research in medicinal plants. Such framework should take care of needs of various stakeholders and also cater to social needs without adverse implications on equity and environment. This would help in harnessing the potential of medicinal plants by facilitating bio-prospecting and equitable sharing of benefits in the post-WTO era.

The major importers of medicinal plants are United States, European Union, Japan, Russia, Latin America and Gulf countries, while the major exporters of medicinal plants or its value added products are China, India, South Korea, Kenya,

Yugoslavia, Argentina, Afghanistan, Japan and Russia. These countries also show strong intellectual property protection. It is noteworthy that active IPR protection has helped countries like China to establish supremacy in the World trade. Forty five percent of all patents on herb or herbal based or related medicine are with China closely followed by Japan with 28 per cent share. Most of the inventors are individuals and only a few are corporate houses and institutions. This fact indicates that in herbal patent system, it is still possible for individual researcher or groups to make the mark and add value in tribal and local community knowledge system. The export market is growing faster than the domestic market. However, the export is being carried out in the form of plants or their parts and not as value added products. Countries that import plant and plant parts are invariably less biodiversity, but are technologically advanced which show a higher capability to develop value added products from the imports.

The Indian exporters face major constraints while exporting medicinal plants. The cyclical nature of medicinal plant market makes it difficult for them to determine supply and demand in the markets. Also, competition from synthetic substitutes, non-compliance with rules and regulation of the importing countries, hoarding and manipulation by large importers, poor quality control, inadequate information on market trends, use of obsolete technology and lack of established methods for cultivation of large number of plants, etc., compound the problem of export of medicinal plants or their products from India.

1.2. NATIONAL SCENARIO OF MEDICINAL PLANTS

India, with approximately eight per cent of World's biodiversity, has the potential of becoming a major global player in screening and developing herbal based formulations, medicines and products. Cosmetic industry as well as aroma therapy are two important areas where Indian medicinal plants and their extracts, and essential oil can contribute globally. About 43,000 plant species are said to exist in India of which 7,500 plant species find mention in the recorded folklore of India. Among these, 1,773 have been documented in Ayurveda, 4,720 in Traditional medicine, 1,122 in Sidda, 751 in Unani and 105 in Allelopathy medical systems, classical healing treatise. Approximately 150 species of them have been commercially exploited. And also, 150 species are found to be endangered (e.g. Sarpagantha) and 36 species (e.g. Mara manjal, Kacholam, Asoka, Red sandal, Neer vaanchi, Kodai paalai) are in extinct in nature.

The collection of these species is mostly done from the forest or other natural ecosystem and only 20 species are under commercial cultivation. This reflects the urgent need for *ex-situ* conservation and cultivation and/or multiplication of medicinal plants through tissue culture or other biotechnological technique. India has also concentrated on exporting the raw material in the form of plant and plant parts/ extracts. In order to be a major player, it is important that we concentrate on exporting value added products rather than the raw material. More than 70 per cent of Indian populations use and rely solely on their health care needs.

Safety parameters

Standardization of herbal drugs is gaining momentum in India and as a result, it is also proposed to include the safety parameters as per International norms. So far as drugs are concerned, one must remember that "Safety comes first and quality afterwards". Therefore, in addition to proper botanical identification and physicochemical parameters, test for heavy metals (Mercury, Lead, Cadmium, Arsenic), microbial contaminants (total viable aerobic count, total Enterobacteri-aceae, total fungal count), specific pathogens (*Escherichia coli, Salmonella* spp., *S. aureus, Pseudomonas aeruginosa*), pesticide residue (organochlorine, organophosphorus, pyrethroids and others) and aflatoxin (B1, B2, G1, G2) should also be an important part of any quality control protocol. State of art facilities have been installed at Quality Control Laboratory of Regional Research Laboratory, Jammu to test such parameters.

Botanical concerns

Similar names have been very frequently used for several related umbelliferous fruits, mint leaves and various other plant materials and thus, different plant species are sold in the market under the same name in different parts of Indian sub-continent. "Ajamoda" (*Trachyspermum roxburghianum*), a reputed drug of the Ayurvedic and Unani systems of medicine, but in the market *Apium graveolens, Apium leptophyllum, Trachyspermum roxburghianum* and *T. ammi* are sold under one trade name *i.e.,* Ajamoda in different regions. The major chemical constituents of all these four plant species are different and can easily be characterized on the basis of chemical markers. Many species of *Mentha* such as *M. spicata, M. longifolia, M. piperata, M. arvensis* and *M. pulegium* are sold under one name Pudina in various parts of the country, but all the species contain different mono-terpenes. In some cases, constituting mono-terpenes are same; they are present in different amounts in different species. Three plant species such as *Carum carvi, Bunium persicum* and *Bunium cylindricum* are sold under one name *i.e.,* Kalazeera and all the three contain entirely different constituents. Fruit of another plant *i.e., Cuminum cyminum* is sold under the name of Safed zeera which have constituents similar to one of Kalazeera *i.e., Bunium persicum.* On the other hand, many time different plant species contain similar chemical markers and same is true with *Mucuna pruriens, M. utilis* and *M. cochinensis,* which all contain L-dopa as marker compound, although these are very much different in size and appearance.

The post-harvest processing of *Crocus sativus* stigma by traditional methods which takes 3-5 days, the amount of total crocins content is 7-9 per cent and if the same stigma is processed at $45^oC \pm 5^oC$ with tray load of $1m^3$ which requires 6-7h of drying, the amount of crocins from the plant increases to more than 14 per cent. Thus, there is a need to remove ambiguities related to proper herbals to be used in Ayurveda, Sidda and Unani (ASU) system of medicine followed by proper taxonomical identification, harvesting time, post-harvest processing, storage conditions etc., as these are of utmost importance in maintaining the quality of an herbal product. Therefore, "Standard Operating Procedures" should be adopted for collection, processing, storage and sampling of herbs and herbal products. Further, ASU drugs should be categorized separately, from the European Union herbal drugs, as these have evidence based documented records and are widely used in Indian sub-continent for centuries.

The medicinal plants which are commonly used for drug preparation are as follows.

Disease / Disorder	Medicinal plants used
Life saving drugs preparation (Cancer, heart disorders and kidney complaints)	*Digitalis, Periwinkle, Ginseng, Toxus, Gingo*
Diabetic problems	Bael, *Gymnema, Stevia, Costus, Andrographis,* Jamun, Aonla
Memory enhancer and blood purifier	*Centella*, Brahmi, Mint
Blood pressure and tension	Sarpagandha, *Coleus, Tinospora,* Safed musli
Laxative	Senna, Isabgol
Jaundice	*Phyllanthus, Eclipta, Andrographis*
Neural problems and muscle strength	Ginseng, Ashwagandha, *Mucuna, Asparagus, Hybanthus*, Aonla
Cosmetics	*Aloe*, Sandal, *Crocus*
Throat infections and cough	*Adhatoda, Solanum trilobatum, Ocimum, Coleus aromaticus, Vitex negundo,* Long pepper
Skin disorders	*Sesbania, Aristolochia*
Kidney stones	*Tribulus, Aerva*
Hair tonic, dyes	*Eclipta, Lippia, Hibiscus*

National Medicinal Plants Board

The National Medicinal Plants Board (NMPB) was set up under the Chairpersonship of Union Health and Family Welfare Minister. Prior to the establishment of NMPB, various activities and programmes related to diverse issues concerning medicinal plants are being pursued independently by numerous ministries, departments and organizations, but there is inadequate coordination and an absence of a focused approach. An appropriate mechanism for coordination and implementation of policies relating to medicinal plants both at the Central and State levels is necessary to facilitate inter-ministry, inter-state and institutional collaboration and to avoid duplication of efforts. The objectives of establishing a board is to establish an agency which would be responsible for coordination of all matters relating to medicinal plants, including drawing up policies and strategies for conservation, proper harvesting, cost effective cultivation, research and development, processing, marketing of raw material in order to protect, sustain and develop this sector. The work would continue to be carried out by the respective departments, organizations, but the board would provide a focus and a direction to the activities. The board is functioning with administrative/technical support of the following committees, which have been carved out of the key departments and have the administrative/technical capability for each area defined below. These committees are:

1. Standing Finance Committee
2. Committee on Demand and Supply
3. Committee on Cultivation of Medicinal Plants including Conservation of rare and endangered species.
4. Committee on Export/ Import
5. Committee on Research
6. Committee on Patents/ Intellectual Property Rights (IPR)

Co-ordination with Ministries/Departments/Organizations/State/Union Territories (UT) Governments for development of medicinal plants in general and specifically in the following fields:

1. Assessment of demand/supply position relating to medicinal plants both within the country and abroad.
2. Advice concerned Ministries/Departments/Organizations/State/UT Governments on policy matters relating to schemes and programmes for development of medicinal plants.
3. Provide guidance in the formulation of proposals, schemes and programmes etc., to be taken up by agencies having access to land for cultivation and infrastructure for collection, storage and transportation of medicinal plants.
4. Identification, invention and quantification of medicinal plants.
5. Promotion of *ex-situ/in-situ* cultivation and conservation of medicinal plants.
6. Promotion of cooperative efforts among collectors and growers and assisting them to store, transport and market their produce effectively.
7. Setting up of data base system for invention, dissemination of information and facilitating the prevention of patents being obtained for medicinal use of plants which is in the public domain.
8. Matters relating to import/export of raw material, as well as value added products either as medicine, food supplements or as herbal cosmetics including adoption of better techniques for marketing of product to increase their reputation for quality and reliability in the country and abroad.
9. Undertaking and awarding scientific, technological research and cost effectiveness studies.
10. Development of protocols for cultivation and quality control.
11. Encouraging the protection of patent rights and IPR issues.

The medicinal plants board at the National level is devoted for overall development of the medicinal plants sector as a whole in the country. However, for the development of the medicinal plants sector at the regional/state level, the National board has initiated action and the respective state/UT governments were requested to constitute the State Medicinal Plants Boards (SMPBs). Statewise natural habitat of the prioritised species is given below.

States	Name of the prioritised species
Andaman & Nicobar Islands	Ashoka
Andhra Pradesh	Chandan, Gudmar, Safed musli
Bihar	Aonla, Ashoka, Bael, Bhumiamalaki, Brahmi, Gudmar, Kalihari, Kalmegh, Makoy, Safed musli, Pippali, Vai vidang, Tulsi, Giloe
Chandigarh	Ashwagandha, Brahmi, Vai vidang, Tulsi, Giloe
Chhattisgarh	Aonla, Ashwagandha, Bael, Bhumiamalaki, Gudmar, Kalmegh
Delhi	Bhumiamalaki, Makoy, Mulethi, Ashwagandha, Bael
Goa	Kokum
Gujarat	Aonla, Bael, Guggal, Isabgol, Safed musli, Senna
Haryana	Aonla, Ashwagandha, Brahmi, Mulethi
Himachal Pradesh	Atees, Chirata, Daruhaldi, Kuth, Kutki, Satawari, Vatsnabh
Jammu & Kashmir	Atees, Chirata, Daruhaldi, Kuth, Kutki, Kesar, Satawari, Vatsnabh
Jharkhand	Aonla, Ashoka, Bael, Gudmar
Karnataka	Brahmi, Chandan, Kokum, Safed musli, Pathar chur, Pippali
Kerala	Brahmi, Kokum, Pippali
Madhya Pradesh	Aonla, Ashwagandha, Bael, Bhumiamalaki, Brahmi, Gudmar, Guggal, Kalihari, Makoy, Safed musli, Pathar chur, Sarpagandha, Senna, Satavari, Giloe
Maharashtra	Aonla, Ashwagandha, Kalihari, Pathar chur, Pippali
Odissa	Aonla, Ashoka, Gudmar, Kalmegh, Kokum, Vai vidang
Punjab	Ashwagandha, Brahmi, Vai vidang
Rajasthan	Aonla, Ashwagandha, Bael, Guggal, Kalihari, Makoy, Mulethi, Pathar chur, Senna, Satawari, Giloe
North East States	Atees, Ashoka, Ashwagandha, Chirata, Giloe, Jatamansi, Kutki, Pippali, Sarpagandha, Vatsnabh, Vai vidang
Tamil Nadu	Brahmi, Chandan, Kalmegh, Safed musli, Pippali Aonla, Bael, Bhumiamalaki, Brahmi, Gudmar, Kalihari
Uttar Pradesh	Kalmegh, Makoy, Mulethi, Safed musli, Sarpagandha, Satavari, Tulsi, Giloe

Uttarakand	Atees, Brahmi, Chirata, Daruhaldi, Jatamansi, Kesar, Kuth, Kutki, Satavari, Vatsnabh, Tulsi, Giloe
West Bengal	Ashoka, Brahmi, Kokum, Pippali, Sarpagandha

Kitchen medicinal plants garden

The non-poisonous medicinal plants like *Solanum trilobatum, Coleus aromaticus, Eclipta alba, Tridax procumbens, Centella asiatica, Piper longum, Ocimum* spp., *Aloe vera, Adhatoda vesica* and brahmi can be recommended for kitchen medicinal plants garden.

Constraints in medicinal plants cultivation in India

The major constraints of medicinal plants cultivation in India are as follows.

1. Lack of quality planting materials
2. Lack of research on post-harvest handling and storage
3. Lack of high yielding varieties and technologies
4. Lack of commercial coordination
5. Lack of patenting for commercial medicinal plants
6. Dependence on middle man
7. Lack of adequate technology and machinery for extraction of active principle.

Suggestions for improvement

At present, the available production of medicinal plants is not sufficient to fulfill global demand. Therefore, the production system has to be intensified by utilizing existing cropping programmes. The production systems to be adopted are as follows.

1. Intercropping under perennial systems
   a. Development of coconut based medicinal plants intercropping system
   b. Development of fruit crops based medicinal plants intercropping system.
   c. Adoption of tree crops/ tree medicinal plants based intercropping system.
2. Promotion of intensive cultivation
3. Contract farming
4. Cooperative farming
5. Self help group approach
6. Research and development for yield and quality improvement.

The medicinal plant species suitable for intercropping under perennial systems are kacholam, lesser galangal and basil. There is a demand for Senna, Periwinkle, *Coleus* and Ashwagandha grown under organic farming systems. The medicinal plant species *viz.,* Isabgol, Brahmi, Gogum, *Phyllanthus, Centella,* Safed musli, *Stevia,* Asthma plant, *Gymnema, Coleus,* Sarpagandha, *Artemisia,* Guggal, Turmeric, Ashwagandha and Sandal wood are having good scope in near future.

1.3. INDUSTRIAL USES OF MEDICINAL PLANTS

In the recent times, the use of traditional knowledge and traditional systems of healthcare has gained wide attention. Ayurveda evolved in the Indian sub-continent thousands of years back and the treatment was more of a service than a profession. Over 50 per cent of prescription drugs are derived from chemicals first identified in plants. Based on observation and experiential wisdom, Charaka, Sushruta and Vagbhat described 700 herbal drugs with their properties and clinical effects. Based on these, they described 50 categories of drugs.

Today, there are about 20 well established ayurvedic companies and trusts besides numerous small manufacturers in India, who make thousands of generic and proprietary products. Currently, the total trade of ayurveda products in India is about Rs 50,000 million. India exports ayurveda products worth about Rs. 6000 million. When the trade of ayurvedic products picked up, the demand for the raw material mainly the medicinal herbs also grew proportionately. Some of the initiatives taken by the government include the establishment of National Medicinal Plants Board (NMPB) at the centre level as well as State Medicinal Plant Boards (SMPB) in almost all the states of India to promote the cultivation of medicinal plants through financial assistance and technology dissemination.

A couple of good industries have started their own cultivation of selected plants, while there are others who get them cultivated through contract cultivation programmes with farmers. Due to the widening gap in supply and demand of the medicinal plants, farmers adopted them to replace a part of their regular crops to get better returns. Currently, there are about 25 plants which were originally procured only from the wild, but they are being cultivated at some level by the farmers. The Ministry of Health and Family Welfare has published five volumes of Ayurvedic Pharmacopoeia. One of the leading industry associations in collaboration with few of the reputed scientific institutions in India, have brought out Indian Herbal Pharmacopoeia containing 52 monographs of widely used herbs.

Many reputed scientific institutions in India *viz.,* ICMR, ICAR, CSIR, DST, DBT, Universities are continuously working on medicinal plants. One such example is Guggul (*Commiphora mukul*), which was extensively studied by Indian Council of Medical Research (ICMR). It helped its introduction in Indian Allopathic Pharmacopoeia. ICMR has also collaborated with five other research institutions in India and came out with scientific monographs of about 60 medicinal plants. These are standardized scientifically using modern chromatographic and analytical tools.

Indian Government's efforts to share the documented scientific data of ayurveda resulted in the acceptance of it in various regions outside India. House of the Lords of United Kingdom Parliament was convinced to move ayurveda from proposed category 3 (which contained traditional systems devoid of any scientifically validated claims) to category 1 (which includes those with all the scientific data with modern allopathic system). Based on the efforts by the Government of India, South Africa has decided to enact a regulation, which will oversee the import of ayurvedic and unani products. The research work in ayurveda, shared by the Government of India with National Centre for Alternate and Complimentary Medicine, USA, led to the introduction of capsule courses of ayurveda in US medical schools. The Government of Hungary, after getting convinced with the scientific validation of ayurveda, has shown their willingness to develop ayurveda in Hungary. It has also recognized ayurveda and about 50 ayurvedic products are registered and are being sold there.

Today, there are 121 pure chemical substances extracted from about 130 species of higher plants used in the modern pharmacopoeia throughout the World. Out of these, 89 plant derived drugs, currently used in modern medicine, were originally discovered through the study of traditional cures and folk knowledge of indigenous people.

Chemical standardization and chemo-profiling of ASU drugs

There are always chances of wide variations with respect of their chemical contents in crude drugs/ raw materials of plant origin due to varied reasons such as climatic conditions, geographical distribution, source and season of collection and lack of scientific methods of post-harvest processing, storage and preservation. Keeping these facts in view for acceptance or rejection criteria, minimum - maximum limits for marker compounds, based on chemical analysis of large number of samples for each drug/ herbal collected from different agro-climatic zones with passport data has to be worked out, to have reproducible results. Therefore, the only solution to place ASU system of medicines on global market, is to chromatographically standardize the herbal products and wherever possible quantify them for active phyto-chemical marker(s). Thin Layer Chromatography (TLC) is a powerful and simple analytical tool used for this purpose. Quantitative and qualitative HPLC, HPLC/MS, HPLC/MS/MS, GC, GC/MS, GC/MS/MS, HPLC-UV–DAD and HPTLC are well suited analytical methods of choice to control the quality of phyto-pharmaceuticals.

According to World statistics, development of one new drug needs screening of 10,000 molecules with 10-15 years of time and an expenditure of more than $ 900 million. In the present global market scenario related to traditional and herbal drugs, People Republic of China is leading with 50 per cent market share followed by Japan (20%) and Russia (16%). Unfortunately India's share is only 2.5 per cent although the internal market has crossed over Rs. 4000 crores per year. The reason is very simple and to ensure fair share in World market, the quality control and scientific evaluation data for Indian products has to be generated / presented to global community.

Source of plant material

Wild harvesting is still prominent mode of obtaining medicinal plants and most of the requirement of medicinal plants for the industries is still met through wild collection. Though many medicinal plants are commonly available in the wild and can be freely harvested, collection and sale of large quantities of plant material from the forest can lead to destruction of many forest plants especially the endemic species that have a restricted geographical distribution. For example, medicinal plants like *Curcuma caesia, Rauwolfia serpentina* were reported to occur abundantly in central India. However, due to their growing economic importance and rampant harvesting, these plants have now been categorized as critically endangered. Cultivation of medicinal plants would seem as a commercially attractive option to companies because they have greater control over supply of the plant material and it is easier to control post-harvest treatment. Moreover, cultivation can reduce the dependence on collection of plants from wild and thus, have the potential to save wild populations and conserve their genetic diversity.

Selection of plants

As per World Health Organization guidelines, the plant selected for collection should be taxonomically same as recommended by the National pharmacopoeia or other related documents. The voucher specimens of the plant material should be submitted to Regional or National herbaria for authentication. Complete taxonomical identification including varieties of the medicinal plants is an important factor during selection as taxonomy of the plant species can play an important role in their biological activity.

Collection of medicinal plants

Good collection practices are necessary for the long term survival of wild populations and their habitats. Prior to initiating collection, essential information regarding the target plant species should be obtained. The botanical identity, scientific name including genus, species, sub-species or variety and family of the plant should be recorded. If available, the local name should also be verified. Information regarding environmental conditions, such as topography, geology, soil, climate and vegetation at the collection site should be obtained. Information such as the geographical distribution of the plant, its abundance, whether it is threatened or endangered, shrub/fast growing tree etc., should also be obtained. It is of immense importance that a voucher specimen be deposited in a National or Regional herbarium for identification and further consultation by other researchers.

Medicinal plant materials should be collected in the proper season so as to ensure the best possible quality of both the starting material as well as the finished product. Seasonal variations can affect the chemical composition of the plants and thus, its biological activity. In most cases, maximum accumulation of chemical constituents occurs at the time of flowering which then declines at the beginning of the fruiting stage. The time of harvest should also depend on the plant part to be used, since it is well known that depending on the plant species, the level of biologically active constituents can vary in different parts at different stages of the plant growth and development. Climatic conditions, e.g. light, rainfall and temperature (including day and night temperature differences) also influence the physical, chemical and biological qualities of medicinal plants. The water and temperature stress related increase in the content of active constituents such as the total phenolic compounds. Information such as the correct plant parts that are used (roots, leaves, fruits etc.) and whether these parts are seasonal or replenishable should be obtained. The collection levels and the collection practices should also be known before initiating collection. It is necessary that the collection practices employed should be non-destructive. For example, while collecting roots, the main root should not be cut or dug up or while collecting the bark, the tree should not be girdled or completely stripped of its bark. Parts that are not required or decomposed and any foreign matter such as soil or toxic weeds should be removed during collection. Collection of medicinal plants should not be done from places that are prone to or close to sources of contamination such as areas where high levels of pesticides or other possible contaminants are used or found e.g. roadsides, drainages, mine tailings, garbage dumps and industrial facilities which may produce toxic chemicals or active pastures that may lead to microbial contamination. Quality control ensures that the plant material is not contaminated with microbes, pesticides, heavy metals or other toxic and that the final product is of consistent high standard.

Processing of plant materials and their preparation

Preliminary processing of the plant material that can be done include elimination of undesirable materials and contaminants, washing to remove soil, sorting and cutting. It would be advisable to dry the plant materials prior to transportation, if the processing facilities are located away from the collection sites. Cross contamination of the different collected plants or plant parts should be avoided during transportation. The plant materials should be protected from conditions that may cause deterioration such as rain, moisture etc. during or after transportation till the processing begins. The plant materials that need to be used fresh should be delivered as quickly as possible to the processing facility to prevent microbial fermentation or thermal degradation. Specific processing methods are often required, to reduce drying time, to detoxify the inherent toxic constituents, to reduce side effects or to enhance therapeutic effects. The medicinal property of plants is closely related to the different classes of phyto-constituents (such as essential oils, alkaloids, acids, steroids, tannins, saponins etc.) present in the plant, each of which would have a preferred effective method of extraction, facilitating maximum yield in preparation.

Storage

Storage can also influence the physical appearance and chemical quality of plant materials and hence, it is necessary to maintain appropriate storage conditions so as to increase their shelf life. It is customary to store the plant material in dried form, since preparations like decoctions/ infusions can only be stored for a few days. Dried plant materials can be stored in whole, crushed or powdered forms in storage conditions that include use of cloth bags, clear glass bottles and plastic. Plant materials that are used fresh should be stored under refrigeration, in jars or sandboxes, or using enzymatic or other appropriate conservation methods. However, they should be used as quickly as possible to avoid microbial contamination. Shelf life of plant material is usually ignored due to the general belief that the plant materials do not have an expiry date, however, dried plant materials usually retain their activity for about six months only. It is observed that the powdered plant material degrades faster than the whole or crushed plant material. Different types of plastics can be used which prevent absorption of moisture and oxidation of the plant material by preventing the exchange of gases to increase the shelf life of the plant material.

Phyto-chemical studies

Medicinal plant preparations are chemically complex and may contain one or many structurally related active compounds that produce a combined effect. Phyto-chemical studies help in standardizing the herbal preparations so as to get the optimal concentrations of these active constituents, as well as in preserving their activities. The aim of phyto-chemical studies is to identify the bioactive constituents in the plants, devise best methods for their extraction, understand their side effects and calculate appropriate dosages. Standardization can be carried out by obtaining a chemical fingerprint/profile or through bioactivity guided fractionation. Chemical fingerprints through chromatographic techniques are more commonly used for standardization and are obtained in terms of one or more marker compounds. It would be ideal to use the active constituent in the plant as the marker compound, however in cases where active constituents are not known, the marker compound can be independent of the therapeutic activity. Furthermore, the plant extracts can also be standardized to class of compounds e.g. ginsenosides in ginseng, kava lactones in kava, or oxindole alkaloids in cat's claw. Such an approach would be suited to situations where though the active constituents are not known, but are expected to belong to a particular class of compounds. Common analytical methods such as high performance liquid chromatography (HPLC) or high performance thin layer chromatography (HPTLC) be sufficiently stable; and be commercially available or able to be isolated by the company in its own laboratory.

TLC and HPLC are the most commonly used methods for obtaining chemical fingerprints and identification of the crude plant extracts. However, there are several possibilities that may arise while using these techniques for standardizing the crude extracts. It is possible that the plant materials collected from the same plant in two different seasons can show different phyto-chemical fingerprint and therefore, different biological activity or two plants with identical taxonomy collected under same environmental conditions can show different phyto-chemical fingerprint, but similar biological activity. In such situations, comparisons of the phyto-chemical profiles as an indicator of important constituents can act as a shortcut for identifying biologically active constituents. Another possibility that may arise is when two different plants showing similar phyto-chemical fingerprints show different biological activity. In such situations, bioassay guided fractionation or any other suitable method is the only option in identifying the biologically active constituents. DNA fingerprinting is another technique, which though still in its early years, seems to be of immense potential in identification of medicinal plants, particularly when profiling the genotypic differences. Apart from identifying these genetic variations, it can also aid in identification of germplasm of important or endangered plants for future cultivation or conservation. Use of isolated compounds can result in better biological activity due to higher concentrations, but it can also lead to potential side effects. The quality of a phyto-medicine is defined by several parameters in sequence like quality of herbal resource and herb, manufacturing of the drug preparations from the herb, properties of the finished product etc., thus, demanding special quality attention on each individual herbal species in accordance with Good Manufacturing Practice (GMP) standards at each step.

Many drugs of modern medicine have their origin in traditional medicine. Some common examples include the discovery of the alkaloid diosgenin in *Dioscorea deltoidea* used as source for the partial synthesis of cortisone and steroid hormones in the forties, the discovery of the hypotensive alkaloid reserpine in *Rauvolfia serpentina* and the analgesic alkaloid aspirin in *Filipendula ulmaria* in the fifties, the discovery of anti-asthmatic alkaloid ephedrine in *Ephedra sinica* and the anti-cancer alkaloid podophyllotoxin in *Podophyllum hexandrum* in the sixties, to mention a few (Table 1).

Table 1. Important plant derived drugs used in modern medicine

Plant species	**Family**	**Drug**	**Medicinal use**
Ammi majus	Apiaceae	Xanthotoxin	Vitiligo
Ammi visnaga	Apiaceae	Khellin	Vascodilator
Anamirta cocculus	Menispermaceae	Picrotixon Barbiturate	Antidote
Artemisia annua	Asteraceae	Artemisine	Antimalarial
Atropa belladonna	Solanaceae	Atropin	Ophthalmology
Cassia angustifolia	Caesalpiniaceae	Sennosides	Laxative
Catharanthus roseus	Apocynaceae	Vinblastine Vincristine	Hodgkin's disease Pediatric leukemia
Cephaelis ipecacuanha	Rubiaceae	Ipecac, Emetine	Emetic, Amoebic dysentery
Colchicum autumnale	Liliaceae	Colchicine, Demecolcine	Gout, Leukemia
Datura stramonium	Solanaceae	Scopolamine	Motion sickness
Digitalis lanata	Scrophulariaceae	Digoxin	Cardiotonic
Digitalis purpurea	Scrophulariaceae	Digitoxin	Atrial fibrillation
Dioscorea deltoidea	Dioscoreaceae	Diosgenin	Induces sterilization
Drimia maritima	Liliaceae	Proscillaridin	Cardiac malfunction
Ephedra sinica	Ephedraceae	Pseudoephedrine	Central nervous system stimulant, Rhinitis
Erythroxylum coca	Erythroxylaceae	Cocaine	Ophthalmologic, Anaesthetic
Filipendula ulmaria	Apocynaceae	Asprin	Analgesic, Anti inflammation
Gossypium herbaceum	Malvaceae	Gossypol	Male contraceptive
Hamamelis virginiana	Hamamelidaceae	Gallotanins	Hemorrhoid suppository
Marsilea minuta	Marsileaceae	Marsilin	Sedative, Anticonvulsant
Melilotus officinalis	Fabaceae	Dicoumarol	Thrombosis
Mucuna pruriens	Papilionaceae	L-DOPA	Antiparkinsonian
Papaver somniferum	Papaveraceae	Codeine, Morphine, Noscapine, Papaverine	Analgesic, Antitussive, Antispasmodic
Physostigma venenosum	Fabaceae	Physotigmine Stigmasterol	Glaucoma Steroidal precursor

Podophyllum hexandrum, P. peltatum	Berberidaceae	Podophyllotoxin Teniposide	Vermifuge, Cancer, Bladder neoplasms
Psoralea corylifolia	Fabaceae	Psoralen	Vitiligo
Rauvolfia canescens	Apocynaceae	Deserpidine, Rescinnamine, Reserpine	Hypertension
Rauvolfia serpentina	Apocynaceae	Ajmalin, Rescinnamine Reserpine	Heart arrhythmia Antihypertensive, Tranquilizer
Rhamnus purshiana	Rhamnaceae	Cascara	Purgative
Ruta graveolens	Rutaceae	Rutine	Decreases capillary fragility
Strophanthus gratus	Apocynaceae	Strophanthin	Congestive heart failure
Strychnos guianensis	Loganiaceae	Toxiferine	Surgery relaxant
Veratrum album	Liliaceae	Protoveratrine	Hypertension

PART - II

ANNUAL MEDICINAL PLANTS

2.1. *Abelmoschus moschatus*

Common names: Ambrette seeds, Musk mallow, Mushkdana, Kasturibendi

Botanical name: *Abelmoschus moschatus* Medik.

Family : Malvaceae

Vernacular names : Kasturidana, Mushkdana (Hindi and Bengali); Kasuridana (Gujarati); Kadukasturi (Kannada); Kattukasthuri (Malayalam); Mushkbhendibij, Kasturibhenda (Marathi); Kattukasturi, Vartlaikasturi (Tamil); Kasturi bhenda, Kasturi bhendvettulu (Telugu); Gandapura, Kasurilatika, Latakasturikam (Sanskrit).

Origin and distribution : *A. moschatus* is native to India and distributed in India, southern China, peninsular Indochina, the south-western Pacific islands, New Guinea and northern Australia.

Habitat : The plant grows wild all over the Deccan regions, and the hilly regions of Karnataka. It grows in a range of habitats, from marshy areas to forest edges, at elevations of upto 450m. Commonly found in disturbed, open areas, as well as gardens, plantations and rice fields. It is able to grow on salt affected wastelands. Soils ranging from sandy to clayey types are suitable for growing this species. Loamy soils with a neutral pH of 7.0 with plenty of organic matter are ideal for its cultivation. It cannot withstand waterlogged conditions. As this crop is very susceptible to frost, its cultivation in winter season with severe cold/winter should be avoided. Tropical to sub-tropical conditions are suitable for growing this crop. An optimum rainfall ranging between 1200 and 1400mm is required for its cultivation.

Habit : It is a stout herb or under shrub, upto about 2m tall, minutely stellate - hairy, with a long slender taproot (Fig. 1).

Leaf : Leaves are polymorphous, 15cm long, the lower ovate, acute or roundish angled, the upper palmately 3-7 lobed, hairy on both surfaces.

Flower : The flower bud appears in the axil of each leaf above the third to eighth leaf, depending upon the cultivar. The flowers are large, corolla bell shaped, yellow with purple centre usually solitary and axillary, sometimes in few flowered raceme (Fig. 2).

Flowering, pollination and fruiting : Flower bud initiation occurs 25-30 days after sowing and it takes another 25 days to reach full blossom. Flowering continues for 45-80 days. Flowering period from July to September and the seeds ripen from August to October. Anthesis normally occurs in the morning during 8-10.30A.M. The stigma is receptive only on the day of anthesis. Because the flower structure is homogamous, it is a self pollinated crop. The fruits start ripening by the end of November and continue doing so till the end of April. The pods mature in two months period.

Fruit : Fruit is capsule fulvous hairy, 8cm long, oblong lanceolate, acute. The pods possess irritating hairs which cause itching (Fig. 3a and 3b).

Seed : Seeds are sub-reniform and blackish. A delicate musk like odour is produced by the seed coat (Fig. 4).

Propagation : It can be propagated by seeds, small tubers and stem. Non-leafy semi-hardwood cuttings are also used for propagation. These cuttings are planted directly in the main field.

Seed collection : Fruits should be plucked before they split and shed the seeds. The pods are ripe for plucking when three quarters of their bodies have turned blackish brown. Beyond this stage, the fruits split and seeds scatter which reduces yield. Since all the fruits do not mature at one time, the pods have to be plucked as and when they mature. Plucking once in a week may be desirable. After shade drying of pods, thrashing may be done by beating the pods with sticks and winnowing them to remove the husk. Similarly the seeds can be obtained by splitting the dry fruits by hand. The seeds have also to be shade dried before sowing. A normal crop may yield 7.5-10 quintals of seed per hectare.

Presowing treatments : Seeds treated with carbofuran at 4 or 5g/100g of seed, grew vigorously, flowered earlier by 2-3 days and produced more seed yield. For controlling fungal diseases, the seeds should be treated with bavistin or carbendazim before sowing. Pink bollworms are controlled by heat treatment of the seeds (Srivastava, 1995). Presoaking the seeds for 24h in water improves the germination (Farooqi and Sreeramu, 2001).

Seed germination : Dibble 2-3 seeds to a depth of 1 to 2cm and cover with soil. Sow the seeds at least a foot apart.

Seeds may also be dibbled 1cm deep at a distance of 1m. Give a spacing of 60cm between rows. Sow the seeds during early summer or rainy season.

Seeds take at least 15 days to germinate, even though early germination is seen after 8-10 days.

Economic part used : Seeds, roots, leaves, seed pods.

Fig. 1. Plant

Fig. 2. Flower

Fig. 3a. Immature pod

Fig. 3b. Pod with seeds

Fig. 4. Seeds

Alkaloids : The seeds are more valuable because of the volatile oil present in the seed coat. Its aromatic constituents have long been used in perfumery industry. Seed analysis report 11.1% moisture, 31.5% crude fibre, 14.5% lipids, 13.4% starch, 2.3% protein and volatile oil (0.2-0.6%).

The seed constituents include myricetin-3-glucoside and a glycoside of cyanidin in flowers, an aromatic constituent in seeds, beta-sitosterol and its beta-D-glucoside, myricetin and its glucoside in leaves and petals, and beta-sitosterol from dry fruit husk.

Medicinal uses : This plant is used in treating spasms of digestive tract, muscle cramps, poor circulation and aching joints. The seeds are much more valuable medicinally due to their diuretic, demulcent, stomachic, stimulant, cooling, tonic, carminative, aphrodisiac and antiseptic properties. When seeds are chewed they act as nerve tonic, stomachic and sweeten the breath. A paste of leaves is applied on the cuts, wounds and sprains. The essential oil from the seeds is used in aromatherapy for the treatment of depression and anxiety, cramps, poor circulation and aching joints. Other people add seeds to coffee, whilst leaves and new shoots are eaten as vegetables in some cultures.

Past work done on Seed Science and technology

Seed technological studies were undertaken by Sivakumar (2005) in ambrette and findings are given below.

a. The pods and seeds attained physiological maturity on 35 days after anthesis as evidenced by the maximum dry weight of pod (2.69g) and seed (12.63mg) combined with minimum moisture content of pod (23.45%) and seed (21.64%). The change in colour of pod and seed from green to brown could be considered as a visible index of maturity.

b. The seed quality parameters *viz.,* germination (41%), 100 seed weight (1.47g), seedling length (16.3cm) and drymatter (8.0mg), vigour index (668), protein (9.85%), oil (21.70%) and starch (13.32%) contents were also maximum at 35 days after anthesis.

c. Application of fertilizers at the rate of 150:60:60kg NPK ha^{-1} had significantly increased the plant height (129cm) and also advanced the events of flowering such as days to first flowering (86.5 days) and fifty per cent flowering (95.8 days), irrespective of spacing levels. Significantly higher pod and seed yield of 2172kg and 1177kg ha^{-1}, respectively were obtained by nourishing the plants with 150:60:60kg NPK ha^{-1} with the plants spacing of 60 x 45cm.

d. The seed quality in terms of 100 seed weight, germination, seedling length and drymatter, vigour index, protein, oil and starch contents were found to be higher due to application of 150:60:60kg NPK ha^{-1} with 60 x 45cm as well as in 60 x 60cm plant spacing. However, the yield of pod (1688kg) and seed (975kg) was significantly low in 60 x 60cm plant spacing. Similarly, though the yield of pod and seed was maximum in 60 x 30cm plant spacing, the quality of seeds were inferior in terms of the above quality parameters when compared to higher plant spacing.

e. Studies conducted to fix optimum number of pickings revealed that the seed yield parameters *viz.*, pod and seed weight, number of seeds pod^{-1}, pod and seed yield and the seed quality parameters *viz.*, seed recovery, 100 seed weight, germination, seedling length and drymatter, vigour index, protein, oil and starch contents were higher upto fifth picking. Hence, pods could be harvested upto fifth picking for seed purpose.

f. The results of the dormancy breaking treatments revealed that the soaking of seeds in hot water at 100°C for 30 minutes effectively improved the germination (83%) and seedling length (18.8cm), drymatter (14.20mg) and vigour (1560) accompanied with minimum abnormal seedling (6%) production.

g. Size grading the seeds with BSS 8 x 8 wire mesh sieve registered the maximum recovery (89.8%) of good quality seeds with higher germination of more than 82 per cent with higher protein, starch and oil contents and seedling vigour. Separation of seeds based on the density by specific gravity separator had further enhanced the recovery (91.8%) of good quality seeds in first four grades with higher germination, seedling length, drymatter production and vigour index.

Geetharani (2006) found that the seeds treated with hot water at 50°C for 30 minutes recorded maximum germination (98%) and vigour followed by soaking of seeds in hot water for 15 minutes (90%). The germination of untreated seeds was found to be very low (16%) (Table 1).

Table 1. Effect of dormancy breaking treatments on germinability of musk seeds

S.No.	Treatments	Germination percentage	Dry matter production (mg 10 $seedlings^{-1}$)	Vigour index
1.	Control	16	0.082	1.31
2.	Soaking in water			
	(i) 6 hours	16	0.070	1.12
	(ii) 12 hours	20	0.090	1.80
3.	Soaking in hot water (50°C)			
	(i) 15 minutes	90	0.094	8.46
	(ii) 30 minutes	98	0.097	9.51
4.	Soaking in conc. H_2SO_4			
	(i) 2 minutes	84	0.092	7.73
	(ii) 5 minutes	82	0.093	7.63
	(iii) 10 minutes	60	0.071	4.26
	(iv) 15 minutes	54	0.062	3.35
	(v) 20 minutes	30	0.086	2.58

	(vi) 25 minutes	24	0.065	1.56
5.	Rubbing on cement surface	60	0.082	4.92
6.	Rubbing on sand paper	64	0.084	5.38
	CD (0.05)	**15.18**	**NS**	**2.04**

The results of the seed development and maturation studies in ambrette revealed that the pods and seeds attained physiological maturity on 35 days after anthesis as evidenced by the maximum dry weight of pod (2.694g) and seed (12.63mg) combined with minimum moisture content of pod (23.45%) (Table 2 and 3). The change in colour of pod and seed from green to brown could be considered as a visible index of maturity. The seed quality parameters *viz.*, germination (41%), 100 seed weight (1.468g), seedling length (16.3cm) and dry matter (8.0mg) and vigour index (668) were also maximum at 35 days after anthesis (Table 4) (Sundareswaran *et al.*, 2011).

Table 2. Changes in ambrette pod characters during development and maturation

Stages of pod development (days after anthesis)	Length of pod (cm)	Girth of pod (cm)	Volume of pod (cc pod^{-1})	Fresh weight of pod (g)	Dry weight of pod (g)	Moisture content of pod (%)
7	4.54	6.58	12.10	3.364	1.160	83.13 (65.65)
14	6.10	8.67	15.40	6.732	1.959	79.43 (63.01)
21	8.15	9.77	18.51	7.725	2.236	63.41 (52.77)
28	8.47	9.83	16.00	3.896	2.587	41.36 (40.16)
35	8.40	9.70	15.40	2.830	2.694	23.45 (28.93)
42	8.38	9.65	14.41	2.810	2.583	17.46 (24.65)
SEd	0.082	0.163	0.249	0.163	0.082	0.938
CD at 5%	0.178	0.356	0.546	0.356	0.178	2.044

(Figures in parentheses indicate transpformed values)

Table 3. Changes during seed development and maturation in ambrette

Stages of pod development (days after anthesis)	No of seeds pod^{-1}	Fresh weight of seed (mg $seed^{-1}$)	Dry weight of seed (mg $seed^{-1}$)	100 seed weight (g)	Seed moisture content (%)
7	103	10.20	2.55	0.289	75.29 (60.13)
14	105	25.15	7.69	0.846	69.54 (56.48)
21	103	31.33	10.34	1.037	63.90 (53.07)
28	102	22.13	11.12	1.233	44.36 (42.15)
35	101	16.12	12.63	1.468	21.64 (27.69)

42	89	15.04	12.65	1.472	19.33 (25.06)
SEd	2.40	1.59	0.61	0.070	2.638
CD at 5%	5.24	3.46	1.32	0.153	5.749

(Figures in parentheses indicate transpformed values)

Table 4. Changes in ambrette seed attributes in development and maturation

Stages of pod development (days after anthesis)	Germination (%)	Root length (cm)	Shoot length (cm)	Hard seed content (%)	Dry matter production (mg seedling^{-1})	Vigour index
7	0 (0.00)	0.0	0.0	0 (0.00)	0.0	0
14	0 (0.00)	0.0	0.0	0 (0.00)	0.0	0
21	0 (0.00)	0.0	0.0	0 (0.00)	0.0	0
28	20 (26.57)	7.0	8.6	0 (0.00)	7.2	312
35	41 (39.82)	7.3	9.0	46 (42.71)	8.0	668
42	42 (40.40)	7.4	9.1	51 (45.57)	8.1	693
SEd	1.732	0.158	0.115	0.942	0.092	12.440
CD at 5%	3.774	0.343	0.252	2.054	0.201	27.095

(Figures in parentheses indicate transpformed values)

REFERENCES

Farooqi, A. A. and B. S. Sreeramu 2001. Ambrette. In: Cultivation of medicinal and aromatic crops. Universities Press Ltd., Hyderabad. pp. 295-299.

Geetharani, P. 2006. Studies on seed maturation, germination and storability of musk (*Abelmoschus moschatus* L.). In: Horticulture Crop Scientist Meet. Dept. of Floriculture and Landscaping, TNAU, Coimbatore.

Sivakumar, D. 2005. Seed technological studies in Ambrette (*Abelmoschus moschatus* Medic.). M.Sc. (Ag.) Thesis, TNAU, Coimbatore.

Sundareswaran, S., S. Sathish and A. Vijayakumar 2011. Seed development and maturation studies in ambrette (*Abelmoschus moschatus* Medic.). *Madras Agric. J.,* 98 (1-3): 29-32.

Srivastava, U. C. 1995. Ambrette seed. Advances in Horticulture Vol. 11 - Medicinal and Aromatic Plants (Eds. K.L. Chadha and Rajendra Gupta). Malhotra Publishing House, New Delhi, India. pp.887-897.

2.2. *Abutilon indicum*

Common names: Country Mallow, Indian Mallow

Botanical name: *Abutilon indicum* (L.) Sw.

Family : Malvaceae

Vernacular names : Potari (Bengali); Kanghi (Hindi); Tutti (Kannada); Velluram (Malayalam); Petari (Marathi); Atibalaa (Sanskrit); Thuththi, Paniyaratutti, Perunthuththi (Tamil); Duvvena Kayalu, Tuturabenda (Telugu).

Origin and distribution : It is native to Southeast Asia and grows as a weed on road side and waste places in Africa, Afghanistan, Israel, China, Taiwan, Bhutan, Nepal, Pakistan, Sri Lanka, Malaysia and India.

Habitat : It is occurring in disturbed sites at low elevations near the ocean.

Habit : A diffusely much branched shrub; growing to a height of 60 to 150cm (Fig.1).

Stem : Round with smooth close tomentum often tinged with purple.

Leaf : Leaves ovate, cordate irregularly toothed, covered on both surfaces with white down.

Flower : Solitary, axillary, yellow sometimes with a purple centre; floral parts as in other members of Malvaceae; carpels 10-20, syncarpous, individual carpels pointed or mucronate with thin tomentum and scattered tuft of still hairs; Pedicel 5cm long and jointed at the top, calyx is tomentose, cupular below with five lobes, corolla is yellow sometimes with purple centre, petals five, obovate and rounded with prominent veins, stamina tube hairy with stellate hairs.

Flowering, pollination and fruiting : Flowers are slightly protandrous, opening and closing times of flower are temperature and light dependent. Flowering and fruiting occurs during September - April. If the pollen from the same flower falls onto the stigma, it will not germinate; some times the flower may wilt. It is known as pollen pre-potency. It is pollinated by honeybees (*Apis* sp.) and *Bombix* sp (Abid *et al.*, 2010).

During the months of February, March, September, October and November, the plants exhibited maximum number of flowers/plant, pollen production, pollen viability, fruit set and seedset. On the other hand, with the either reduction in temperature in the months of December and January or considerable increase in the months of May and June, the plants exhibited considerable reduction in the number of flowers/plant, number of pollen/flower, pollen viability, fruitset and seedset (Kumar *et al.,* 2012).

Fruit : Fruit is schizocarpic capsule called carcerulus, quite interesting - it is circular in shape, consisting of 11-20 radiating hairy carpels with 2-3 seeds/ chamber, brown when dry; each carpel flattened, somewhat boat shaped (Fig. 2 and 3).

Fig. 1. Plant

Fig. 2. Immature fruit

Fig. 3. Dry fruit

Fig. 4. Seeds

Seed : The seeds are kidney shaped, light brown in color, but turn black on ripening with stellate hairs. They get dispersed after the dehiscence of fruits (Fig. 4).

Seed dispersal : Seeds are mainly distributed through winds.

Propagation : By seeds

Seed treatment : Gupta *et al.* (2001) showed that different treatments like hot water treatment at 70°C for 10 minutes followed by concentrated H_2SO_4 scarification for five minutes was most effective in breaking the seed coat dormancy.

Economic part used : Whole plant.

Alkaloids

Leaves and other aerial parts contain mucilage, tannin, alkanol, flavonoids, saponins, cyanidin-3-rutinoside, asparagin, eudesmol, geraniol, caryophylline, sesquiterpene lactones, tocopherol oil (0.3%), essential oil and eugenol.

Seeds yield raffinose and a semi-drying oil consisting of linoleic, oleic, palmitic and stearic acids.

Plant also contains gossypetin -8 & 7-glucosides, cyanidin-3-rutnoside.

Medicinal uses

In traditional medicine, it is used as a demulcent, aphrodisiac, laxative, diuretic, pulmonary and sedative (leaves). The bark is astringent and diuretic; laxative, expectorant and demulcent (seeds); laxative and tonic, anti-inflammatory and anthelmintic (plant); analgesic (fixed oil); diuretic and for leprosy (roots). The whole plant is uprooted, dried and is powdered. In ancient days, maidens were made to consume a spoonful of this powder with a spoonful of honey, once in a day, for 6 months until the day of marriage, for safe and quick pregnancy. The leaves can also be used to treat ulcers, headaches, gonorrhoea and bladder infection. The plant is very much used in Siddha medicines. In fact, the root, bark, flowers, leaves and seeds are all used for medicinal purposes by Tamils. The leaves are used as adjunct to medicines used for pile complaints. The flowers are used to increase the semen in men. A methanol extract had some antimicrobial properties. A chemical compound, ß-sitosterol, which has been identified as the active ingredient in many medicinal plants, is present in *A. indicum* and a petroleum ether extract provided larvicidal properties against the mosquito larvae (*Culex quinquefasciatus*).

REFERENCES

Abid, R., J. Alam and M. Qaiser 2012. Pollination mechanism and role of insects in *Abutilon indicum* (L.) Sweet. *Pak. J. Bot.*, 42(3): 1395-1399.

Gupta, V., A. Kak, V. Mitter and B. B. Singh 2001. Techniques to remove hard seededness in the wild medicinal plant *Abutilon indicum*. *J. Med. Arom. Pl. Sci.,* 23 : 369-371.

Kumar, P., S. Chauhan and A. Rana 2012. Impact of temperature on reproductive biology of *Abutilon indicum* (L.) Sweet (Malvaceae), *International J. Plant Reproductive Biology* 4(2): 127–131.

2.3. *Acalypha indica*

Common names: Indian Acalypha, Indian nettle. Three seeded mercury, Indian copper leaf.

Botanical name: *Acalypha indica* L. Syn. *A. australis* L.

Family : Euphorbiaceae

Vernacular names : Muktajhuri, Swetbasanta, Biralhatchi (Bengali); Khokali, Kuppi (Hindi); Kapi, Khokoli (Marathi); Kuppaimeni, Kuppivani (Tamil).

Origin and distribution: It occurs throughout tropical Africa and South Africa, India and Sri Lanka, as well as in Yemen and Pakistan. In West and East Africa, the plant is used as a medicinal plant. In West Africa, the leaves are cooked and eaten as a vegetable. This plant is held in high esteem in traditional Tamil Siddha medicine as it is believed to rejuvenate the body.

Habitat: It is a very common weed found growing wild on the side of the road, lawn, or on the slopes. Distributed profusely in wastelands and pastures particularly soon after rains. In cultivated areas, grown with crops requiring frequent inter cultivation.

Habit: Erect, annual, pubescent herb growing to a height of 20 to 60cm (Fig. 1).

Stem: Round or angular and densely shortly pubescent, branched with coarse elongated lines.

Leaf : Leaves are single, long stemmed, spreading. Leaf blade is oval to lanceolate, thin, tapered tip and base, serrated edge, 2.5 to 8cm long, 1.5 to 3.5cm wide, green. Petiole longer than the blade.

Flower: Minute, green, unisexual, monoecious, in axillary erect spikes; bracteate, bracts cuneiform, many nerved, dentate, folded, alternate; male flowers few and terminal; calyx lobes four, membraneous, valvate; petals; stamens many on a convex receptacle; female flowers basal: calyx 3-4 lobed, lobes imbricate, petals 0; ovary superior, hispid, syncarpous, three celled, ovules solitary in each cell, style filiform, often very long (Fig. 2).

Flowering and fruiting: Flowering and fruiting occurs during June-November.

Fruit : Capsules covered by the bract, often one seeded, black in colour.

Seed : Seeds ovoid, acute, smooth and brown in colour (Fig. 3).

Propagation : By seeds.

Effect on crop: It is a host of several plant viruses, e.g. pumpkin yellow vein mosaic virus, okra yellow vein mosaic virus, leaf curl virus and tobacco virus, often resulting in very severe and recurrent outbreaks in crops. Roots may suffer from nematode infestation (*Meloidogyne* spp.). In India, *Acalypha indica* is a host of the spiralling whitefly (*Aleurodicus dispersus*).

Fig. 1. Plant

Fig. 2. Flower

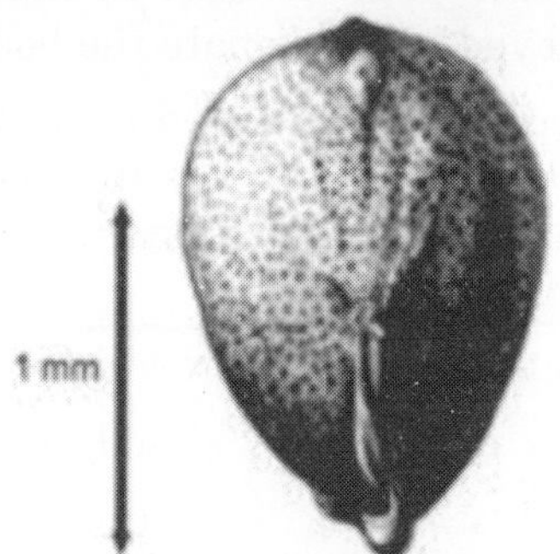

Fig. 3. Seed

Economic part used : Whole plant, roots and leaves.

Alkaloids: The plant contains kaempferol, a cyanogenic glucoside, a base, triacetonamine and an alkaloid, acalyphine. It also contains the amide, acalyph-amide and some other amides, 2-methyl anthraquinone, β-sitosterol, β-sitosterol glucoside, stigmasterol, n-octacosanol, quinine, tannin, resin and essential oil.

Medicinal uses : It is an important medicinal plant for its expectorant properties. It also has significant antibacterial and antifungal activities, both against human and plant pathogens. In East Africa, the leaf sap is used as eye drops to treat eye infections. Leaf powder is applied to maggot infested wounds. In Comoros, a leaf decoction is used as a massage cream to treat pain of the joints. In the Seychelles and Réunion, a root infusion or decoction is taken to treat asthma, and also to clean the liver and kidneys. In the Seychelles, a root decoction is also taken to treat intestinal worms and stomachache. A poultice made from the whole plant is applied to treat headache. The leaf sap is taken as an emetic and the root decoction as a laxative.

Agricultural uses: The leaf extracts of *Acalypha indica* and *Catharanthus roseus* effectively inhibited the mycelia growth, sporulation, spore germination and germ tube elongation of *Sarocladium oryzae* (Yesu Raja and Syamala, 2012).

REFERENCE

Yesu Raja, I. and M. Syamala 2012. Biopotency of promising plant extracts against rice sheath rot disease. *Madras Agric. J.*, 99 (4-6): 344-348.

2.4. *Acorus calamus*

Common names: Sweet flag, Rat Root, Flag Root, Calamus

Botanical name: *Acorus calamus* L.

Family : Acoraceae

Vernacular names: Bajai, Gora bach, Vasa Bach (Hindi); Vayambu (Kannada and Malayalam); Vekhand (Marathi); Vacha, Bhutanashini, Jatil, Haimavati (Sanskrit); Vasambu (Tamil); Vasa, Vadaja (Telugu).

Origin and distribution: Likely indigenous to India or Arabia. *Acorus calamus* is now found across Europe, in southern Russia, northern Asia Minor, southern Siberia, China, Indonesia, Japan, Myanmar, Sri Lanka, Australia, as well as southern Canada and the northern United States. This has been cultivated in Karnataka, Kerala, Tamil Nadu and other regions.

Habitat: This herb is grown throughout India upto an altitude of about 200m, marshy or moist situations of northern and eastern Himalayas. It's a hardy plant found growing from tropical to sub-tropical climates. Plenty of sunshine should be made available to the plant during its growth and after harvesting for drying the rhizomes. Temperature ranging from 10°C to 38°C and annual rainfall between 70 and 250cm is best suited. It prefers moist clayey loams, sandy loams and alluvial soils.

Habit: It is a tall perennial (1m height) wetland monocot with scented leaves and more strongly scented rhizomes (Fig. 1).

Rhizome: Rhizome is cylindrical, light brown or white and spongy, long, creeping and much branched aromatic rhizomes.

Stem: The stem is tall, erect; with the radical leaves clustered round it, and the plant has a grass like habit. Single prominent mid-vein and then, on both sides slightly raised secondary veins (with a diameter less than half the mid-vein) and many fine tertiary veins.

Leaf: The plant is erect with lanceolate leaves (0.75 to 1.5m long) and 2-4cm breadth that are erectophilic and green in colour when young and turn yellow when rhizomes mature. The leaves are bright green possessing thick midrib and wavy margins and sweet scented when bruised.

Flower: Flowering shoots supported by a large leaf like structure called spathe. The flower stalk is compact spike. Flowers hermaphrodite, small, pale green, in 5-10cm long spikes called spadix. The flowers are longer too, between 3 and 4mm. There are six stamens, fixed on the base of the segments, with flattened anther-stalks. The six stamens with anthers are in pairs, the stigmas sessile and very small. The flowers are sweet scented. The flowers are infertile and show an abortive ovary with a shrivelled appearance (Fig. 2).

Flowering, pollination and fruiting: Flowers from May to July and the seeds ripen from July to August. It is pollinated by insects.

Fruit: The fruit is several seeded (1-9, most frequently 5-7). Three locular, dry berry. It is 4-5mm long, roughly wedge shaped, narrowing toward the base, and somewhat four angled as a result of crowding on the spadix. The scarious pericarp is lustrous, light brown to straw coloured, and marked with longitudinal dark brown streaks. The septa are thin and delicate. The seeds are borne in a cluster on an axile placenta near the summit of each locule (Fig. 3).

Seed : The seeds are oblong and contain both endosperm and perisperm and the testa is foveolate (Fig. 4).

Fig. 1. Plant

Fig. 2. Flower

Fig. 3. Head with fruit

Fig. 4. Seed

Seed dispersal : Pieces of rhizome are easily dispersed by water along rivers. In particular, the sterile triploid cytotype of *Acorus calamus* has been dispersed by this mean. The seeds are also dispersed by water along stream margins. It was widely dispersed in Europe and parts of North America by human agency due to its medicinal importance and later became naturalised in many regions (Kubitzki, 1998).

Propagation

This aquatic herb can be propagated through rhizomes. Take the rhizomes from earlier planting and keep them preserved in the soil and constantly moist.

Once the sprouts emerge, cut these rhizomes into smaller pieces consisting of two to three nodes and plant in the nursery.

In the month of July-August, these sprouted rhizome pieces may be planted at half a finger depth at a distance of one foot from each other. Second fortnight of June is the best time for planting.

As the growth is fast, one can see the sprouts on the very second day of planting. Ninety per cent of the rhizomes sprout and successfully establish in the field. For transplanting, pick the sprouted rhizomes from the nursery and plant directly in marshy open field.

Rhizome can be harvested one year after planting. During the harvesting stage, the leaves are yellow in colour.

Economic part used : Dried rhizome

Alkaloids : Alpha-asarone, Beta-asarone, eugenol

Medicinal uses : The oils can be used as carminative, appetizer, expectorant, and nerve stimulant. Tannins in this plant can be used to cure diarrhoea and dysentery. Its powdered roots are used as vermifuge.

Agricultural uses

Anandi (2001) reported that, the cowpea seeds dried to 8.5 per cent moisture treated with vasambu (*Acorus calamus*) rhizome powder @ 50g kg^{-1} of seed and stored in 300 gauge polylined cloth bag showed higher germinability, longer roots and shoots, higher dry matter production, better cell membrane integrity and protein content as compared with other treatments. The seeds could be stored safely for 9 months. The seeds treated with vasambu rhizome powder were completely free from bruchid infestation.

Subramanian and Manikandan (2002) reported that 500ml of sweet flag (*Acorus calamus*) extract with 1 litre of cows urine and 100ml of khadi soap solution per tank controlled *Fusarium* wilt attack in chilli crop. Malarkodi (2003) found that the seed treatment with vasambu (*Acorus calamus*) rhizome powder @ 100g/kg of seed was effective to maintain insect free seed storage upto 21 months in greengram. Studies on storage of castor hybrid seeds with prestorage treatment revealed that slurry treatment with carbendazim @ 2g/kg of seed and dry dressing with *Acorus calamus* rhizome powder @ 1:100 ratio preserved the seed quality upto 12 months with 92 per cent germination (Shete, 2004). The rhizome extracts can be used to control the aphids infested in vegetables (Sridhar *et al.,* 2006).

Sudha (2007) observed that the organically produced cotton seeds treated with vasambu rhizome powder @ 100g/kg of seed before storage recorded better shelf life maintenance upto 6 months.

REFERENCES

Anandi, R. 2001. Seed technological studies in cowpea (*Vigna unguiculata* (L.) Walp) cv. CO 5. M.Sc. (Ag.) Thesis, TNAU, Coimbatore.

Kubitzki, K. 1998. The families and genera of vascular plants. Vol (IV) flowering plants monocotyledons. Springer verlog berlin Heidelberg, New York.

Malarkodi, K. 2003. Integrated management techniques for seed storage in greengram (*Vigna radiata* L. Wilczek). Ph.D. Thesis, TNAU, Coimbatore.

Shete, D. M. 2004. Seed production techniques in castor hybrid TMVCH1. Ph.D. Thesis, TNAU, Coimbatore.

Sridhar, S., S. Arumugasamy, H. Saraswathy and K. Vijayalakshmi 2006. Organic vegetable gardening. Centre for Indian Knowledge Systems, Chennai. p.46.

Subramanian, K. and S. Manikandan 2002. Diseases attacking chilli crop. *Indigenous Agriculture News*, 1(1-3):11.

Sudha, P. 2007. Organic seed production in cotton (*Gossypium hirsutum* L.) cv. LRA 5166. M.Sc. (Ag.) Thesis, TNAU, Coimbatore.

2.5. *Andrographis paniculata*

Common names: Creat, Green Chirayta, King of bitters.

Botanical name: *Andrographis paniculata* (Burm. F.) Wall. Ex Nees Syn. *Justicia paniculata* Vahl. CL.

Family : Acanthaceae

Vernacular names: Kalmeg (Assamese, Bengali); Kiriyata, Olikiriyat (Gujarati); Kalmegh, Kalpanath, Kiryat, Mahatit (Hindi); Nelaberu (Kannada); Vhadlem Kiratyem (Konkani); Nilaveppu, Kiriyatta (Malayalam); Vubati (Manipuri); Olen Kirayat, Kalpa (Marathi); Hnakhapui (Mizo); Bhuinimba (Oriya); Bhuinimb, Kirata, Mahateet (Sanskrit); Nilavempu (Tamil); Nela Vemu (Telugu); Naine havandi (Urdu).

Origin and distribution: It is native to India. It is widely distributed in Indian plains (Andhra Pradesh, Assam, Bihar, Karnataka, Kerala, Madhya Pradesh, Chattisgarh, West Bengal and Himachal Pradesh), Sri Lanka, parts of China, Thailand, Bangladesh and Taiwan.

Habitat: Grows well in moist and shaded places, but it prefers sunny situations. It can be cultivated on wide range of soils from loam to lateritic soils with moderate fertility. It can be cultivated on shady wastelands also.

Habit: It is small, erect and branched, herbaceous to semi woody annual. It is thickened at lower nodes, quadrangular and bearing long, divaricate branches (Fig. 1).

Stem: The stem is dark green, 0.3-1.0m in height, 2-6mm in diameter, quadrangular with longitudinal furrows and wings on the angles of the younger parts, slightly enlarged at the nodes.

Leaf: Leaves are opposite, glabrous, upto 8.0cm long and 2.5cm broad, entire, lanceolate, pinnate, dark green above and pale beneath.

Flower: Flowers are numerous, distantly arranged in a much branched cymose panicle with small bracts. Calyx has five linear segments, corolla is rose coloured and two lipped, upper lip arched and bifid and lower lip divided into three short acute lobes. Stamens are two in number, inserted in the throat of corolla with flattened filaments, with a large tuft of hairs, beneath the two celled anthers. Ovary is small, laterally compressed with an annual disc two celled with a few ovules in each cell, style is long and stigma, slightly bifid (Fig. 2).

Flowering, pollination and fruiting : With the onset of monsoon, plant grows luxuriantly and starts flowering with the moderation in temperature during September. Flowering and fruiting continues upto December until temperature drops drastically in Northern plains. It is self-pollinated, self-compatible and habitual inbreeder (Lattoo *et al.*, 2006).

Fruit : Capsules are linear - oblong, acute at both ends, 1.9cm x 0.3cm, compressed and two celled, seeds about 12 in number (Fig. 3a and 3b).

Seed : Seeds are numerous, round to ovoid, slightly compressed and yellowish brown (Fig.4).

Propagation : It can be propagated by seeds and stem cuttings. Vegetative propagation is also possible in certain special cases through layering as each node is capable of producing enough roots.

Seed propagation

Its propagation is through shattered seeds in nature. Harvest mature pods and dry under sun. Collect the dried seeds, discard the chaffy seeds. Healthy seeds are viable for six months. Since the seeds are too small, mix them with sufficient quantity of fine sand and sow the seeds on raised beds. The seeds are to be soaked in water for six hours before sowing. Sowing is to be done during the months of July- November.

Eighty per cent of the seeds would germinate within 10 days. Young seedlings may be later transplanted to nursery beds. Avoid deep sowing. Maintain the seedlings in the nursery for 2-3 months.

Fig. 1. Plant

Fig. 2. Flower

Fig. 3a. Immature pod

Fig. 3b. Pod with seeds

Fig. 4. Seed

Vegetative propagation: For vegetative propagation, semi-mature stem cuttings 10-15cm long with three nodes are used. These cuttings have to be planted in either polybags or raised beds to a depth of at least 9cm with one node exposed. Eighty per cent of them would sprout within a week. These cuttings may be transplanted to the main field after 1-2 months.

Economic part used : Except roots, all other parts are used

Alkaloids : It contains bitter principles andrographolide, a bicyclic diterpenoid lactone and kalmeghin (up to 2.5%). The leaves contain the maximum active principle content, while in the stem it is in lesser amount.

Medicinal uses : Immune booster, bitter tonic, antityphoid, anti inflammatory, anti hepatitic, antibiotic, antidiabetic, used in curing fevers, worms, dysentery, general weakness and excessive gas formation in stomach. The plant is useful in acidity and liver complaints. It is also useful in the treatment of sannipata type of fever, difficulty in breathing. etc.

Agricultural uses

The occurrence of Epilacna beetle in brinjal and army worm in beans were controlled by *Andrographis paniculata* extract at 45 per cent concentration (Thambidurai and Arumugasamy, 2001). Jayashankar *et al.* (2002) reported that the incidence of pests in the field bean such as pod borer, pod sucking bug and aphids were controlled by spraying 5% extract of *Andrographis paniculata*, which also helped in complete control of the aphids.

Past work done on Seed Science and Technology

Gangare *et al.* (1998) recommended top of paper method for testing seed germination. Seed germination is a problem in this crop and information available on seed quality is very limited. Hot water treatment of seeds at 50°C for five minutes gives highest germination percentage 84 per cent followed by 3% KH_2PO_4 (Saraswathy *et al.*, 2003). It could be inferred that seeds possess the combined dormancy of physical and innate nature. Hence, the hot water treatment increased the germination potential due to softening of outer layer of hard seed coat (Table 1).

Table 1. Effect of seed treatments on germination and vigour parameters in kalmegh

Treatments	Germination (%)	DMP (mg/10 seedlings)	Vigour index	Abnormal seedlings (%)
T_1 - Thiourea 0.5%	50	17.50	204	8
T_2 – Thiourea 1.0%	64	16.50	342	12
T_3 – Thiourea 1.5%	56	16.35	180	10
T_4 – Thiourea 2%	38	17.20	181	10
T_5 – Thiourea 3.0%	66	16.95	238	4
T_6 – KCl – 0.5%	74	17.45	495	-
T_7 – KCl – 1.0%	46	17.70	217	-
T_8 – KCl – 1.5%	70	17.60	395	12
T_9 – KCl – 2.0%	70	17.45	349	8
T_{10} – KNO_3 –0.5%	40	16.20	158	-
T_{11} – KNO_3 –1.0%	60	16.25	349	10
T_{12} – KH_2PO_4 –0.5%	62	16.90	184	-

T_{13} – KH_2PO_4 –1.0%	62	16.25	292	12
T_{14} – KH_2PO_4 –1.5%	60	17.05	213	16
T_{15} – KH_2PO_4 –2.0%	56	17.35	199	12
T_{16} – KH_2PO_4 –3.0%	76	18.05	390	-
T_{17} – IAA 250 ppm	64	16.65	366	8
T_{18} – IAA 500 ppm	70	16.70	402	12
T_{19} – GA_3 100 ppm	68	16.70	375	8
T_{20} – GA_3 250 ppm	74	17.05	405	10
T_{21} – GA_3 500 ppm	74	17.15	407	14
T_{22} – H_2SO_4 (one min.)	04	16.85	5	-
T_{23} – Hot water (5 min. at 50°C)	84	18.70	540	-
T_{24} – Water soaking (12 h.)	30	16.05	143	-
T_{25} – Leaching (12 h.)	12	15.00	40	4
T_{26} – Sand scarification (5 min.)	74	17.50	393	4
T_{27} – Control	62	16.30	368	12
CD (0.05)	15.59	0.482	8.67	4.027

Note: Treatments T_1 – T_{21} – soaking the seeds for three hours.

Manimohan (2002) carried out the experiment in kalmegh to evolve suitable seed production technologies and the findings are ;

(a) The seeds attained physiological maturity at 30 days after anthesis associated with change of colour of pods from light green to light brown with brown seeds. Beyond this stage, splitting of pods occur and resulted in shattering of seeds.

(b) Foliar application of combination of panchakavya 2 per cent, humic acid 0.2 per cent and moringa leaf extract 2 per cent at vegetative, flowering and pod maturity stages significantly increased the yield parameters like number of pods and seed yield as well as seed quality parameters like germination and vigour index.

(c) Standardization of quick viability test revealed that soaking of embryos either in 0.05 (or) 0.1 per cent tetrazolium chloride solution for two hours was optimum for assessing the viability with a proper staining intensity.

(d) The dormancy of kalmegh could be broken by accelerated ageing for three days followed by soaking of kalmegh seeds with GA_3 @ 200ppm for three hours. The kalmegh seeds treated with halogen mixture @ 3g kg^{-1} and stored in 700 gauge polythene bag proved to be the best practices for storing seeds with minimum loss in viability and vigour.

Bharath (2008) carried out the studies on standardization of seed testing procedures and storage studies and his findings are given below.

- For germination, the best results were obtained in the top of paper method in presence of light for 24 hours at 20°C. The germination percentage (78.80), root length (8.20cm), shoot length (18.90 cm), seedling vigour index (2134), seedling dry weight (15.80mg) and rate of germination was faster (0.359).
- The first and final counts were fixed at seventh and thirteenth day, respectively.
- The lower values of these parameters were found at alternate temperature of 20/30°C in dark without KNO_3 in sand media.
- The extent of decrease in the germination was 37.00 per cent in the storage period. KNO_3 treated seeds recorded the highest values in all the parameters compared to all other treatments.
- The lowest values were observed in control at the initial months of storage. At the later months, cow's urine treated seeds recorded lowest values for all the seed quality parameters.

REFERENCES

Bharath, V. B. L. K. 2008. Standardization of seed testing procedures and storage studies in selected medicinal crops. M.Sc. (Agri.) thesis, University of Agricultural Sciences, Dharwad.

Gangare, K. C., Patil and Y. B. Suryawanshi 1998. Seed dormancy and germination studies in some medicinal plant species. *Seed Tech. News*. 28: 59.

Jayashankar, M., S. Manikandan and S. Thambidurai 2002. Management of pest and diseases in field Bean. *Indigenous Agriculture news*.1 (1-3):4.

Lattoo, S. K., S. Khan, A. K. Dhar, D. K. Choudhary, K. K. Gupta and P. R. Sharma 2006. Genetics and mechanism of induced male sterility in *Andrographis paniculata* (Burm. f.) Nees and its significance. *Current Science*, 91(4): 515–519.

Manimohan, M. 2008. Seed technological studies in kalmegh *Andrographis paniculata* (Burm. F.) Wall. Ex Nees M.Sc. (Ag.) thesis, TNAU, Coimbatore

Saraswathy, S., R. S. Azhakia Manavalan, E. Vadivel, K. Manian and S. Subramanian 2003. Studies on seed germination in kalmegh (*Andrographis paniculata* Nees.) *South Indian Hort.*, 52(1-6): 286-290.

Thambidurai, S. and S. Arumugasamy 2001. Plant extract in pest control. *Pesticide Post*. 9(5-6): 4.

2.6. *Asteracantha longifolia*

Common names: Kulikhara, kokilaksah, long leaves barleria, Hygrophila, Marsh Barbel.

Botanical name: *Asteracantha longifolia* Nees.

Family : Acanthaceae

Vernacular names: Shulamardan (Bengali); Ekhro (Gujarati); Talimkhana (Hindi); Kalavankabija (Kannada); Voyal-chullai (Malayalam); Talimkhana, Vikhra, Kolsunda (Marathi); Kokilaksha (Sanskrit); Neermulli, Kullichedi (Tamil); Kokilakshi (Telugu).

Origin and distribution: This plant is widely distributed throughout tropical and sub-tropical regions of India and other parts of the world including Phillippines, Srilanka, Myanmar, Malaya and Nepal.

Habitat: Generally the plant is found on wet places.

Habit: This is an annual herb or under shrub, it grows to a height of about 60cm (Fig. 1).

Stem: The plant has a number of fasciculated, usually unbranched sub glandular stems, each 60 to 120cm tall. The stems have nodules hispid with long hairs.

Leaf: Leaves are sessile, oblong-lanceolate or linear lanceolate, spines yellowish brown and 2 to 3cm (sometimes more) long. These are found in whorls of six at each node. The two outer leaves of the whorl remain much larger than the four inner ones. Each leaf is greenish brown in colour. These are acute, entire and hairy.

Flower: Flowers are yellowish brown to blue in colour. These usually occur in apparent whorls of eight in four pairs at each node. Bracts occur on nodes. These are usually 2.5cm long with long and white hairs. Corolla is about 3cm long, widely two lipped, tube about 1.6cm long, abruptly swollen at top. Stamens are four, didynamous with the second pair larger. Calyxes are four- partite with upper sepal 1.6 to 2cm long, broader than the upper three. Calyxes are linear, lanceolate and coarsely hairy on the back and with hyline ciliated margin (Fig. 2).

Flowering and fruiting : Submerged plants do not flower, but flowering and fruiting take place when plants emerge out of water. It appears that different temperatures and light conditions might be playing some role in the ecophysiology of this plant species.

Fruit : Fruits are two celled, linear, and oblong, compressed about 8cm long, and pointed 4 to 8 seeded capsules. These have single layer of epidermis covered with striated cuticle, followed by 5 to 10 layered, thick walled, oval to hexagonal, lignified sclerenchymatous cells (Fig. 3).

Seed : Seeds like the eye of a cuckoo. Seeds are ovate, flat or compressed, 0.20 to 0.25cm long and 0.10 to 0.15cm wide, hairy, appearing smooth when soaked in water immediately get coated with mucilage, light brown, taste slightly bitter and odour not distinct (The Ayurvedic Pharmacoepia of India, 1999) (Fig. 4).

Fig. 1. Plant

Fig. 2. Flower

Fig. 3. Fruit

Fig. 4. Seed

Propagation : By cuttings and seeds.

Life cycle : *Asteracantha longifolia* grows well and completes its life cycle mostly when water recedes and then, it takes terrestrial mode of life.

Seed dispersal: Dispersed by water and wind.

Seed dormancy and germination : The seeds of *Asteracantha longifolia* prefer germinating in light. Germination was also favoured in blue and red lights, whereas total darkness delayed this process. The optimum temperature for germination of seeds was 29°C in continuous white light. The seeds did germinate in total darkness as well, but the percentage of germination remained poor and with high temperatures beyond 30°C, the pace of germination became slow. Higher temperatures suppressed the seedling growth both in continuous white light and total darkness. Blue and red lights promoted hypocotyls growth, whereas radicle was inhibited (Sen and Chawan, 1970).

Economic part used : Roots, leaves and seeds

Alkaloids: It contains phytosterol alkaloid, potassium, mucilage and essential oil.

Medicinal uses : In Ayurveda, the leaves, seeds and the roots are used for treatment of rheumatism, urinary tract infection, inflammation, jaundice, hepatic obstruction, pain, etc. Powdered seeds taken with milk is said to increase libido.

REFERENCES

Sen, D.N. and D.D. Chawan 1970. Role of light and temperature in relation to seed germination and seedling growth of *Asteracantha longifolia* Nees. *Osterr. Bot. Z.* 118(3): 226-232.

The Ayurvedic Pharmacoepia of India 1999. Government of India, Ministry of Health and Family Welfare, Department of Indian System of Medicine & Homeopathy, New Delhi.

2.7. *Bacopa monnieri*

Common names: Thyme leaved Gratiola, Water hyssop

Botanical name: *Bacopa monnieri* (L.) Pannell. Syn. *Herpestis monniera* (L.) H.B. & K., *Bramia indica, Bacopa monnieria*

Family: Scrophulariaceae

Vernacular names: Brahmi (Assamese, Hindi and Kannada); Brahmi sak (Bengali and Manipuri); Jalanevari (Gujarati); Bam (Marathi); Brahmi, Gundala, Indravalli, Jalasaya (Sanskrit); Nirbrahmi, Sambrane elaye (Tamil); Neeri sambraani mokka, Sambrani chettu (Telugu).

Origin and distribution: It is originated from Pan Tropic and commonly found in all parts of India. It grows in marshy land and can be found throughout the United States, Australia and Asia.

Habitat: The plant is a short duration annual herb and is found in moist or wet places, such as on borders of water channels, wells, irrigated fields etc.

Habit: It is a succulent, glabrous, perennial, creeping herb, with rooting at nodes (Fig. 1).

Root : Shows a single layer of epidermis, cortex having large air cavities; endodermis single layered; pericycle not distinct.

Stem: The leaf and flower bearing stems are 10-30cm long and arise from creeping stems that form roots at the nodes.

Leaf: The leaves are ovate oblong, sessile, opposite, obtuse apex approximately 2 x 1cm with entire margin. Small, succulent, shows a single layer of upper and lower epidermis covered with thin cuticle; glandular hairs sessile, subsidiary cells present on both surfaces; a few prismatic crystals of calcium oxalate occasionally found distributed in mesophyll cells; mesophyll traversed by small veins surrounded by bundle sheath; no distinct midrib present.

Flower: Arise in the axils of the leaves and are borne on short pedicels with two bracteoles. Sepals are five in number, 0.4-0.9cm long. One of the five sepals is larger than others. The corolla is bluish white in colour and about 1cm across. Corolla tube is cylindrical with spreading lips, twice as long as sepals (Fig. 2).

Flowering, pollination and fruiting: Flowers and fruit appear in summer (Chopra, 1958; Bone, 1996). Butterflies pollinate the flowers and the resulting fruits are small and inconspicuous. This is the nectar and larval food plant for the white peacock butterfly (*Anartia jatrophae*).

Fruit: The fruit is an upto 5mm capsule which develops in the persistent calyx. Capsule is ovoid in shape, acute, two grooved and two valve with numerous seeds (Fig. 3).

Seed : Seed ellipsoid/oblong, pitted or ribbed or spiny, very minute and pale (Fig. 4).

Propagation : This herbaceous creeper is propagated mainly by stem cuttings. Sometimes, it can also be propagated by seeds.

Vegetative propagation: For vegetative propagation, cut the whole plant into small units with 4-6 nodes each. The plant shows luxurious growth during rainy season when the propagules multiply fast. Dip the cuttings (5-10 cm length) in cow dung slurry to ensure better rooting. Plant these stem cuttings directly in the field. Use marshy open fields for the cultivation of the plant. The plantlets sprout and establish as independent plants within a few days.

Economic part used : Whole plant

Alkaloids: Bramhine and herpestine

Medicinal uses : It has been reported to reduce oxidation of fats in the blood stream, which is a risk factor for cardiovascular diseases. It has been used for centuries to help benefit epilepsy, memory capacity, increase concentration and reduce stress induced anxiety. According to Ayurveda, it is bitter, pungent, heating, emetic, laxative and useful in bad ulcers, tumours, ascites, enlargement of spleen, indigestion, inflammations, leprosy, anaemia, biliousness etc. According to Unani system of medicine, it is bitter, aphrodisiac, good in scabies, leucoderma, syphilis etc. It is promising blood purifier and useful in diarrhoea and fever.

Fig. 1. Plant

Fig. 2. Flower

Fig. 3. Fruit

Fig. 4. Seed

REFERENCES

Bone, K. 1996. Clinical applications of Ayurvedic and Chinese herbs: Monographs for the Western herbal practitioner. Phytotherapy Press, Warwick, Queensland.

Chopra, R. N. 1958. Indigenous drugs of India. 2nd ed. U.N. Dhur and Sons, Calcutta, India. P.341.

2.8. *Centella asiatica*

Common names: American coinwort, Asiatic coin wort, Asiatic pennywort, Coinwort, Indian pennywort, Marsh pennywort, Water pennywort, Penny weed, Spade leaf

Botanical name: *Centella asiatica* (L.) Urb. Syn. *Hydrocotyle asiatica*

Family : Apiaceae

Vernacular names: Aghinya, Bor-mani-muni (Assamese); Bora, Thulkari Thankuria (Bengali); Barmi (Gujarati); Brihmi, Brahmmanduki, Budhbrahmani, Valari (Hindi); Vondelega (Kannada); Muthil, Kudangal (Malayalam); Peruk (Manipuri); Brahmi, Karinga (Marathi); Brahmibuti (Punjabi); Mandukaparni (Sanskrit); Vallarai (Tamil); Bokkudu, Nacley, Saraswati Aku (Telugu).

Origin and distribution: It is a native of Madagascar. It is found in swampy areas of India, commonly found as a weed in crop fields and other waste places throughout India upto an altitude of 600m, in tropical and subtropical regions. The plant is reported to occur also at higher altitudes of 1550m in Sikkim and 1200m in Mount Abu. It grows widely in regions of East India, China, Japan and Australia, Sri Lanka and South Africa.

Habitat: It is commonly found in rice fields along the bunds of irrigation channels. It can be commonly found in marshy banks of streams, ponds, rivers and irrigated lawns, where there is moisturized a soil upto 1500m Elevation.

Habit : It is a small creeping herb with shovel shaped leaves emerging alternately in clusters at the stem nodes. The runners lie along the ground and the inch long leaves with their scalloped edges rise above on long reddish petioles (Fig. 1).

Root : The plant consists of vertical root stalk.

Stem : Stem is prostrate, creeping with long stolen and nearly glabrous or hairy on young parts.

Leaf: Leaves are cordate or hastate or orbicular or reniform or sub-entire or palmately lobed consisting of long petiole and small stipules. Crowded leaves can be seen at nodes, consisting of very long stalks and sheathing leaf bases. Leaf blades are dentate, crenate with thick radiate veins and dark green in colour. Leaves are glabrous on both surfaces and having a diameter of 2-10cm.

Flower: Flowers are small, sessile and dark pink in colour. They arise as simple umbels of 3-6 flowers at the end of slender peduncles arising from the axils of leaves and much shorter than petioles supported below by an involucre of two boat shaped membraneous persistent bracts (Fig. 2).

Flowering, pollination and fruiting: Flowering in June to October; fruiting in July to November. The flowers are pollinated by insects and the plant is self fertile.

Fruit : Fruits are small look like barley grains, clusted at joints. Carpels oblong, sub-cylindric curved and less in length much laterally compressed, readily separating into two indehiscent halves (mericarps) united by a very narrow plane of junction. Vittae are seen in the furrows of each mericarp. The fruits when bruised have an aromatic odour, pungent, nauseous and bitter taste (Fig. 3).

Seed : Solitary in each mericarp, pendulous embryo, laterally compressed (Fig. 4). It has 15 seeds per ramet which are brownish in colour and slightly kidney shaped. The average length and breadth of seeds are 2.84 and 1.81 mm, respectively. The 100 seeds weigh 0.13g (Devkota and Jha, 2010).

Seed dispersal: Through water.

Fig. 1. Plant

Fig. 2. Flower

Fig. 3. Fruit

Fig. 4. Seed

Propagation : It reproduces both vegetative and sexually. Severe competition exists between ramets and seedlings. Seedling mortality is high in natural settings. Ramets play an important role in population maintenance in nature (Singh *et al.*, 1999).

Seed propagation : The fresh seeds are collected during February-March. The seeds are sown in nursery beds or directly in the fields. Sixty per cent germination would be obtained within 10 days.

Seed storage : The tetrazolium test indicated 82 per cent viability in freshly harvested seeds. Seeds viability deteriorated as duration of storage increased and they became non-viable after storage for thirty months. Seeds offer genetic variation and ease of portability and storage as compared to vegetative propagules apart of resilience to drought (Brock *et al.*, 2003).

Vegetative propagation: The stem divisions are used as the planting material. The selected division should have 3-4 nodes. One hundred per cent sprouting will be seen in 7-10 days.

Economic part used : Fresh and dry leaves.

Alkaloids: Brahminoside, asiaticoside, madecassoside, centelloside, thankuniside.

Medicinal uses

It has been used externally for burns, psoriasis, prevention of scar formation following surgery, recovery from an episiotomy following vaginal delivery of a newborn and treatment of external fistulas.

Extract exerted anti-inflammatory and antifilarial effects and is used as memory enhancer and wound healer (Shukla *et al.*, 1999).

Past work done in Seed Science and Technology

According to Devkota and Jha (2010), freshly harvested seeds did not germinate even after GA_3 treatment and incubation under different light qualities, whereas two-three months old seeds exhibited germination (82%) without pre-treatment at warm environment (25-30°C). Germination of seeds pre-treated with 60°C water were significantly lower than germination of seeds pre-treated with 10% HNO_3 and seeds treated with different concentrations of GA_3. Among the different treatments of GA_3, soaking seeds in 10ppm resulted in the highest germination followed by 20ppm (Fig. 5).

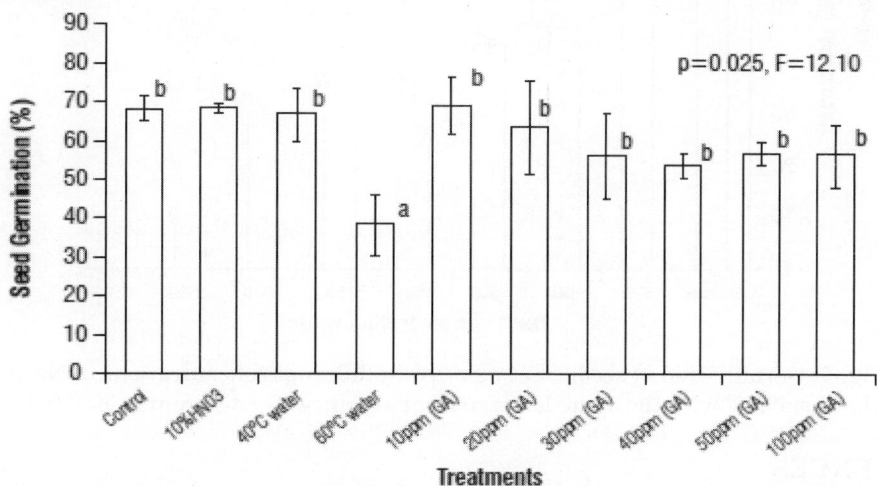

Fig. 5. Effects of different pre-treatments on seed germination of *C. asiatica* after six weeks (bars marked with the same letter are not significantly different from the control, p>0.05).

The germination was slightly higher in red light and inhibited by blue and far red lights. The seeds were defective on germination in dark conditions (Fig. 6).

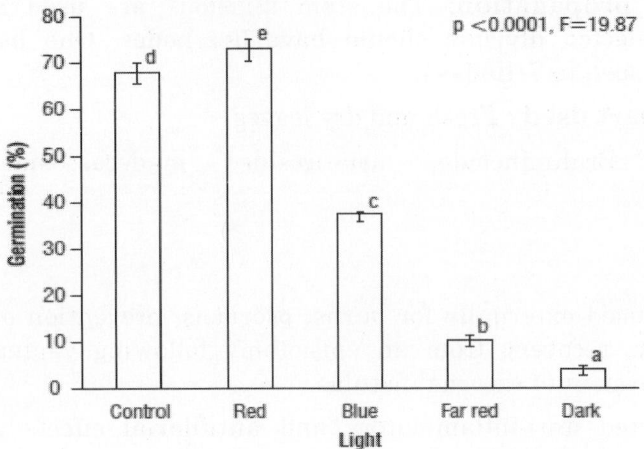

Fig. 6. Final germination percentages of *C. asiatica* seeds in different coloured light (observation at sixth week).

At salinities upto 3500ppm, NaCl germination rate ranged 70-73 per cent. The germination was inhibited by 57 and 89 per cent at 4500ppm and 5500ppm NaCl, respectively and completely abolished at 6500ppm salinity (Fig. 7).

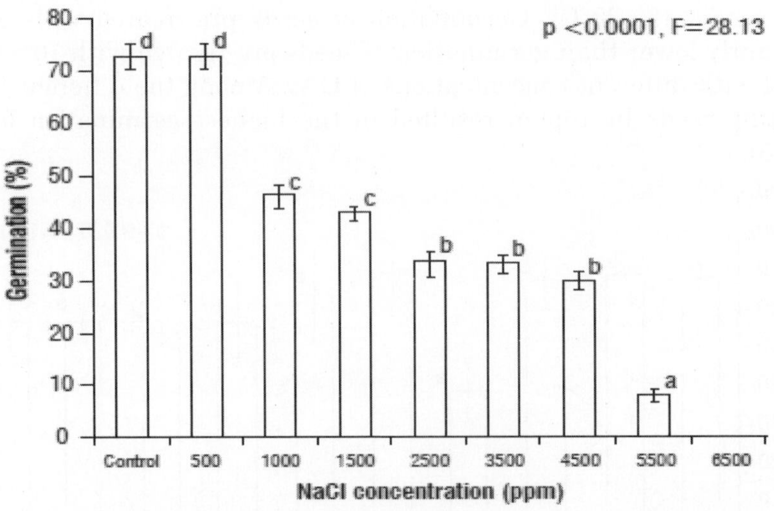

Fig. 7. Germination of seeds of *C. asiatica* in different concentrations of NaCl (bars marked with the same letter are not significantly different at p>0.05).

REFERENCES

Devkota, A. and P. K. Jha 2010. Seed germination responses of the medicinal herb *Centella asiatica*. *Brazilian J. Plant Physiology*, 22(1): 143-150.

Brock, M. A., D. L. Nielsen, R. J. Shiel, J. D. Green and J. D. Langley 2003. Drought and aquatic community resilience: the role of eggs and seeds in sediments of temporary wetlands. *Freshwater Biol.*, 8: 1207-1218.

Singh Pratibha, U. P. Singh and J. S. Singh 1999. The effect of leaf extracts of *Centella asiatica* and *Andrographis paniculata* on spore germination of some fungi. *J. Plant Protection Tropics*, 12 (2): 106-112.

Shukla, A., A. M. Rasik and B. N. Dhavan 1999. Asiaticoside induced elevation of antioxidant levels in healing wounds. *Phytotherapic Res.*, 13: 50-54.

2.9. *Chlorophytum borivilianum*

Common names: India Spider Plant, White Musale

Botanical name: *Chlorophytum borivilianum* Santapau & R. R. Fern.

Family: Liliaceae

Vernacular names: Dholi Musli (Gujarati); Khiruva, Safed Musli (Hindi); Shedheveli (Malayalam); Safed Musli (Marathi); Swetha Musli (Sanskrit); Taniravi Thang (Tamil); Swetha Musli (Telugu).

Origin and distribution: It is native to India and distributed in Rajasthan, Gujarat, Maharashtra and Madhya Pradesh.

Habitat: It grows successfully in the wide range of temperatures and rainfall. It was originally grown in thick forests of India. It prefers well drained loamy to sandy soil rich in organic matter.

Habit: It is small tuberous plant measuring 30-50cm has life span of 3-8 months and it surfaces once in a year during the rainy season (Fig. 1).

Root: Tuberous root, root stock stout, short or elongate, more or less cylindrical, 4 to 15cm long, average being 7.4cm, 0.5- 1cm wide. The upper portion of the root stock remains clothed with the withered leaf bases and with copious lateral roots which are long, almost whitish or yellowish gray, shrieked. Externally the root stock is yellowish brown, thin. Fracture is starchy, odour indistinct; taste slightly bitter and mucilaginous (Fig. 2).

Stem: Safed Musli can grow upto a maximum height of 45cm. Tubers can grow upto a depth of 25cm.

Leaf: Sessile or short petiolate, with sheathing bases, 15 - 45cm. by 1.5-2.5cm, linear or linear lanceolate, membraneous, glabrous or sparsely softly hairy. The leaf tip in contact with soil produces adventitious roots.

Flower: Axillary, scapose, racemose, scape short, clavate flattened with the pedicels, bracts and ovary concealed in the leaf sheaths. The lower flowers on the scape are mostly bisexual and the upper staminate flowers; epigynous; bright yellow, bisexual or unisexual with anceolate membraneous bracts. Perianth;

Gamophyllous, rotate, six partite, perianth segments 12 to 16mm long, elliptic oblong acute, hairy on the back. Stamens; six filaments very small, adnate to the base of the perianth lobes, anthers linear erect. Ovary: inferior, tricarpellary, syncarpous with a fairly long, slender beak and stripe, ovules 6-8 with long finical styles one short, stigma 3 cleft with the lobes oblong, erect and apprised (Fig. 3).

Fig. 1. Plant

Fig. 2. Root

Fig. 3. Flower

Flowering, pollination and fruiting: The plant exhibits shy flowering and poor seed setting. Flowering and fruiting are seen during June-September. It is pollinated by insects.

Fruit: Capsule, four seeded with a slender beak and spongy septa.

Seed: Two month old saplings begin to produce small black colored seeds resembling the seeds of onion and the seeds are oblong, black and shiny with crustaceans' testa.

Propagation: By seeds and root divisions.

Seed propagation

Seeds are reported to have a dormancy period upto 10 months. Seed propagation is not recommended as the plants raised this way take long time to mature (about two years), have a poor germination per cent, high variability and are less vigorous compared to the vegetative propagated plants.

Seeds are to be sown in very well prepared seedbeds, which need to be heavily manured with farmyard manure or well rotten leaf manure. First or second week of June is ideal for sowing. The germination extends upto 16 days with 11 to 24 per cent germination.

Seed collection

Since the fruits dry and split open to disperse the seeds, it is not easy to collect them. Cover the inflorescence with a bag after the fruits are developed. This would ensure the collection of the seeds otherwise dispersed. Collection of seed should be made before capsule color turns fully brown. If the seeds are turn brown, as soon as the seeds will dehisce.

Vegetative propagation

More vigorous plants are produced when the root divisions are used. These root divisions are to be taken from the root system before the plant begins new growth. The root bunches taken from the field are spread under shade for 15 to 20 days so that moisture content of the root bunches is reduced. This planting material can be stored in a dry place for the next seven to eight months by covering the root at 15cm depth below the ground under the shade. Another storage method is to pack the root bunches in about 1/2 to 1kg bundles in perforated plastic bags having an equal amount of soil, tie at the mouth and keep in a cool dry room. Care should be taken not to include the damaged and fleshy roots without the skin.

Collect the sprouted seedlings between 10 to 30 days after the rains or take out the fleshy root bunches from the ground during May and store in polythene bags. Tubers weighing 5-10g can be used as a whole for planting. To ensure good yield, certified material from approved and established sources should be used for the first planting. Around 2,00,000 propagules can be planted in an area of one hectare. Tubers treated with bavistin for two minutes should be planted and depth not exceeding 2.5-3cm. Even a small (1cm), slightly shrunken fleshy root has the capacity to reproduce a new plant. In case of roots that could not be taken out and have already started sprouting, separate them with a sharp blade so that a part of the stem disc remains attached to the fleshy root. These divisions may be used for planting. The first or second week of June is the appropriate planting season. The seedlings can emerge out of the ground within four to six days after rains. Degeneration of tuberous roots during storage affects their growth rate. Hence, special care should be taken while storing the fleshy roots for propagation.

Economic part used: Dry root

Alkaloids: The roots (tubers) are rich in alkaloids, vitamins, minerals, proteins, carbohydrates, saponins, polysaccharides and steroids (Anon., 2001).

Medicinal uses: It is used for increasing male potency and to fight fatigue. The roots of this herb have been found to strengthen the immune system. It is being used as an antiarthritic and anticancer drug.

REFERENCE

Anonymous 2001. Medicinal plants more on safed musli. *Agriculture Industry survey (May)*, pp: 38-39.

2.10. *Coleus aromaticus*

Common names: Aromatic coleus, Country borage, Indian borage

Botanical name: *Coleus aromaticus* L. Syn. *Plectranthus aromaticus* Benth.

Family: Lamiaceae

Vernacular names: Patharchur (Hindi); Doddapatre (Kannada); Panikku-rukka (Malayalam); Pan Ova (Marathi); Pashanabhedi (Sanskrit); Karpooravalli, Omavalli, Oma elai (Tamil).

Origin and distribution: It is native to Seychelles, India and south east Asia. It is widely cultivated in Rajasthan, Maharashtra, Karnataka and Tamil Nadu.

Habitat: The herb grows easily in a well-drained, semi-shaded position. It is frost tender and grows well in sub-tropical and tropical locations, but will do well in cooler climates, if grown in a pot and brought indoors, or moved to a warm sheltered position in winter.

Habit: A succulent herb which has the typical four cornered stem of the Lamiaceae family, it is known as the five seasons herb (Fig.1).

Leaf: Leaves are simple, opposite, broadly ovate, crenate and fleshy. The leaves are very thick and succulent, grey-green and hairy and can grow to around 50cm tall. The leaves are highly aromatic with a strong flavour of mixed herbs.

Fig. 1. Plant

Fig. 2. Flower

Flower: The plant bears terminal spikes of small, bluish and bi-lipped flowers. Pale purplish in dense whorls at distant intervals in a long slender raceme (Fig. 2).

Flowering and pollination: It blooms in summer. Pollination is entomophilous.

Fruit: Orbicular or ovoid nut lets.

Propagation: Cuttings

Economic part used: Leaves

Alkaloids: Thymol, cymine, terpine, carvacrol

Medicinal uses: Expectorant, diuretic, liver tonic, cough, chronic asthma, bronchitis.

2.11. *Coleus forskholii*

Common name: Indian Coleus

Botanical name: *Coleus forskholii* Briq. Syn. *Coleus barbatus, Plectranthus forskohlii*

Family: Lamiaceae

Vernacular names: Pashanbhedi, Pathatchur (Hindi); Garmalu (Gujarati); Makandiberu (Kannda); Maimnul (Marathi); Makandi (Sanskrit); Koorkan Kilangu, Marunthu koorkan (Tamil).

Origin and distribution: It is native to India, Myanmar and Thailand. It is widely distributed in the tropical regions of India, Pakistan, Sri Lanka, tropical east Africa, Brazil, Egypt, Arabia and Ethiopia. It is cultivated in parts of Gujarat, Maharashtra, Rajasthan, Karnataka (Belgaum district) and Tamil Nadu.

Habitat: It is a crop of the tropics growing on the hills upto an altitude of 2400m. The annual rainfall requirement is 100-160cm mainly during June-September. Warm and moist climate with a relative humidity of 85-95 per cent and a temperature of 10-25°C is best suited for it, but under irrigated conditions it can be grown in areas with lesser rainfall. It can be easily grown in well drained, fertile, sandy loam soil having a pH of 5.5 to 7. Red or lateritic soils are very good for its cultivation. It does not require very fertile soils and can be economically grown on soil with marginal fertility.

Habit: It is a perennial aromatic herb with annual square stems, 50cm tall, branched, the nodes are often hairy (Fig. 1).

Root: The roots are tuberous, fasciculate, upto 20cm long and 0.5 – 2.5cm thick; they are conical, fusiform, straight, orange and strongly aromatic.

Leaf: Leaves are tear drop shaped, shimmering green framing a bright purple center. The leaf color varies depending on the amount of shade. Pubescent leaves are narrowed into petioles.

Flower: Snowy bluish to pale lavender coloured flowers, arranged in whorls on a long spike like racemes. The flowers, borne on racemes are perfect; the calyx is fine toothed and deflexed in the front. The pale blue corolla is bilabiate, the lower lobes are elongated and concave. The ovary is four parted and the stigma is bilobed. The plants have four didynamous with declinate stamens where filaments unite as a sheath at base (Fig. 2 and Fig. 3).

Fig. 1. Plant

Fig. 2. Flower

Fig. 3. Seed spike

Flowering and pollination: Flowering occurs during August–October. The flowers are cross pollinated by wind or insects.

Fruit: Fruits are seedless.

Propagation: Coleus can be propagated by seeds as well as by stem cuttings.

Seed propagation: Propagation through seeds is little difficult and slow and should be used only for the breeding of new varieties. Coleus seeds are small. They should be sown onto a layer of moistened, sterile potting soil in a shallow tray and then, covered with a thin layer of fine soil. In about 15 to 20 days, the germination is completed. When the seedlings are 45 days old and have attained about 8-10cm height, they are ready for transplanting.

The viability of the seeds being very poor (8-10%), a sufficient quantity of fresh seeds has to be sown in well prepared nursery beds to obtain good germination.

Vegetative propagation: Propagation by rooted terminal cuttings is easy and economical to raise this crop on a large scale. Normally, 10 to 12cm long cuttings, comprising of three to four pairs of leaves, are sown in already prepared nursery beds and regular care about shading and watering has to be taken. The cuttings establish well in the nurseries and there is no problem in their rooting. After about a month's time, when the cuttings have produced sufficient roots, they are transplanted to the main field.

Economic part used: Fresh roots

Alkaloids: Chief alkaloid found is Forskohlin. Many other diterpenoids as deacetyl forskolin, 9-deoxy forskolin, 1, 9-deoxy forskolin, 1, 9-dideoxy- 7-deacetyl forskolin have been isolated. Other minor phytochemicals are allylroyleanone, barbatusin, plectrin, plectirinon A, acetoxycoleosol, coleol, coleonone, coleosol, deoxycoleonol, crocetin dialehyde, napthopyrones.

Medicinal uses: Plant pacifies vitiated pitta, fever, burning sensation, inflammation, muscular spasm, hypertension, diabetes, cardiac debility, allergy, anaphylaxis, high cholesterol and bronchial asthma.

2.12. *Dioscorea composita*

Common name: Medicinal Yam

Botanical name: *Dioscorea composita* Hemsl.

Family: Dioscoreaceae

Vernacular names : Chupri alu, Khamalu (Hindi); Peru valli kilangu, Kappan Kaccil (Tamil).

Origin and distribution : It is native to tropical Asia, Africa and South America.

Habitat : This species commonly prefer to grow on low lying areas. The crop can be grown as an irrigated crop in Andhra Pradesh in light and medium textured, deep soil rich in organic matter.

Habit : Vines are robust, perennial, branched, climbers, nearly glabrous and right twining.

Rhizome: Tubers are large, white and deep rooted.

Leaf : The vines produce alternate leaves having long petiole, membraneous or coricaceous lamina measuring upto 2.0 x 18.0cm, abruptly acute or cuspidate - acuminate, shallowly or deep cordate, 7-9 nerved.

Flower : The fasciculate glomerate inflorescence is single or branched with two or three sessile male flowers having fertile stamens. Male fascicle is 15-30cm long. The female flowers have bifid stigma.

Propagation : By seeds, cuttings and rhizomes.

Seed propagation : Anamariaviana and Felippe (1990) observed the effects of storage on germination seeds. They observed that the initiation of germination was delayed in freshly harvested seeds for about nine months.

Vegetative propagation

Single node leaf cuttings: This consists of a single leaf with petiole and about 0.8cm of the stem. The cuttings are quick dipped in 5000ppm IBA and planted in mist chamber for rooting. After about 8-10 weeks, the plants are transferred to polythene bags. After 5-6 months, they are planted at a spacing of 45 x 30cm during June- July or September-October.

Rhizomes: Tubers give three kinds of pieces namely crown (low alkaloid content), medium and tips (high in alkaloids). Crown (stem ends) are used for planting as they sprout early. Rhizomes each piece weighing 50-60g with one or two buds can also be used for planting. A pre-planting dip in carbendazim 2g/ lit for 10 minutes is essential. After 30 days, the crowns sprout and are ready for planting.

Economic part used: Fresh tubers

Alkaloids: Diosgenin, yamogenin, botogenin and kryptogenin.

Medicinal uses: Anti-inflamatory, anabolic, antifertility, cures arthritis, rheumatism and asthma.

REFERENCES

Anamariaviana and Felippe, G. M. 1990. Effect of storage on germination of *Dioscorea composita* seeds. *Econ. Botany*, 44(3):311-317.

2.13. *Dioscorea deltoidea*

Common name: Medicinal Yam

Botanical name: *Dioscorea deltoidea* Wall. Ex. Kunth.

Family: Dioscoreaceae

Vernacular names: Chupri Alu, Khamalu (Hindi); Peru valli kilangu, Kappan kaccil (Tamil).

Origin and distribution : It is native to tropical Asia, Africa and South America.

Habitat : This species commonly prefer to grow on low lying areas. The crop can be grown as an irrigated crop in Andhra Pradesh in light and medium textured, deep soil rich in organic matter.

Habit : Glabrous robust, perennial, branched and leaf twining vines bear alternate petiolate leaves. The petioles are 5-10cm long.

Rhizome: Rhizome is lodged superficial, horizontal, tuberous, digitate and chestnut brown in color.

Leaf : The lamina is 5 -15cm long and 4 -12cm wide, widely cordate.

Flower : Flowers are borne on axillary spikes, male spikes 8 - 40cm long and stamens are six. Female spikes are 15cm long, 3-5cm broad and 4-6 seeded.

Seed : Seeds are winged all round.

Propagation: By seeds, cuttings and tuber pieces

Economic part used: Fresh tubers

Alkaloids: Diosgenin, yamogenin, botogenin and kryptogenin.

Medicinal uses: It is an anti-inflammatory, anabolic, antifertility, cures arthritis, rheumatism and asthma. It also cures leprosy and piles (Gogoi and Islam, 2009).

REFERENCE

Gogoi, M. and M. Islam. 2009. Certain endangered and threatened ethnobotanically important plants of Sivasagar district, Assam. *Ad. Plant Sci.* 22(1): 205-207.

2.14. *Dioscorea florifunda*

Common name: Medicinal Yam

Botanical name: *Dioscorea florifunda* Mart and Gal.

Family: Dioscoreaceae

Vernacular names : Chupri Alu, Khamalu (Hindi); Perum valli kilangu, Kappan kaccil (Tamil).

Origin and distribution: The origin of this plant is Central America. It is grown in states like Karnataka, Assam, Meghalaya, Andaman and Goa.

Habitat : This species commonly prefer to grow on low lying areas. Well drained clay loam soil is suitable. Can be grown in tropical and subtropical climatic conditions.

Habit: It is a dioecious robust, perennial, branched herb. The vines are glabrous and left twining (Fig.1).

Rhizome: Tubers are thick with yellow coloured flesh, branched and grows upto a depth of 30cm.

Leaf : Leaves are borne alternatively on slender stem. They are broadly ovate or triangular ovate, shallowly or deeply cordate, coriaceous lamina with nine nerves. The petioles are 5-7cm long, thick and firm.

Flower : Male flowers are solitary and rarely in pairs. Female flowers have divericate stigma which is bifid at apex (Fig. 2).

Fruit : Capsule is obovate and seed winged all round (Fig. 3).

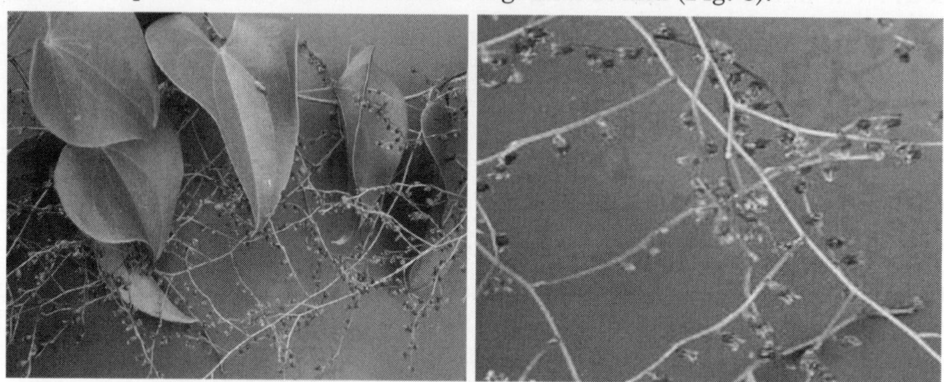

Fig. 1. Plant Fig. 2. Flower

Fig. 3. Fruit

Propagation : By tuber pieces, stem cuttings and seeds.

Micro-propagation: Segments of male inflorescences cultured on Murashige and Skoog (MS) medium supplemented with 13.94mM kinetin (Kn) resulted in the conversion of floral buds into vegetative buds and these later developed into plantlets. Growth and multiplication of shoots could be obtained by culturing individual shoots in MS modified basal medium, replacing the MS standard three vitamins with 10.0mg thiamine in addition to 13.94mM Kn. Root induction was also obtained simultaneously from the base of the shoots in the same medium. Such plantlets have been successfully transferred to potted soil, where they grew normally (Fig. 4). Plantlets were also made to develop tubers on MS medium with 18.91mM abscisic acid (ABA) and also with 2.68mM α-naphthalene acetic acid (NAA) and 40–50g sucrose (Borthakur and Singh, 2002).

Economic part used: Fresh tubers

Alkaloids: Diosgenin (3.0 to 3.5%), yamogenin, botogenin and kryptogenin.

Medicinal uses: It is used as a anti-inflammatory, anabolic, antifertility and also cures arthritis, rheumatism and asthma.

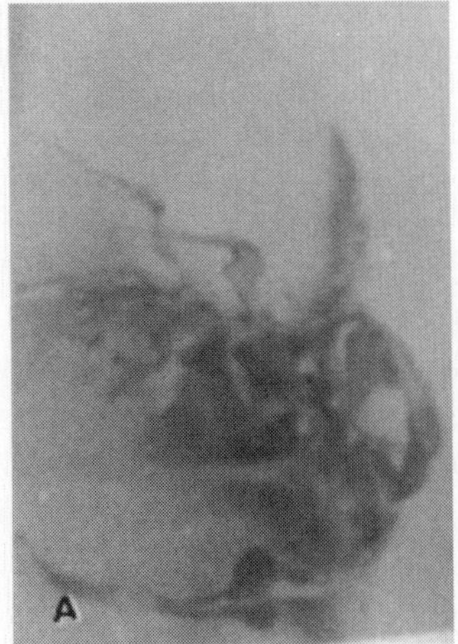

A. Vegetative shoots emerging from the axil of bracts after six weeks of culture

B. Formation of well developed shoots

C. Regenerated plants established in potted soil

D. Formation of tubers at the bases of rooted shoots

Fig. 4. Regeneration of plantlets in *Dioscorea floribunda*.

REFERENCE

Borthakur, M. and R. S. Singh 2002. Direct plantlet regeneration from male inflorescences of medicinal yam (*Dioscorea floribunda* Mart. & Gal.). *In Vitro Cell. Dev. Biol. - Plant* 38: 183–185.

2.15. *Eclipta alba*

Common names: Trailing eclipta, False daisy

Botanical name: *Eclipta alba* (L.) Hassk. Syn. *Eclipta prostrata* L., *Eclipta erecta, Eclipta punctata, Verbesina prostrata*

Family : Asteraceae

Vernacular names: Kehraj (Assamese); Bhringaraj, Kesharaj, Bhangra (Hindi); Ajagara, Garuga (Kannada); Kannuni (Malayalam); Uchi sumbal (Manipuri); Maka (Marathi); Kesarda (Oriya); Bhringraj (Sanskrit); Karisilanganni, Karisalai, Kaikassi, Karippan, Kavanthakara, Thegaraasam (Tamil); Galagara (Telugu).

Origin and distribution: It is native to United States and tropical areas worldwide. It is widely distributed throughout India, China, Thailand and Brazil.

Habitat: It is a hardy crop and can be grown on varied types of soil. Soils with high moisture content are preferred. Red loamy soils rich in organic matter are best for its cultivation. The crop comes up well in tropical, sub-tropical and temperate regions. However, it prefers warm climate with a temperature range of 25°C to 35°C for its good growth and yield. In paddy growing areas of India, it occurs as common weed.

Habit: It is an annual herb (Fig. 1).

Root: Root well developed, cylindrical, greyish

Stem: Stem is 50cm tall, single from base, but with many spreading branches, from fibrous roots, strigose, herbaceous, sub-succulent, erect or ascending, often rooting at lowest nodes, terete, purplish in strong sun.

Leaf: Opposite, sessile, lanceolate, shallow serrate, 13cm long, 3cm broad, strigose, acuminate.

Flower: Axillary and terminal clusters of 1-3 flower heads. Peduncles 2cm long, densely antrorse strigose. The flowers are protrandrous in nature.

Involucral bracts - Typically a single series of 8-10 unequal phyllaries. Phyllaries to 7mm long, 2.3mm broad, antrorse strigose externally, glabrous internally, green, acute (Fig. 2).

Ray flowers - Fertile, pistillate. Ligules 2mm long, 5mm broad, notched or rounded at apex.

Disc flowers - Disc 4-10mm broad. Corolla tube 0.9mm long, four lobed, glabrous, white. Lobes acute, 0.3mm long. Stamens four. Anthers purple, 0.5-0.6mm long included connate around style. Style whitish at apex (stigmas). Receptacle flat. Chaff to 3mm long, translucent yellow, antrorsely barbellate (Fig. 3).

Flowering, pollination and fruiting: Flowering starts about two months after the seedlings appear and the plant completes its life in about three to four months. Under Indian conditions, the flowering is throughout year and fruiting freely. The flowers are showy and are borne in aggregates of capitular heads for pollinator attraction. The ray florets open before the disc florets to enhance the chances of cross pollination. The flower opening is asynchronous which ensures pollen availability for a longer time to affect pollination and possibilities of higher fruit and seed set. The anthesis starts at morning and florets remained open till the stigma shrivels. The outer ray florets opens first so as to enhance the chances of cross pollination from pollens of disc florets of other flowers than the pollens from disc florets of the same capitulum. The white flowers are not visited by many insects except thrips and ants and these may carry out the pollination suggesting entomophilous pollination. Small *Thrips tabaci* insects are often seen visiting all florets of a head one by one going down to the corolla tube and comes out to visit the next one.

Fruit : Fruit is achene, four angled, white, 2.5mm long, warty glabrous. Pappus is absent to a minute crown. Fruits are many seeded (Fig. 4).

Fig. 1. Plant

Fig. 2. Flower

Fig. 3. Disc flower

Fig. 4. Seed

Seed : Seeds are black and resemble cumin seeds.

Propagation : The crop can be propagated either through seeds, root suckers and cuttings.

Seed propagation: A single plant can produce as many as 17,000 seeds; germination affected by light, moisture level, pH and temperature, but seeds have no dormancy. Matured seeds can be collected from well grown plants. Then, the seeds are dried and soaked in water for 24 hours. The seeds are sown in raised beds during rainy season, since water logging would be injurious to the seedling. In other seasons, the seeds are sown in sunken beds also. In 6-7 days, 80-90 per cent germination is obtained.

Seed dispersal : Ridley (1930) mentions the following method of dispersal of seeds for the species. The surface of the achene is papillose and the fruit is sticky due to mucilagenous exudation. Thus, the dispersal is possible by its adherence to the plumage and feet of birds. The human agency is also helping its dispersal either by adhesion to feet or cloths or to importation with other plants. He further mentions of seed drifts in water during floods though not quit satisfactorily as in case of buoyant seeds.

Seed dormancy and germination

According to Dhawan Rupa (2007), *Eclipta alba* seeds germinated immediately after harvest and showed no dormancy. Light is essential and 25°C was seen to be the optimal temperature for germination. Seeds did not germinate beyond a water stress of −0.4 MPa. Low levels of salinity (0–120mM NaCl) did not affect germination. Acidic pH completely inhibited germination. Only seeds at the surface layers of soil germinated. Those at a depth of even 2.5cm did not germinate.

Seeds were strongly photoblastic with none germinating in the dark. Seeds germinated in light intensities between 6 and 100 per cent of full light. Seeds germinated 78 to 88 per cent in water potentials of 0.0, -0.1, and -0.2MPa. In distilled water (pH 5.5) and a buffered solution pH 6 and 7, seeds germinated 85 to 89 per cent; however, some seeds germinated in a buffered solution pH range of 5 to 8.

According to Ferreira *et al.* (2001), the optimum temperature for germination was 30°C and light was must for germination.

Seed germination was completely inhibited in the dark, whereas in the light/ dark it was 76, 93, and 87 per cent at 25/15, 30/20, and 35/25°C alternating day/ night temperatures, respectively. Germination was greater than 80 per cent upto a temperature of 140°C, when seed were placed in an oven for 5min. followed by incubation at 30/20°C for 14d, but declined progressively with a further increase in exposure temperature with no germination at 200°C. Seed germination was tolerant of salt stress, but highly sensitive to water stress. Seed germinated (87 to 93%) over a pH range of 4 to 10. Seedling emergence was greatest (83%) for the seed placed on the soil surface, but declined thereafter and no seedlings emerged from a depth of 0.5cm (Chauhan and Johnson, 2008).

Ramakrishnan (1960) reported that the seeds are found to have no dormancy period, but the germination percentage increases with storage (Table 1).

Table 1. Germination of seeds of *E. alba* after dry storage for different periods

Month of storage	Percentage germination
0	4
2	16
5	62
12	62

The maximum germination of 82 per cent is obtained in continuous light which may be partly due to the warm atmosphere around the electric bulb and also due to the direct effect of light itself. Very poor germination is obtained in the seeds that are kept in continuous darkness. Fresh seeds buried under mud for four months gave a germination of 72 per cent which is greater than that obtained from dry stored seeds.

Vegetative propagation

For vegetative propagation, the root suckers are planted in small pits, cover with loose soil and allow for sprouting. Then, the seedlings at four leaves stage are to be planted.

This can be propagated by using terminal cuttings with 5-6 nodes and 10-15cm length. They are planted in well prepared nursery beds or polythene bags. In about 4-6 weeks time the rooting is complete and they are ready for transplanting into the main field.

Economic part used : Leaves and roots

Alkaloids: It contains mainly coumestans *i.e.*, wedelolactone (I) and demethyl wedelolactone (II), polypeptides, polyacetylenes, thiophene derivatives, steroids, triterpenes and flavonoids. Coumestans are known to possess estrogenic activity.

Medicinal uses : The plant has a better sharp dry taste, good for the complexion of the hair, the eyes, the teeth; cures inflammation, eye diseases, jaundice, bronchitis, asthma, leucoderma, anemia, diseases of heart and the skin, itching, night blindness, syphilis, used to prevent abortion, miscarriage and uterine pain after delivery.

References

Chauhan B. S. and D. E. Johnson 2008. Influence of environmental factors on seed germination and seedling emergence of Eclipta (*Eclipta prostrata*) in a tropical environment. *Weed Sci.,* 56(3): 383-388.

Dhawan Rupa, S. 2007. Germination potential and growth behaviour of *Eclipta alba. Indian J. Weed Sci.,* 39(1&2): 116-119.

Ferreira, A. G., B. Cassol, S. G. T. Da Rossa, T. S. Da Silveira, A. L. Stival and A. A. Silva 2001. Germination of seeds of *Asteraceae* native of Rio Grande do Sul, Brazil. - *Acta Bot. Bras.* 15: 231-242.

Ramakrishnan, P. S. 1960. Ecology of *Eclipta alba* hassk. *Proc. Nat. Inst. Sci. India*, 26, B(4): 191-204.

Ridley, H. N. 1930. The dispersal of plants throughout the world, L. Reeve and Co., Ltd., Ashford, Kent.

2.16. *Euphorbia hirta*

Common names: Asthma weed, Cats hair, Common spurge, Pill-bearing spruce, Snake weed

Botanical name: *Euphorbia hirta* L. Syn. *Euphorbia pilulifera*

Family: Euphorbiaceae

Vernacular names: Barokarni (Bengali); Bara dudhi, Lal Dudhi (Hindi); Achchedida (Kannada); Dudurli (Konkani); Nelapalai (Malayalam); Pakhamba maton (Manipuri); Dudhi (Marathi); Amman Pacharisi (Tamil); Nanabalu (Telugu).

Origin and distribution: It is native to Central America and a very common weed of the tropics and subtropics; it occurs throughout tropical Africa and also in South Africa.

Habitat: Most common dry land weed, flowering and fruiting all round the year on roadsides and waste lands.

Habit: It is a slender stemmed, annual hairy plant with many branches, growing upto 40cm tall, reddish or purplish in color (Fig. 1).

Stem: Round and covered yellow crisp hairs; it is usually solitary and bent backwards at the top in characteristic way.

Leaf: Leaves are opposite, elliptic-oblong to oblong lance like, 1-2.5cm long, blotched with purple in the middle, toothed at the edge.

Flower: Flowers, purplish to greenish in color, dense, axillary, short stalked clusters or crowded cymes, about 1mm in length. Flowers are monoecious; male several surrounding a single female in each cyathium; limb of stalked glands minute; stamen one; ovary superior, tricarpellary syncarpous; cocci appressed, pubescent (Fig. 2).

Flowering, pollination and fruiting: Flowering and fruiting occurs during August to April. Cross pollination is necessary because of unisexual flowers. Certain characters (coloured bracts, well developed nectarines and glands) also favour insect pollination in some plants (Sharma, 1993).

Fruit: Fruit is berry and broadly ovoid, hairy, three angled, about 1.5cm.

Seed : Seeds reddish with shallow furrows. Seeds are minute, quadrangular bluntly pointed with 5-6 shallow transverse furrows. The average weight of the seed is 0.13g. It produces upto 3000 seeds per plant.

Fig. 1. Plant Fig. 2. Flower

Seed dispersal: It is observed that maximum number of seeds are dispersed to a distance of about 5 to 25cm around the plant. In nature, the seeds are dispersed by the bursting of the capsules and further be carried away by wind currents. Ants also may be dispersal agents of seeds.

Propagation : It is propagated by seeds.

Seed dormancy and germination : Maximum germination of 48 per cent is obtained in diffused daylight. In continuous darkness, germination percentage is low and in continuous light, it is very poor. It shows a germination rate of upto 92 per cent. Seeds can be dried without damage to low moisture contents, usually much lower levels than normally reached in nature. Over a wide range of storage environments, their longevity increases with reduction in both moisture content and temperature.

Economic part used: Seeds, leaves

Alkaloids: Sterols, alkaloids, tannins, glycosides, triterpenoids, alkenes, phenolic acids, choline, shikimic acid.

Medicinal uses: Relaxation of bronchioles, depressant action on heart and respiration, removal of worms in children in bowel complaints, asthma and cough, promotes formation and flow of milk in women.

Past work done on Seed Science and Technology

Development of embryo and seed

Kajale (1954) reported that fertilization is porogamous. The pollen tube passes in between the wall of the embryo sac and of the synergids. Generally one and sometimes both the synergids are destroyed. The polar nuclei increase in size prior to triple fusion.

The pro-embryo consists of three to four cells and is T shaped. The small suspensor is multiseriate towards the embryo side and persists in the mature seed.

The primary endosperm nucleus divides earlier than the oospore. The free nuclear endosperm later becomes cellular. Wall formation starts at the periphery and extends towards the centre. One to four nucleoli are present in the nuclei of the endosperm cells. The endosperm surrounds the embryo except at the tip of the root cap.

The hypostase persists in the mature seed as perisperm and serves as an organ of storage.

The testa consists of four layers of cells. The inner most layer representing the outer epidermis of the inner integuments becomes brittle and stony. Next come three more layers derived from the outer integument. Granular deposits are present in the epidermal cells of the testa which have a finely warty appearance. There is a difference in the size and shape of the cells of the different layers.

REFERENCES

Kajale, L. B. 1954. A contribution to the embryology of the Phytolaccaceae II. Fertilization and the development of embryo, seed and fruit in *Rivina humilis* Linnaeus and *Phytolacca dioica* Linnaeus. *J. Indian Bot. Soc.*, 33: 206-225.

Sharma, O. P. 1993. Plant taxonomy. Tata McGrow-Hill Publishing Company Ltd., New Delhi. pp. 483.

2.17. *Lippia nodiflora*

Common names: Rog fruit, sawtooth fogfruit, turkey tangle.

Botanical name: *Lippia nodiflora*

Family : Verbenaceae

Vernacular names: Bukkan (Hindi); Poduthalai, Poorsatham (Tamil).

Origin and distribution: Native to tropical America and Europe.

Habitat: It grows best on bared ground in periodically moist sites (heavy rainfall, flooding or opening of springs) that also undergo periods of water stress

(Richardson, 1994) and with apparently poor soil structure. It is found mostly on high moisture holding clay to clayloam soils in floodplains or wetlands.

Habit: Prostrate herb widely creeping and rooting, minutely hairy (Fig. 1).

Roots : Roots are produced from leaf axils along stems and consist of a central taproot, usually 50–60cm in length, but thought to extend upto 2m (NWWCC 2002). An extensive fibrous root system extends from the taproot. Lippia also produces fibrous roots from some stem nodes.

Stem: Stems lay along ground and root.

Leaf: Green to blue-green; oval, 2cm long; jagged margin.

Inflorescence: The inflorescence consists of a purple coloured centre encircled by small white to pink flowers (Fig. 2). The flower takes on a match like look, which is why the plant is sometimes called matchweed.

Flower: Bracts broadly obovate, apiculate; calyx membraneous two to four lobed later two valved; corolla small salver shaped two tiped upper lip, bifid lower three lobed; stamens four, didynamous; ovary bilocular, single ovule in each locule (Fig. 3).

Flowering, fruiting and pollination: Lippia flowers anytime in the spring-summer-autumn period given favourable soil moisture conditions (Lucy *et al.*, 1995; Dellow *et al.* 2001). Flowering takes place during May-October. Flowering can be triggered by rain or flooding at any time during this period (Lucy *et al.*, 1995). Lippia can self- and cross-pollinated (McCosker, 1994) resulting in a high percentage seed set and maintenance of sufficient genetic variation to adapt and rapidly colonise new environments (Kumar and Dutt, 1989).

Fruit: Capsule 1.5-2mm long, breaking into two segments.

Seed: Dehiscent fruits (1–1.5mm in diameter) release two tiny brown, oval, flattened seeds at maturity (Fig. 4) (Lucy *et al.*, 1995). Seeds are barely visible to the naked eye and exalbuminous.

Propagation: Through seed. Vegetative reproduction occurs *via* production of roots at stem nodes, and can occur while the plant grows under water. When inundated with shallow flood waters, lippia stems grow towards the water surface and become shortened in their stem sections, which then break off as the nodes weaken. The detached fragments float and can remain dormant before becoming established on suitable soils again when floodwaters recede (McCosker, 1994).

Seed dispersal: Seeds are initially dispersed below the parent plant with secondary dispersal *via* floodwaters, probably by floating on the water surface and germinate when they are deposited in silt by the declining waters (McCosker, 1994). Other dispersal agents include ants (McCosker, 1994) and possibly birds (SQDNR 2000). Vegetative dispersal may be assisted *via* the displacement of plant material (e.g. muddy stem fragments stuck on the feet of animals) by machinery, cattle, sheep and birds and is especially likely to occur in wet conditions (McCosker, 1994).

Fig. 1. Plant

Fig. 2. Inflorescence

Fig. 3. Flower

Fig. 4. Seed

Seed dormancy and germination: It has the ability to remain dormant when moisture reserves become depleted and can withstand long dry periods. It can also regrow quickly after rain events, allowing it to compete with other species. Germination can occur under a wide range of conditions. Estimations of the viable period of stem fragment dormancy vary from at least six to eight weeks (Lucy *et al.*, 1995) upto three months (Dellow *et al.,* 2001).

Soil seed bank: Viable seed bank remains around a parent plant even after the plant's removal (McCosker, 1994; Lucy *et al.*, 1995). Seeds of lippia may possess an effective drought survival mechanism involving enhanced germination from the seed bank as a result of alternate drying and wetting (McCosker, 1994).

Agricultural importance: Lippia may reduce seed set in commercial crops *via* pollinator competition. Lippia has allelopathic effects on other plant species (McCosker, 1994). One study showed that lippia leachate inhibited the growth of lettuce seedlings and the researchers correlated this by the presence of potential allelopathic compounds that have been isolated from the plant (Elakovich and Stevens, 1985).

Economic part used: Leaves and pods

Uses

It is used as a lawn and garden ornamental in the nursery industry and it is sold as shade-loving lawn replacement species.

It produces many flowers that are very attractive to bees. Bees are able to extract large quantities of nectar from lippia flowers, which is said to produce a pleasant tasting honey. Because of this and the fact that lippia is very easy to grow, the plant is favoured by honey producers (Richardson 1994).

REFERENCES

Dellow, J. J., K. Motley, A. Storrie and J. Spenceley 2001. *Lippia*, Agfact P7.6.52, NSW Agriculture.

The State Queensland Department of Natural Resources 2000. Better management practices manual, Department of Natural Resources, Upper Condamine Floodplain, Natural Heritage Trust and Landcare Queensland, avai-lable online at <http://www.nrm.qld.gov.au/community/condaminefp/bmp-manual.htm>.

Elakovich, S. D. and K. L. Stevens 1985. Volatile constituents of *Lippia nodiflora, J. Natural Products*, 48; 504–506.

Kumar, P. and B. Dutt 1989. Cytogenic basis of breeding system in some Verbenaceous species, *Cytologia*, 54; 347–353.

Lucy, M., E. Powell, R. McCosker, G. Inglis and R. Richardson 1995. *Lippia (Phyla canescens):* A review of its economic and environmental impact on floodplain ecosystems in the Murray Darling Basin, Department of Primary Industries, Brisbane.

McCosker, R. O. 1994. *Lippia* (*Phyla nodiflora*): An invasive plant of floodplain systems in the Murray Darling Basin, a report on the distribution and ecology of lippia in the lower Gwydir Valley and the Murray Darling Basin prepared for the Gingham Watercourse Landcare Group, Department of Ecosystem Management, University of New England, Armidale, NSW.

North West Weeds County Council (NWWCC). 2002. *Lippia*, available online at <http://www.northwestweeds.nsw.gov.au/Lippia.htm>.

Richardson, R. 1994. *Lippia* (*Phyla nodiflora*): its distribution and impact throughout the eastern mainland states of Australia, a postgraduate report, Faculty of Applied Science, The University of Queensland, Gatton College, Lawes.

2.18. *Mucuna pruriens*

Common names: Cow itch, Itch bean, Cowage

Botanical name: *Mucuna pruriens* (L.) DC. Syn. *M. prurita*

Family: Fabaceae

Vernacular names: Kiwach, Atmagupta, Cowhage (Hindi); Poonai kaali, Kanduthi, Markadi (Tamil); Tita-kogila, Pilliadugu (Telugu).

Origin and distribution: It is native to India and widely distributed all over India including Andaman & Nicobar Islands.

Habitat: Tropical and sub-tropical climates are suitable. It is sensitive to frost and frost prone areas and can withstand drought. A wide range of soils are suited for this species, however well drained sandy to clay loam soils are best suited.

Habit: Herbaceous twining annual with long slender branches covered with short hair (Fig.1).

Leaf: The leaves are tripinnate, ovate, reverse ovate, rhombus shaped or widely ovate. The sides of the leaves are often heavily grooved and the tips are pointy. In young plants, both sides of the leaves have hairs. The stems of the leaflets are two to three millimeters long. Additional adjacent leaves are present and are about 5mm long.

Flower: Large, short stalked, purple corolla, flowers are elongate, 6-30 flowered racemes of 15-30cm long. The flowers are solitary or 2-3 together along a slender silky rachis. Pedicel 3-6mm long, hairy; bracts 1-2cm long, lanceolate, hairy; bracteoles 8mm long, hairy, calyx 1cm long, silky and with a few irritant bristles outside. Upper teeth completely connate into a triangular lip equaling the tube. Lateral teeth lanceolate as long as the tube. Lower tooth lanceolate and slightly longer (Fig. 2).

Flowering, pollination and fruiting: Flowering occurs from November to January followed by fruiting. It is self pollinated.

Fruit: Pods are 5-7.5/1.2cm, turgid with a longitudinal rib running the length of each valve, not winged on the margins nor transversely plaited on the faces, falcate curved on both ends, somewhat like the letter S, densely clothed with persistent irritant bristles which are at first pale brown, afterwards steel grey. The husk is very hairy and carries upto seven seeds (Fig. 3).

Seed: The funicular hilum is about 4mm long and is surrounded by a distinctive white aril. The seed coat is hard, thick, black and glossy. The dry weight of the seeds is 55 to 85g/100 seeds (Fig. 4).

Fig. 1. Plant

Fig. 2. Flower

Fig. 3. Pods

Fig. 4. Seed

Propagation: This can easily be propagated through seeds.

Seed propagation: The fresh seeds are collected during winter and soak them in water for 24 hours. The seeds are sown in raised beds or polybags in the nursery. Sowing is done preferably during May-June or at the onset of monsoon for better growth. In about 4 to 12 days, 90 per cent of the seeds germinate.

Pre-sowing seed treatment: Dormancy in the freshly harvested seeds was reported by Yogesha and Shivananda (2001). They suggested that the seeds soaked in conc. H_2SO_4 for 90 minutes resulted in very high germination (74%) and seed coat abraded by rubbing on hard rough cement surface or using sand paper also resulted in high germination (82 and 86 per cent respectively) against the control with 10 per cent germination.

Economic part used: Seeds and roots.

Alkaloids: L-DOPA (L-3,4 di hydroxyl phenyl alanine), glutathion.

Medicinal uses: It cures Parkinsons disease, roots are used as diuretic, tonic, stimulant, purgative, emmenagogue, anthelmintic.

Past work done in Seed Science and Technology

Jeeva (2003) carried out the experiments to evolve a suitable seed production technology and to develop an effective seed storage method in *Mucuna pruriens* L. The findings of this study are as follows.

(a) Seeds attained physiological maturity on 45 days after anthesis with maximum seed dry weight, germination and vigour with minimum seed moisture.

(b) The seeds collected from first two pickings recorded higher seed quality, germination and vigour than third and fourth pickings. But there was not much reduction in seed quality in the late pickings. Hence, seed collection from all the four pickings could be used for seed purpose.

(c) The results of seed positional polymorphism indicated that middle position seeds gave the highest values for 100 seed weight, germination and vigour than those of proximal and distal position of seeds. The seed quality from the vertical pods was better than the seeds from horizontal pods.

(d) Among the orientations of seed sowing, seeds sown micropylar end sideward at 3cm depth registered higher germination and vigour than micropylar end upward, downward and flat orientations.

(e) The seeds treated with captan @ 2g kg^{-1} of seed and stored in 700 gauge polyethylene bag maintained better seed quality of 87 per cent germination upto 10 months of storage, under ambient conditions.

REFERENCES

Jeeva, B. 2003. Seed development and maturation, picking and storage studies in *Mucuna pruriens* L. M.Sc (Ag.) Thesis, TNAU, Coimbatore.

Yogesha, H. S and Shivanandha 2001. Seed dormancy studies in *Mucuna pruriens* (L.) DC. Proc. of National seminar on Seed Science and Technology, p. 42-43

2.19. *Nervilia aragoana*

Common name: Tall Shield Orchid

Botanical name: *Nervilia aragoana* Gaudich. Syn. *Aplostellis aragoana* (Gaudichaud-Beaupré) Ridley; *Aplostellis flabelliformis* (Lindley) Ridley; *Epipactis carinata* Roxburgh; *Nervilia carinata* (Roxburgh) Schlechter; *Nervilia flabelliformis* (Lindley) Tang & Wang; *Nervilia scottii* (Reichenbach f.) Schlechter; *Nervilia tibetensis* Rolfe; *Nervilia yaeyamensis* Hayata; *Pogonia carinata* (Roxburgh) Lindley; *Pogonia gracilis* Blume; *Pogonia flabelliformis* Lindley; *Pogonia nervilia* Blume; *Pogonia scottii* Reichenbach f.

Family: Orchidaceae

Vernacular names: Sthala padma, Sthala kamal (Hindi); Orila tamara, Nila tamara, Kal tamara (Malayalam); Padma karini (Sanskrit); Orilai thamarai (Tamil).

Origin and distribution: Global: Indo Malaysia, Pacific and Australia. National: Common in the Himalayas, Gujarat and occasionally found in the higher altitudes of Western Ghats of Karnataka, Kerala and Tamil Nadu.

Habitat: It is commonly grown in humid shady localities of dense forests and rarely in wet grassy slopes of moist deciduous forests.

Habit: Terrestrial single leaved herbs, with underground tubers.

Tuber: Spherical to egg-shaped, 1-2cm across, white, fleshy, 3-5 noded

Leaf: Leaf stalk is 10-20cm long, greenish purple, with a distinct groove in the middle. Leaves are single, round to slightly kidney shaped, 10-25cm diameter, base deeply cordate, apex rounded to acute, margin wavy and coarsely toothed, glabrous, green with purple shade on both surface and with prominent 9-15 ribs from the base.

Flower: Inflorescence 15-45cm, 4-15 flowered; peduncle green or purple, with 3-5 tubular sheaths; floral bracts refluxed, linear-lanceolate, 10-16x1-1.5mm, exceeding ovary. Flowers nodding, resupinate, usually opening widely, emitting an unpleasant odor; pedicel and ovary 7-13mm, ridged. Sepals are yellowish green, linear-oblong, 15-22 × 2.5-3mm, apex acuminate or acute. Petals yellowish green, linear-lanceolate, 12-20 × 2-2.5mm, apex acute; lip pale green, white, pinkish white, or pale green with purplish brown venation, variable in shape, ovate-oblong, obovate, or rhombic, 12-17×5-8mm, base cuneate, spurless, three lobed above middle; lateral lobes erect and loosely embracing column, triangular, 2-4×2-4mm, apices acute; mid-lobe ovate-elliptic, ovate-triangular, or sub-obovate square, 3-5×3-5mm, apical margin undulate, apex obtuse or acute; disk villous primarily on veins. Column white, 6-10mm, apex dilated; stigma sub-orbicular (Anon. 2009).

Flowering, pollination and fruiting: Flowering and fruiting during March to July. The leaves and flowers do not appear at the same time. The flowers are short lived and the inflorescence lengthens as the capsules develop. Leaves appear following fruiting when the seeds have dispersed and the flowering stem has collapsed. It is self pollinated crop.

Fruit: Fruit usually a capsule, opening laterally by slits.

Seed: Seeds are very numerous, minute, lacking endosperm, the embryo not differentiated

Propagation: Tubers

Economic part used: Tubers

Medicinal uses: Tubers used in postnatal treatments.

REFERENCE

Anonymous. 2009. Nervilia, *Flora of China* 25: 197–201.

2.20. *Nigella sativa*

Common names: Fennel flower, nutmeg flower, Roman coriander, black seed or black caraway, black cumin

Botanical name: *Nigella sativa* L.

Family: Ranunculaceae

Vernacular names: Kaljeera, Kolajeera (Assamese); Mota Kalajira, Kalajira, Kalo jeeray (Bengali); Kalonji jeeru, Kalounji (Gujarati); Kalonji (Hindi and Urdu); Krishna Jeerige (Kannada); Karim jeerakam (Malayalam); Kalaunji jire, Kalejire (Marathi); Kalvanji (Punjabi); Susavi (Sanskrit); Karum cheerakam (Tamil); Peddajila karra (Telugu).

Origin and distribution: It is native to south and southwest Asia.

Habitat: The plant is indigenous to the Mediterranean region, but now found widely in Jammu and Kashmir, Himachal Pradesh, Bihar, Assam and Punjab. The herb is also cultivated in Bengal and north-east India.

Habit: It is an annual flowering plant. It grows to 20 - 30cm tall (Fig.1).

Leaf: Finely divided, linear leaves.

Flower: The flowers are delicate and usually coloured pale blue and white with 5-10cm petals (Fig. 2).

Flowering, pollination and fruiting: Flowers appear in early winters and fruits in winters. Four orders of insect pollinators were captured by the insect sweep net technique. These orders were Hemiptera, Coleoptera, Diptera and Hymenoptera. Most abundant species were related to orders Diptera and Hymenoptera. Daily peak activity of caught insect species on flowers was detected at 12 noon and 2pm. Visits of honeybees increased the number of seeds set and yield production (Abd El-Wahab and Ebadah, 2011).

Fruit: The fruit is a large and inflated capsule composed of three to seven united follicles, each containing numerous black seeds (Fig.3a and 3b).

Seed : Seeds trigonous, regulose-tubercular and black in colour (Kirtikar and Basu, 2006) (Fig.4).

Propagation : Seed

Economic part used : Seed

Alkaloids: Carvone, Alfa - pinene, Beta - pinene, P - cymene

Fig. 1. Plant

Fig. 2. Flower

Fig. 3a. Fruit (Mature capsule)

Fig. 3b. Fruit (Dry capsule)

Fig. 4. Seed

Medicinal uses : Since ancient days, black seed is used in folk medicine as diuretic and hypotensive agent. Another important factor associated with black seeds, its effect on human immune system which combats diseases that are product of defective immune system. It contains essential oils which provide health benefit. One such important activity is its antioxidant property which makes it useful in treating cardiovascular disorders. The nutritional value of black seed is mainly due to rich constituents of proteins (21%), carbohydrates (35%) and fats (35-38%). The carbohydrates present in black seeds are mainly monosaccharides in the form of glucose, rhamnose, xylose and arabinose which are useful as source of dietary fibre.

REFERENCES

Abd El-Wahab, T. E. and Ebadah, I. M. A. 2011. Impact of honeybee and other insect pollinators on the seed setting and yield production of black cumin *Nigella sativa* L. *J. Basic. Appl. Sci. Res.*, 1(7): 622-626.

Kirtikar, K. R. and B. D. Basu 2006. Indian Medicinal Plants Vol. I. International Book Distributors, Dehradun.

2.21. *Papaver somniferum*

Common name: Opium poppy, Joy plant

Botanical name: *Papaver somniferum* L.

Family: Papaveraceae

Vernacular names : Khas khas, Afim (Hindi); Kasa kasa (Tamil).

Origin and distribution: It is native to East Mediterranean region and introduced into India in the early sixteenth century. Since ancient times, its cultivation is going on in Italy, Greece and Minor Asia. At present, it is being cultivated mainly in India, Russia and Turkey. It is also grown in Yugoslavia, Bulgaria, Afghanistan and Japan in a limited scale. Central Europe grows this crop for seed production only, but since 1930, it is being cultivated for production of seedless capsules from which opium is obtained. In India, it is cultivated under a license issued by the excise department as such, its cultivation is restricted to about 24,000ha in the districts of Neemuch, Mandsaur and Ratlam in Madhya Pradesh; Faizabad, Barabanki, Bareilly and Shahjahanpur in Uttar Pradesh and Chittoor, Jhalawar and Kota in Rajasthan (Gupta, 1997).

Habitat: It is cultivated on large scale at altitude of 1500m - 2100m. It prefers moderately cool weather and open sunny locations; severely cold spell, frost, dull cloudy weather, high winds and very heavy rainfall during the lancing period adversely affect the yield. The crop is grown on well drained, fertile, clayey loam to rich sandy loam soils. Well drained loam and sandy loam soils are best for its cultivation. Soil having pH of seven are best for its cultivation. Soil should also have adequate quantities of organic matter and plant nutrients.

Habit: Erect, annual herb, rarely branches, growing upto a height of 60-120cm (Fig. 1).

Leaf : Leaves are ovate- oblong or linear to oblong, dentate or serrate.

Flower: Large, gorgeous, white to pink or purple flowers, flower buds are ovate and drooping, but the flowers are erect, solitary and large; sepals two, green, broad, smooth, disarticulate and are pushed away as the flower opens; petals four, decussate, the outer two are wider and much overlapping, the inner ones are slightly narrower, concave, undulated with numerous closely placed veins radiating from the stiff thick wedge shaped base, pure snow white and glossy; stamens are numerous, hypogynous, inserted in two to three rows on the under surface of the dilated thalamus; filaments are long, flat and ribbon shaped, slightly dilated at the top; anthers are linear and attached by a narrow base to the filaments, cream coloured wavy and are twisted after dehiscence; ovary is large, globular, but contracted below into a neck which again dilates to form the receptacle and also narrows off below into the pedicel. The ovary is unilocular and contains large spongy parietal placenta which bulges out nearly to the center. The placenta are almost always equal to the number of stigmatic rays and bear numerous ovules over all parts of their surface; the stigma is sessile, peltate, spreading over the top of the ovary with 8-13 short obtuse oblong rays (Fig. 2).

Flowering, pollination and fruiting

After about 90 to 100 days of sowing, the plants which are waist high begin to flower, *i.e.*, flowering will take place during first week of March, if the crop was sown during the second fortnight of November. Usually after three days of flowering, the petals fall off and after another 10-14 days the capsules are ready for lancing. The intensity of the sun's heat, especially in the flowering period and while the fruit pods are forming is of major importance for the accumulation of alkaloids in the milky juice. Under the effect of abundant moisture, the poppy sickens and the seeds grow in the pod during ripening and risk becoming completely spoiled.

It is a self pollinated crop, but a degree of out-crossing has been reported. The bees are less effective in cool at cloudy days while prefer white colour more than the purple one (Singh *et al.*, 1995).

Fruit: Capsule, large, 2.5cm in diameter, globose, stalked, the pericarp is hard, smooth, dry and brittle and brownish yellow, containing small white or black kidney shaped seeds (Fig. 3).

Seed: The seeds are numerous, very small; kidney shaped, whitish yellow, grey brown, reddish brown or black, testa with a raised reticulated network, its embryo is slightly curved in the axis of oily endosperm. One hundred seed weight is 0.25 - 0.5g (Fig. 4).

Propagation: Seeds

Seed treatment: It is necessary to protect the crop from seed borne fungal diseases. Before sowing, the seed should be treated with fungicides like dithane M45 or brassicol at about 20g of the fungicide for 5 kg of seed (4g/kg of seeds).

Economic part used: Seeds, Latex

Fig. 1. Plant

Fig. 2. Flower

Fig. 3b. Dry fruit

Fig. 3a. Mature fruit

Fig. 4. Seed

Alkaloids: There are 25 types of alkaloids identified in raw opium; of them, morphine, codeine, narcotine, narcein, thebaine, papaverin and hydrochloride are important. Morphine is the most important alkaloid.

Medicinal uses: Traditionally, the dry opium was considered an astringent, antispasmodic, aphrodisiac, diaphoretic, expectorant, hypnotic, narcotic and sedative. Poppy has been used against toothaches and coughs. The ability of opium from poppy to serve as an analgesic is well known. Opium and derivatives of opium are used in the pharmaceutical industry as narcotic analgesics, hypnotics, and sedatives. Opium and the drugs derived from opium are addictive and can have toxicological effects.

REFERENCES

Gupta, R. 1997. Medicinal and aromatic plants. **In:** Handbook of Agriculture, ICAR, New Delhi.

Singh, S. P., S. Shukla and K. R. Khanna 1995. Opium poppy. **In:** Advances in Horticulture, Vol. (11) Medicinal and aromatic plants (Eds. K. L. Chadha and Rajendra Gupta), Malhotra Publishing House, New Delhi. p. 535-574.

2.22. *Phyllanthus amarus*

Common names: Country goose berry, Stone breaker

Botanical name: *Phyllanthus amarus* Schum. & Thonn. Syn. *P. fraternus*

Family : Euphorbiaceae

Vernacular names: Bhuin amla (Assamese); Bhumamla, Bhumi amalaki (Bengali); Bhoi Amali, Bhony amari, Bhonyamali (Gujarati); Bhui amala, Jangli amli, Jar amla (Hindi); Nelanelli (Kannada); Kizanelli, Keezhanelli, Ajjhada (Malayalam); Bhuia vali (Marathi); Bhuin amla (Oriya); Lodhar (Punjabi); Bhumyamalaki (Sanskrit); Keezha nelli, Keezhkaai nelli, Keezhvaai nelli (Tamil); Nelausirika (Telugu).

Origin and distribution: Native to Asia and has spread as a weed throughout the tropics. It also occurs in tropical Africa and the Indian Ocean islands.

Habitat: It is well adapted to variety of soils, at soil pH ranging from alkaline to neutral and acidic soil. Plants have shown preference for calcareous well drained and light textured soils. Plants do not grow properly under shades. It grows under semi-temperate to tropical conditions and under high rainfall. It however, rarely survives under dry or very low temperature conditions. Water logging does not show any lethal effects on this crop. It is found mainly as a weed in waste lands, agricultural lands and riverbanks. It grows abundantly during the rainy season.

Habit: It is an annual, glabrous herb grows upto 15-60cm height (Fig. 1).

Stem: It has smooth erect cylindrical stem of 1.5 to 2mm thick and deciduous horizontal branchlets, 4 to 12cm long and about 0.5cm thick, with 15 to 30 leaves.

Leaf: Alternate, distichous and crowded along lateral branchlets, simple, glabrous, sessile; stipules ovate to lanceolate, long-acuminate, blade oblong to elliptical oblong, 7-12mm × 3-6mm, base cuneate to rounded and slightly unequal, apex rounded, often pointed, margins finely toothed.

Flower: Solitary minute flowers are unisexual. Female flower axillaries in the base of twigs, the male flowers regrouped by 2-4 in the axils of leaves in the top of twigs. Female flowers in peduncle 2mm long, with five sepals. The disc is flattened, small with five lobes. The ovary is spherical surmounted of a short trifid style. Male flowers in hail peduncle, 0.5mm long, with five elliptic sepals and three welded stamens (Fig. 2).

Flowering, pollination and fruiting: Flowering and fruiting occur twice a year *i.e.*, March - May and November - January. Cross pollination is necessary because of unisexual flowers. Certain characters (coloured bracts, well developed nectarines and glands) also favour insect pollination in some plants (Sharma, 1993).

Fruit : Capsules are dehiscent in three cells with two seeds each. The capsule is globose, depressed, and smooth, 2mm in diameter that appears to the face lower of twigs (Fig. 3).

Seed : Wedge shaped, one concave and two flat sides, 1mm in height. Seeds on the back longitudinally ribbed, on the sides concentrically ribbed, in between and over the ribs finely transversely ribbed.

Propagation : Through seeds

Seed propagation: The seeds are sown in well prepared nursery beds. Sowing in April-May will help to obtain higher germination rate. The minute seeds are mixed with sufficient quantity of fine sand to ensure uniform distribution of seeds on the nursery bed. Later, thin layer of soil is spread to cover the seeds in the nursery beds. Appropriate moisture is maintained in the beds till the seeds have germinated. Sowing in April-May gives higher seed germination and good yield.

Seed collection: The whole plant can be pulled out after maturation of the fruits. Then, the foliage is allowed for drying and seeds are separated.

Seed dispersal : The seeds are dispersed by water and animals.

Seed dormancy and germination : *Phyllanthus* is a prolific seed producer. The seeds can remain dormant for an extended period of time. They germinate massively during the rainy season and the seedlings are very fast growing. Seed requires light to germinate (Unander *et al.*, 1993; Webstar, 1957). A germination rate of upto 80 per cent was observed at temperatures of 25–35°C, but germination was poor at 20°C or 40°C. Germination of seed is also poor under moisture stress conditions (osmotic potential higher than 300kPa). A 1000 seed weight of 150mg has been reported for *Phyllanthus amarus.* Stratification significantly improved the germination of the seeds (Unander *et al.*, 1995).

Seed storage: The seeds are dried to eight per cent moisture content and treated with turmeric rhizome powder under ambient conditions maintained viability for upto six months; untreated seeds stored poorly. According to Unander *et al.* (1995), the freezing did not significantly affect the germination of dry seeds compared to seeds stored at ambient room temperature or higher, but storage at 10°C was deleterious.

Fig. 1. Plant Fig. 2. Flower

Fig. 3. Fruit

Economic part used : Dry herbage

Alkaloids: Hypo phyllanthin, phyllanthin

Medicinal uses : It is an important plant of Indian ayurvedic system of medicine which is used for stomach problems and for the genitourinary system, liver, kidney and spleen. It possess antiviral activity against hepatitis B and related hepadna viruses, and its wide usage in traditional medicine, suggesting it might possess other biological effects as well (Unander and Blumberg 1991; Unander *et al.*, 1991).

Past work done on Seed Science Technology

Unander *et al.* (1995) reported that seed color was associated with time of dehiscence. All seeds from capsules which dehisced by 30h after harvesting plants from the field were dark green. The numbers of light tan seeds increased at subsequent time intervals. Seeds collected at 96h were all light tan. Dark-green, early-to-dehisce seeds had greater and faster germination (Table l).

Table 1. Per cent germination and a germination index (GI) over seed color in Madras derived seeds and four time periods of dehiscence of Loiza derived seeds in *P. amarus*

Madras-derived seeds			Loiza derived seeds		
Seed color categories	**Germination (%)**	**GI**	**Hours after harvest**	**Germination (%)**	**GI**
Dark (early)	56.6 + 11.8	0.39 + 0.07	0-30	22.0 + 34.4	0.31 + 0.21
Light (late)	26.5 + 16.2	0.36 + 0.11	30-48	13.7 + 7.9	0.76 + 0.23
			48-72	7.0 + 1.6	0.50 + 0.16
			72-96	6.0 _a	

[a] 95% confidence intervals not applicable because of small seed sample

Seeds kept continuously in the dark did not germinate even after six weeks. For 40, 160 and 480h of light exposure, the average germination percentages were 0.3% + 0.66, 11.3% + 4.7, and 33.3% + 14.5, respectively. This indicated that a light requirement exists for seed germination in this species (Unander *et al.*, 1995).

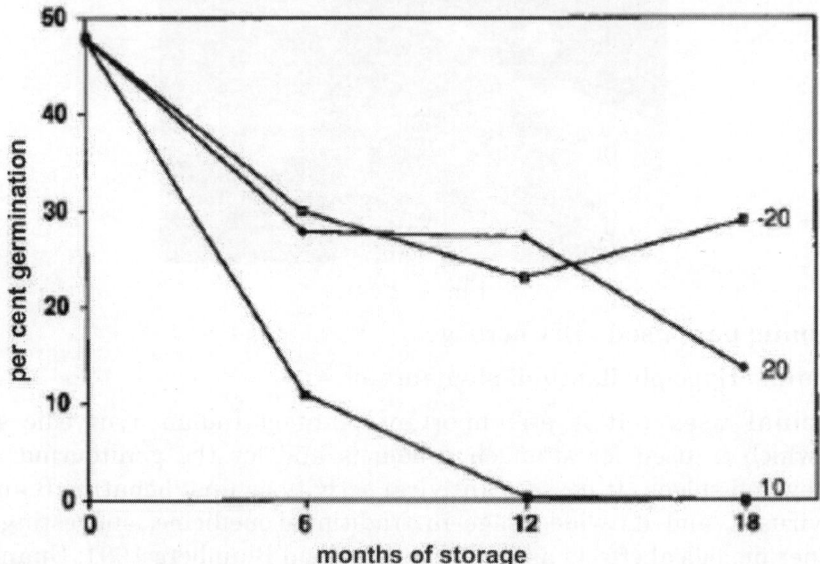

Fig. 5. Mean per cent germination of seeds of *Phyllanthus amarus* at harvest and after storage at -20, 10 or 20°C

According to Unander *et al.* (1995), storage temperature significantly affected the per cent germination of 6, 12 and 18 month old seeds, and the GI of 12 and 18 month old seeds. It is concluded that seeds of *P. amarus* are best kept at room temperature or alternatively in a frozen state, but should not be stored in a conventional seed storage facility at 10°C. No scarified seeds germinated (Table 2). These results suggested that per cent germination would not be increased in older seeds of *P.amarus* through metabolic elimination of putative germination inhibitors, although a slightly more rapid germination rate with time was reflected in GI values for seed stored at 20°C and -20°C (Fig. 5, 6).

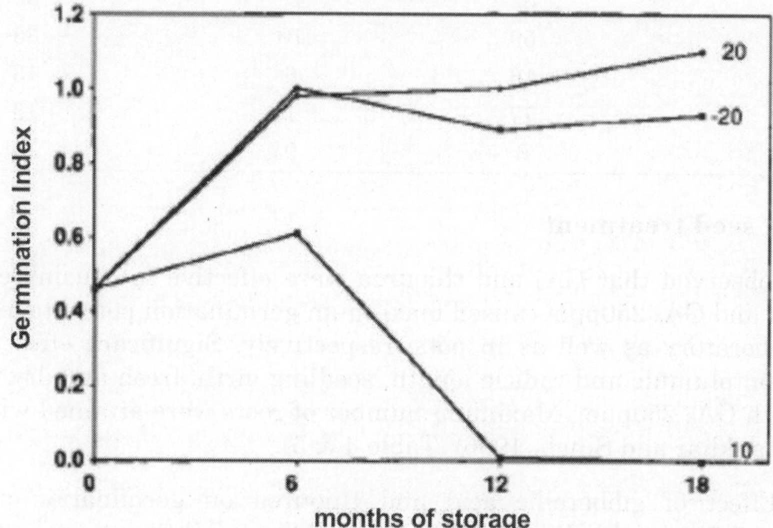

Fig. 6. Mean germination index values of seeds of *Phyllanthus amarus* at harvest and after storage at -20, 10 or 20°C.

Table 2. Means and 95 per cent confidence intervals for per cent germination and a germination index at 0, 6, 12 and 18 months after harvest at three storage temperatures for Madras derived seeds

Storage temperature (°C)	June 1990	Dec. 1990	June 1991	Dec. 1991
		Per cent germination		
20	48.0 + 5.2	28.1 ± 14.6	27.3±5.8	13.7 ± 6.2
10		10.7 ± 8.8	0.3 ± 0.6	0.00
-20		30.2 ± 19.6	23.0 ± 14.7	29.0 ± 5.6
		GI		
20	0.46 ± 0.03	0.99 ± 0.03	2.27 ± 0.46	1.11 ± 0.06
10		0.62 ± 0.45	0.02 ± 0.03	0.00
-20		1.00 ± 0.00	1.58 ± 0.90	0.93 ± 0.16

The *Phyllanthus* seeds germinated faster and higher at a high constant temperature of 25°C or alternating of 25-30°C compared to low temperature (20°C or 20-30°C). The performance of these temperature conditions was the best only when the seeds were exposed to light for 24h (Revathi *et al.*, 2002) (Table 3).

Table 3. Effect of temperature and light on germination of *Phyllanthus.*

Temperature (°C)	Germination (%)		
	Light	Dark	Alternate light and dark
25	82	18	60
30	59	10	35
20-25	16	6	13
25-30	77	1	13
20-30	6	9	10

Pre-sowing seed treatment

It was observed that GA_3 and thiourea were effective in causing early initial germination and GA_3 250ppm caused maxi-mum germination percentage (75.98 and 64.14) in laboratory as well as in pots, respectively. Significant effect over other treatments on plumule and radicle length, seedling girth, fresh and dry weight was observed with GA_3 250ppm. Maximum number of roots were attained with thiourea 750ppm (Dhankhar and Singh, 1996) (Table 4 & 5).

Table 4. Effect of gibberellic acid and thiourea on germination and growth parameters in *Phyllanthus* under laboratory condition.

Treatment	Days taken for initial germination	Germination (%)	35 days after sowing (DAS)		
			Plumule length (cm)	Radicle length (cm)	No.of roots/ plant
Control	2.4	50.36	7.05	2.34	2.57
Distilled water	1.9	58.75	8.03	4.58	2.63
GA_3					
250ppm	1.9	75.98	11.56	5.97	2.78
500ppm	2.4	55.55	9.04	3.09	2.92
750ppm	2.3	59.34	9.32	3.35	2.89
Thiourea					
250ppm	2.0	63.52	10.06	5.89	3.35
500ppm	1.9	62.12	9.99	4.68	3.42
750ppm	2.3	68.26	8.65	3.83	3.60
C.D. (P-0.05)	0.144	4.615	0.502	0.304	0.045

Table 5. Effect of GA_3 and thiourea on germination and growth parameters in *Phyllanthus* in pots

Treatment	Days taken for initial germination	Germination (%)	Fresh weight of seedling (g)	Dry weight of seedling (g)
Control	3.60	47.62	0.683	0.171
Distilled water	3.32	52.69	0.748	0.199
GA_3				
250ppm	3.12	64.14	1.205	0.260
500ppm	3.46	50.17	1.026	0.235
750ppm	3.39	53.47	0.966	0.222
Thiourea				
250ppm	3.16	56.95	1.063	0.243
500ppm	3.16	56.02	1.014	0.233
750ppm	3.35	62.01	0.875	0.211
C.D. (P-0.05)	0.125	3.749	0.021	0.012

Seeds soaking in water for 8h found to be the best for germination improvement (33 per cent) and vigour index. Soaking of seeds in cold water could help in enhancing the seed germination by softening the hard seed coat and facilitating leaching out of the germination inhibitors (Revathi *et al.*, 2002) (Table 6).

Table 6. Effect of water soaking, acid scarification on germination and vigour index in *Phyllanthus amarus*

Treatments	Duration	Germination (%)	Vigour index
Water soaking	6h	31	66
	8h	33	79
HCl scarification	1min	9	10
	2min	8	8
H_2SO_4 scarification	1min	8	9
	2min	6	6
Rubbing on smooth surface	2min	15	27
		18	34
Scarification with pestle and mortar	2min	4	6
Sand paper scarification	2min	4	5
Control		18	34

Table 7. Effect of GA_3, KNO_3 and Thio urea in *Phyllanthus amarus*

Treatment	Germination (%)	Vigour index
Control	33	121
GA_3 (ppm)		
200	72	276
300	62	119
400	36	104
KNO_3 (%)		
3	35	133
4	61	219
5	32	81
Thiourea (ppm)		
100	33	88
150	84	316
200	70	158

From Table 7, it can be seen that, soaking seeds in GA_3 200ppm for 8h improved the germination. This might be due to the involvement of GA_3 on the activation of cytological enzymes. This activation is combining with increasing cell wall plasticity and better water absorption. Also thiourea 150ppm for 1h was found to increase the germination upto 69 per cent. This might be ascribed to the deactivating capacity of thiourea to certain inhibitors present in the seed. The seed soaked for 8h in KNO_3 four per cent solution increased the germination from 25 to 38 per cent. The enhancement in germination may be due to an increase in cytochrome oxidase activity (Revathi *et al.*, 2002).

Seed storage

Revathi *et al.* (2002) revealed the better performance of seeds stored in 700 gauge polythene bag compared to those stored in cloth bag. The seeds store in moisture vapour container retains higher germination (72%) upto six months of storage. This might be due to the ability of the material to prevent the entry of external moisture and the maintenance of moisture content to the original level. Seeds treated with turmeric rhizome powder perform well, as it maintained germination (70%) upto six months of storage. This may be due to some active principle present in the rhizome powder that prevents the loss of viability of seed over long periods. It also acts as an insecticidal repellent as it contains insecticidal activity that prevents the entry of insects (Table 8).

Table 8. Effect of treatment and containers on germination and vigour index of seed during storage (after six months)

Treatments	Germination (%)		Vigour index	
	Cloth bag	**Polythene bag**	**Cloth bag**	**Polythene bag**
Control	16	36	11	29
Bavistin + Carboryl	42	62	37	49
Halogen mixture	22	43	18	37
Turmeric rhizome powder	52	72	71	136
Neem leaf powder	35	62	27	72
Vasambu powder	33	52	26	5
Notchi powder	28	46	19	40

REFERENCES

Dhankhar, D. S. and M. Singh 1996. Seed germination and seedling growth in Phyllanthus as influenced by GA_3 and thiourea. *Crop. Res., 12 (3):* 363-366.

Revathi, R., K. Vanangamudi and G. Mani 2002. Standardization of germination requirements in *Phyllanthus amarus. Seed Tech. News, 32 (6);* 127.

Sharma, O. P. 1993. Plant taxonomy. Tata McGraw-Hill Publishing Company Ltd., New Delhi. pp. 483.

Unander, D. W. and B. S. Blumberg 1991. *In vitro* activity of *Phyllanthus* (Euphorbiaceae) species against the DNA polymerase of hepatitis viruses: Effects of growing environment and inter and intra-specific differences. *Economic Botany*, 45:225-242.

Unander, D. W., G. L. Webster and B. S. Blumberg 1991. A survey of ethnobotanical uses and biological assay results in *Phyllanthus* (Euphorbiaceae). II. The subgenus *Phyllanthus. J. Ethnopharmacology*, 31:97-133.

Unander, D. W., H. H. Bryan, C. J. Lance and R. T. McMillan Jr. 1995. Factors affecting germination and stand establishment of *Phyllanthus amarus* (Euphorbiaceae). *Economic Botany* 49(1): 49-55.

Unander, D. W., H. H. Bryan, C. J. Lance and R. T. McMillan Jr. 1993. Cultivation *of Phyllanthus amarus* and evaluation of variables potentially affecting yield and the inhibition of viral DNA polymerase. *Economic Botany*, 47:79-88.

Webstar, G. L. 1957. A monographic study of the West Indian species of *Phyllanthus* (continued). *J. Arnold Arboretum*, 38: 51-80; 170-198; 295-373.

2.23. *Plantago ovata*

Common names: Desert Indian wheat, Desert Plantain, Blonde Psyllium, Indian Plantago, Isabgol.

Botanical name: *Plantago ovata* Forssk. Syn. *Plantago brunnea, P. fastigiata, P. gooddingii, P. insularis, P. minima*

Family: Plantaginaceae

Vernacular names: Isabgol (Hindi); Iskolvirai, Iskol, Isphagol, Ishappukol-virai, Ishappukol (Tamil).

Origin and distribution: It is native to west Asia and widely distributed in North West India and Pakistan. In India, it is cultivated in Rajasthan, Gujarat, Punjab and Uttar Pradesh.

Habitat: It requires a cool climate and dry sunny weather during maturation, even a mild dew, cloudy weather or light showers cause seed shedding. Temperature 20-25°C is required for seed germination, whereas at time maturing, it requires 30-35°C temperature. It requires 50-125cm annual rainfall. The crop is grown in marginal light, well drained sandy loam to loamy soils having pH between seven and eight.

Habit: Stemless small annual herb, 7-30cm tall, covered with dense or soft hairy growth, tillers arise from the base of the stem, each tiller bears rosette of leaves. Its roots are adventitious. Main root goes straight down in the soil and produces small roots which spread in the soil. Every plant produces 25-100 pseudo stems after 60-70 days of sowing (Fig. 1).

Leaf: Leaves possesses parallel vein structure. Leaves arise in large number from the base of plant. Leaves are 8-20cm long, very narrow, finely acuminate, entire or distinctly toothed, attenuate at the base, usually three nerved, glabrous or pubescent peduncles arise from the leaf axils, longer or shorter than the leaves. These peduncles bear ovoid or cylindrical terminal spikes, 1.3 to 3.8cm long, 10-15 per tiller.

Flower: The inflorescence is a spike with about 60 florets crowded at the top of the fragile, peduncle in about 3cm in length. The flowers are protogynous with floral maturity occurring in acropetal succession. Flowers are petioleless, non-leafy small, bisexual, incomplete and irregular. Minute, in oval or cylindrical spikes, 1.5-4.0cm long, bracts are 4mm long, broadly ovate or sub-orbicular, concave, membranous except the narrow midrib, which is glabrous, bracts and sepals are similar to each other having an herbaceous midrib, bordered by wide membraneous wings. Corolla is nearly colourless, but the mid-veins are often coloured brownish or red, suffusing into the orifice as an eye at the base of the lobes. The lobes are narrowly to broadly ovate, the stamens are adnate to the corolla tube and are exerted. The filaments and style are colourless or pink to dark red (Fig. 2).

Flowering, pollination and fruiting

Usually flowering occurs during spring season. The plant bears the flowering spikes in about 60 days after sowing and matures in the next two months. It is cross pollinated. Anthesis occur during early morning hours but stigma maturity is distributed to both morning and evening hours with a lowest frequency at the noon (Patel *et al.*, 1980). They also reported that 14 florets in a spike base matured stigma before anthesis started in that spike. In a spike, minimum gap between stigma receptivity and anthesis in the same floret was observed to be 14h with variation upto 120h. Pawar (1981) concluded from his studies that

(a) Stigma receptivity as judged by its elongation, continued to occur in acropetal succession in florets of a spike throughout the day and probably at night;

(b) Pollen grains collected in the morning and stored at 0°C without silica gel till evening were most viable as judged by per cent florets which set seed. Pollen grains could, however, remain viable for about 50h, if stored at a temperature range of -3 to 7°C; and

(c) The stigmata remained highly receptive for one day (seed set 92.18%), moderately the receptive for next five days and significantly lost the receptivity in subsequent days. After the 9th day, the receptivity was completely lost.

Fruit: Capsule is ovate, 8mm long, two celled, brown or purple at maturity and the top half lifting up when ripe, releasing the seeds.

Seed: Boat shaped, small, smooth dull ovate seeds, 1.8-3.8mm long, pinkish grey brown or pinkish white with a brown streak on the convex surface. The seeds are covered with a translucent membrane, known as the husk which is odourless and tasteless. The husked seeds are hard and dark red in colour. One gram of the seeds contains 600 seeds (Fig. 3).

Propagation: By seeds

Seed collection: The yellowing of the lower leaves is an indication of maturity, confirmed by pressing a spike between two fingers when the mature seeds come out. The crop is harvested close to the ground in the early morning hours to avoid losses owing to seed shedding. The harvested material is stacked for one or two days, made to be trampled by bullocks, winnowed and the separate seed crop is collected. The average seed yield is 1t/ha. A bold seed crop fetches a better price (Gupta, 1997).

Seed dormancy and germination

Verma *et al.* (1989) recommended top of paper method and 25°C was found to be optimum for conducting germination test. The factors responsible for germination are length of storage period, dormancy and light sensitivity. Under ideal conditions of adequate moisture and low temperature (10°C to 20°C), 30 per cent of the seed germinates in five to eight days. The seeds show some innate dormancy (three months) following harvest. Various treatments including wet and dry heat and cold, scarification, ethylene and CO_2 do not eliminate this dormancy period. Post dormancy seeds show reliable germination in excess of 90 per cent at 29°C and lower rates of germination as temperature is increased (Hanson *et al.,* 1992). Chandler (1953) had reported that dormancy of freshly harvested seed was expressed in their inability to germinate at a temperature as high as 30ºC and in their delay in germination at 20ºC as compared to old seeds. He also reported that the dry seeds stored for ten years at room temperature germinated 99 per cent at 20-30ºC. Fresh seeds too germinated, at this temperature to the some extent.

Fig. 1. Plant

Fig. 2. Flower

Fig. 3. Seed (Husk of isabgol)

Stebbins and Day (1967) reported that seed germination could be raised to almost 100 per cent by scarifying the seed with fine sand paper and then soaking the seed in a solution of 100ppm potassium gibberellate to which was added 0.5 per cent glucose, if these seeds had been recently collected.

Patel *et al.* (1978) concluded that

(a) Freshly harvested seeds were dormant which could be broken by cold shock (by growing seeds at a temperature range of 5-15°C) or by ageing (by storing the seeds under ordinary condition);

(b) Interaction effect of the temperature and age was significant only with the fresh seeds;

(c) Seeds stored for two years lost viability.

The seeds are treated with any fungicides at the rate of $3g.kg^{-1}$ of seed, to protect the seedlings from a possible attack of damping off. Germination starts in 6-10 days after sowing.

Economic part used: Seeds and husk

Alkaloids: Protein, polysaccharides, cellulose, pectin, oil and muscilage

Medicinal uses: Laxative, treatment of amoebic dysentery, intestinal disorders

REFERENCES

Chandler, C. 1953. Seed germination for some species of *Plantago. Contribution from Boyce Thomson Institute,* 17: 265-271.

Gupta, R. 1997. Medicinal and aromatic plants. **In:** Handbook of Agriculture, ICAR, New Delhi.

Hanson, C. V., E. A. Oelke, D. H. Putnam and E. S. Oplinger 1992. Alternative field crops manual. University of Wisconsin, Madison.

Patel, N. H., S. Sriram and K. C. Dalal 1978. Effect of seed age and temperature on germination of Isabgol seed. Third All India workshop on Medicinal and Aromatic plants held at Indore, Madhya Pradesh, India.

Patel, N. H., S. Sriram and K. C. Dalal 1980. Floral biology and stigma - pollen maturation schedule in isabgol (*Plantago ovata* F.). *Curr. Sci.,* 49: 689-692.

Pawar, V. A. 1981. Pollen viability, duration of stigma receptivity and stigma emergence schedule in isabgol (*Plantago ovata* Forsk.), M.Sc. thesis submitted to Gujarat Agricultural University, Anand, India.

Stebbins, G. L. and A. Day 1967. Cytogenetic evidence for long continued stability in the genus *Plantago. Evolution,* 21: 409-428.

Verma, S. S., B. S. Chabra, C. Ram and Urmila Verma 1989. The standardization of the seed testing procedure for Isabgol (*Plantago ovata*). *Intern. J. Trop. Agric.,* **7** (3/4): 199-201.

2.24. *Plectranthus vettiveroides*

Common names: Swedish ivy, Creeping Charlie, Swedish begonia, Prostrate coleus, Spur flower

Botanical name: *Plectranthus vettiveroides* K.C. Jacob. Syn. *Coleus vettiveroides*

Family: Lamiaceae

Vernacular names: Valak (Hindi); Lavanchi and Muchivala (Kannada); Iruveli (Malayalam); Valakam, Udicya (Sanskrit); Kuruver, Vettiver (Tamil); Kuriveru, Vettiveru (Telugu).

Origin and distribution: Possibly endemic. Probably extinct in the wild. Today, it is under cultivation only in four villages near Kollidam in Thanjavur district of Tamil Nadu.

Habitat: In sandy loams along riverbanks in plains.

Habit: A sub-succulent bushy pubescent herb, growing 30-120cm tall (Fig. 1).

Root: Roots fibrous, 30-60cm long, forming a tuft, slender, very thin, easily cut-off, strongly fragrant, straw colour when fresh, changing dark on drying.

Stem: Stems faintly four angled, light brown, sub-succulent, densely minutely hairy.

Fig. 1. Plant

Leaf: Leaves opposite, sub-orbicular to ovate, base obtuse, apex rounded, margin dentate, sub-fleshy to leathery, with dense white minute hairs on both surfaces; lateral nerves 10-13 pairs, less prominent above, prominent below, all nerves arising from the base of the leaves and joining along the margins just below the dentations. Each dentation about 1cm long with one or two small lateral serrations. Leaf stalks 4-13cm long, sub-succulent, minutely hairy, with a central groove.

Propagation : By stem cuttings.

Economic part used : Roots

Alkaloids: It is an aromatic plant containing essential oils and diterpenes in its roots.

Medicinal uses : The roots are used in treating burning eyes, diarrhoea, fever, intrinsic haemorrhage, vomiting, skin diseases, thirst, giddiness, headache and insanity.

2.25. *Rosmarinus officinalis*

Common name: Rosemary

Botanical name: *Rosmarinus officinalis* L.

Family: Lamiaceae

Origin and distribution: It is native to Mediterranean region and distributed in Algeria, China, Middle East, Morocco, Russia, Romania, Serbia, Tunisia, Turkey and to a limited extend in India. It is commercially cultivated in Europe and California.

Habitat: Dry scrub land in the Mediterranean region, but widely cultivated.

Habit: It is a dense, evergreen, hardy, perennial, aromatic herb, 90cm high with small (2-4cm) pointed, sticky, hairy leaves. It can be grown either as a field crop or as an indoor plant (Fig. 1).

Stem: Branches are rigid with fissured bark and stem is square, woody and brown.

Leaf: It has opposite, simple, entire, evergreen leaves upto 5cm long and an eighth of 2.5cm wide. The leaf margins are revolute and the leaves are a shiny green on top and whitish beneath due to a dense collection of very fine hairs (Dirr, 1990). The leaves are resinous.

Flower: Pale blue small flowers appear in cymes inflorescence (Fig. 2).

Flowering and pollination: The plant flowers in spring and summer in temperate climates, but the plants can be in constant bloom in warm climates. Seeds ripen from August to October. It is pollinated by bees (Abrol, 2012).

Fruit: The fruit consists of four dry nut lets (Fig. 3).

Fig. 1. Plant

Fig. 2. Flower

Fig. 3. Fruit

Propagation : By seeds and cuttings

Seed propagation: The ideal season for raising the nursery is between September and November. The seeds are very small and about 0.2 to 2.5g seeds are required to cover 1m^2 area and are sown to a depth of 1-2cm. The seeds are sown in well prepared nursery beds and germinate best at a soil temperature ranging from 14-15°C. When the seedlings are about eight to ten weeks old, they are ready for transplanting into the main field.

Vegetative propagation: Cuttings from healthy mother plants, 10-15cm in length are taken. All leaves about half of the length from bottom should be removed. The cuttings are then planted in nursery beds of sandy soil under partial shade at a depth of about 6 to 10cm. Thereafter, regular watering and weeding is provided to the nursery for about a month. After about 6-8 weeks, the cuttings are ready for transplanting into the main field. They can also be raised in small polybags or seed pans. This helps in easy transportation of the rooted cuttings.

Economic part used : Leaves and tender tops.

Alkaloids: Camosic acid, camosol, rosemarinic acid, rosmanol

Medicinal uses : In medicine, rosemary is credited as carminative, anti-depressant, anti-spasmodic, anti-microbial, anti-inflammatory, carcinogen blocker, liver-detoxifier and anti-rheumatic.

REFERENCES

Abrol, D. P. 2012. Pollination biology, biodiversity conservation and agricultural production. Springer London, p.792.

Dirr, M. A. 1990. Manual of woody landscape plants. 4th ed. Stipes. Champagne, IL.

2.26. *Thymus vulgaris*

Common name : Common thyme, Garden thyme

Botanical name: *Thymus vulgaris*

Family: Lamiaceae

Vernacular names: Banajwain (Hindi)

Origin and distribution: Thyme is a native to southern Europe from Spain to Italy. Apart from Europe, it is grown in Australia, North Asia, North Africa, Canada and USA. In India, it is cultivated in the western temperate Himalayas and Nilgiris.

Habitat

It grows well in a temperate to warm, dry, sunny climate, and where the plants are not shaded. It needs full sun to grow to its best potential. This plant does not like excessive moisture because of its susceptibility to rot diseases. Rainfall in the Mediterranean region where thyme is cultivated most is 500 to 1000mm per year, mainly in winter.

It prefers light, well-drained soils with a pH of 5.0 to 8.0. This species do best in coarse, rough soils that would be unsuitable for many other plants. Although thyme grows easily, especially in calcareous light, dry, stony soils, it can be cultivated in heavy wet soils, but it becomes less aromatic.

Habit: It is a hardy perennial shrub, a semi-evergreen groundcover 20-40cm in height. The flavour is aromatic, warm and pungent. It has both horizontal and upright habits (Fig. 1).

Stem: This plant have ascending, quadrangular, greyish brown to purplish brown lignified and twisted stems.

Leaf: The leaves are very small, usually 2.5 to 5mm in length and vary considerably in shape and hair covering, depending on the cultivar, with each species having a slightly different scent. The leaves are oval to oblong in shape and somewhat fleshy. Leaves are almost stalkless with margins curved inwards and highly aromatic.

Flower: The flowers terminate the branches in whorls. The calyx is tubular, striated, closed at the mouth with small hairs and divided into two lips, the uppermost cut into three teeth and the lower into two. The corolla consists of a tube about the length of the calyx, spreading at the top into two lips of a pale purple colour, the upper lip erect or turned back and notched at the end, the under lip longer and divided into three segments (Fig. 2).

Flowering pollination and fruiting: Flowering and fruiting occur during May to July. The source of seed has to be known as there are possibilities of hybridization (Anon. 2009) due to cross pollination.

Fruit: The fruit consists of four brown ovoid nut lets.

Seed : The seeds are round and very small and retain their germinating power for three years.

Propagation : It is propagated from seeds, stem cuttings and layering.

Seed propagation: Sow seeds in spring to a depth of 6mm or less. Seeds germinate in about two weeks. When planted in seed trays, it will take six to eight weeks to reach transplant readiness.

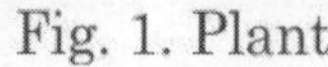

Fig. 1. Plant

Fig. 2. Flower

Vegetative propagation: To have homogeneous plants, it is advisable to make cuttings. Thyme grows easily from 5 to10cm cuttings taken in spring.

Economic part used : Aerial parts are used for essential oil production, mostly by steam distillation. The fresh and dried herb market uses it for culinary purposes.

Alkaloids: Thymol

Medicinal uses : Thyme oil is used in treatment of bronchitis. It has anti-spasmodic and carminative properties. It possesses antioxidant and antimicrobial properties.

Other uses: Thymol is used to control the Varroa mite (*Varroa destructor*), a species that is parasitic to bees. Thymol is volatile and permeates the hive, coming in direct contact with the parasitic mites. The volatilized thymol irritates the mites, causing them to withdraw from the bees and die from starvation.

REFERENCE

Anonymous 2009. Thyme production. Directorate of Agricultural Information Services, Department of Agriculture, Forestry and Fisheries, Pretario, South Africa. p.21.

2.27. *Trianthema portulacastrum*

Common name: Black pig weed, desert horse purslane, giant pig weed, horse purslane

Botanical name: *Trianthema portulacastrum* L.

Family: Aizoaceae

Vernacular names: Santhi, Patharchatta, Lalsabuni (Hindi); Pundhari-ghentui (Marathi); Saaranai (Tamil); Tellagalizeru (Telugu).

Origin and distribution: It is cosmopolitan, with no known centre of origin and widespread in Southeast Asia, tropical America and Africa.

Habitat: A common weed in dry land and garden lands, gregarious causing considerable hindrance to normal cultivation.

Habit: Prostrate, profusely branching, glabrous, annual herb upto 40cm long (Fig. 1).

Root: Tap root white or brown.

Stem: Stem is round, striate, glabrous below and hairy above. There are stipules present with one spine in the centre.

Leaf: Leaves are simple, opposite, petiole sheathing at the base, obovate, entire, apex rounded or emarginated, base acute, unequal pairs and pinnately veined.

Flower: The small purple flowers may only be open for a short time in the mornings. Flowers solitary, bisexual, sessile, sheathed by the base of the petiole; calyx five lobed, lobes obtuse, green above and pink inside; cuspidate. Petals 0; stamens 10-15, inserted near the top of the calyx tube; ovary superior, 1-2 celled, free, style 1, subulaten (Fig. 2).

Flowering, pollination and fruiting: The production of flowers and seeds starts 20-30 days after germination of the seeds. Flowering during June-October and fruiting occurs during July-December. The flowers are pollinated by ants.

Fruit : Capsule scarious below, top mitriform, coriaceous, three to five seeds with concentric mucriculate lines.

Seed : Kidney shaped, in spiral ended by a beak, 2mm in diameter. The tegument is tuberculate, reddish and brown to black.

Fig. 1. Plant

Fig. 2. Flower

Propagation : By seeds and cuttings.

Seed dispersal : Seed dispersal is achieved by several methods: one seed is dispersed in the detached cap of the capsule, which can float and the other seeds are either dispersed individually from the capsule or remain on the annual, parent plant where they will germinate and establish new plants.

Seed dormancy and germination

Seeds do not exhibit dormancy and fresh seeds can germinate when incubated at 35°C to the extent of 54 to 56 per cent (Balyan and Bhan, 1986b). It was found that eighth day was the optimum period for testing seed germination of *Trianthema portulacastrum* (Umarani and Selvaraj Jacqueline, 1994).

Seeds germinated from 20 to 45°C. Maximum germination of 90 per cent was observed at 35°C. Germination was the highest between the 4th and 8th day of incubation. Maximum seedling emergence resulted when seeds were planted at 1cm and gradually declined with the increased seeding depth. Seeds lying on the surface gave very poor seedling emergence (Balyan and Bhan, 1986b).

Seed storage : Germination increases during seven and eight months air dry storage at room temperature (Balyan and Bhan, 1986b). The seeds are orthodox in nature. Seeds stored in soil gave significantly higher germination than stored in the laboratory.

Economic part used : Leaves

Alkaloids: Punarnavine

Medicinal uses : The plant dried out and reduced to powder, is reportedly used against throat troubles and anti-fungal agent. A decoction of the roots is reputedly used as an emmenagogue and in larger doses it is abortive. Recent studies have investigated potential hepatoprotective properties. The leaves are diuretic and applied in the treatment of oedema, jaundice, strangury and dropsy. A decoction of the herb is used as a vermifuge and is useful in rheumatism; it is considered an antidote to alcoholic poisoning. The fleshy nature of leaves makes them suitable for use as a wound dressing.

Past work done on Seed Science and Technology

1. Seed production

A mature horse purslane plant can produce as many as 3330 fruits per plant and 6-10 seeds/fruit (Prakongvongs, 1999).

Balyan and Bhan (1986a) reported that the first three seeding dates resulted in significantly higher seedling emergence than the August 30 date. Leaf number, leaf area, dry matter and number of branches per plant at the peak growth stages of this weed were significantly higher for the July 20 sowing than for the other sowing dates, followed closely by the June 30 sowing. Similarly number of nodes per plant, seeds per plant, and seed weight per plant were also significantly higher for the July 20 sowing, followed closely by the June 30 sowing. Phenological behavior of the plants indicate that: (a) maximum seedling emergence takes place during June and July, (b) rapid and vigorous growth occur during July and August (rainy season) when conditions for growth are optimum, and c) production of flowers and seeds starts 20 to 30 days after sowing.

2. Allelopathic effects germination inhibition

Hussain *et al.* (1987) reported that *Trianthema portulacastrum* contains caffeic, chlorogenic, p-hydroxybenzoic, p-coumaric and ferulic acids which may inhibit or promote seed germination of other plants.

Concentrated sorghum water extract (100%) reduced the germination of *Trianthema portulacastrum* by 15 to 20 per cent (Randhawa *et al.,* 2002). Cheema *et al.* (2002) reported that sorghum water extracts reduced weed growth by 28 per cent, but this inhibition level still not effective for profitable crop production. For better and eco-friendly control of horse purslane, the effectiveness of sorghum water extracts needs to be improved. Duke *et al.* (2000) stated that mixture of two or more allelopathic aqueous extracts caused more inhibition of weeds due to synergistic mode of action. Likewise in another field study, Cheema *et al.* (2003) found that application of sorghum + eucalyptus + sunflower water extracts combination furnished >70 per cent weed reduction than sorghum water extract alone in wheat, probably due to the presence and synergistic effects of different allelochemicals in these species.

Randhawa *et al.* (2002) revealed that sorghum water extract at higher concentration (100%) reduced the germination of *Trianthema portulacastrum* by 15 to 20 per cent. Root and shoot length of *Trianthema portulacastrum* was also significantly suppressed at higher concentrations (*i.e.*, 75 and 100%) of sorghum water extract. While lower concentration (25%) of the extract promoted shoot length of *Trianthema portulacastrum* (Table 1).

Table 1. Allelopathic effect of sorghum water extract on the germination and seedling growth of *Trianthema portulacastrum*

Parameters	**SWE (0%)**	**SWE (25%)**	**SWE (50%)**	**SWE (75%)**	**SWE (100%)**
Germination percentage	75 ab	80a	75ab	70bc	60c
Root length (cm)	3.01 ab	3.15a	2.70b	2.40c	2.31c
Shoot length (cm)	2.24 b	2.82a	2.20b	2.11b	2.02bc

SWE = Sorghum water extract

Mahmood *et al.* (2010) reported that in the laboratory bioassay, sorghum+ sunflower water extract combination at higher concentration (100%) completely inhibited germination of horse purslane. In pot experiment, the foliar application of sorghum + sunflower water extract combination at higher concentration (100%) greatly suppressed growth traits (shoot and root length) of horse purslane seedling and this combination reduced shoot dry weight by 66 per cent over control. Inhibitory effect was proportional to the concentrations of the extracts and higher concentration had the stronger inhibitory effect (Table 2).

Table 2. Allelopathic effect of different plant water extracts on germination, shoot and root length of horse purslane

Treatments	**Germination (%)**	**Shoot length (cm)**	**Root length (cm)**
Plant water extracts			
W0 = Control (Distilled water)	92.08 a †	1.70 a	2.52 a
W1 = Sorghum water extract	35.00 b	0.56 b	0.88 b
W2 = Sorghum + mulberry (WEs)	20.83 d	0.30 e	0.50 e
W3 = Sorghum + brassica (WEs)	21.67 d	0.44 c	0.66 d
W4 = Sorghum + sunflower (WEs)	13.33 e	0.24 f	0.37 f
W5 = Sorghum + maize (WEs)	24.17 cd	0.37 d	0.64 d
W6 = Sorghum + rice (WEs)	27.92	0.45 c	0.70 c

† = means not sharing a letter in common differ significantly at p 0.05;
WEs = Water extracts

Muhammad Saeed *et al.* (2011) revealed that *Helianthus annuus* water extract reduced germination (10.31%), shoot length (0.25cm) and root length (0.28cm) of horse purslane as compared to control which showed germination (86.72%), shoot length (1.81cm) and root length (2.43cm). The degree of per cent germination inhibition of horse purslane was in the order of *Helianthus annuus* > *Parthenium hysterophorus* > *Datura alba* > *Sorghum bicolor* > *Eucalyptus camadulensis* > *Trianthema portulacastrum* > Control (Table 3).

Table 3. Effect of different aqueous extracts on horse purslane.

Treatments	**Germination (%)**	**Shoot length (cm)**	**Root length (cm)**
Control	86.72 a	1.81	2.43 a
Trianthema portulacastrum L.	31.95 b	0.63 b	0.79 b
Parthenium hysterophorus L.	17.82 d	0.38 d	0.41 e
Datura alba L.	18.67 d	0.39 d	0.55 d
Helianthus annuus L.	10.31 e	0.25 e	0.28 f
Sorghum bicolor L.	21.15 d	0.49 c	0.57 cd
Eucalyptus camadulensis L.	25.00 c	0.50 c	0.61 c
LSD (P 0.05)	3.47	0.04	0.05

Means not sharing a letter in common are differ significantly at $P_{0.05}$

2. Seed dormancy and germination

According to Lee *et al.* (2011), germination response of both populations (IRRI: IR population; university of the Philippines; UP population) was greater at 30/20°C and 35/25°C day/night temperatures than they were at 25/15°C alternating day/night temperatures. Germination of both populations was greater in the light/dark regime than in darkness. In dark, depending on the temperature, seed germination of the UP population ranged from 37 to 62 per cent, whereas seed germination of the IR population was <20 per cent . Exposure to 5min. at 117 and 119°C for the IR and UP popu-lations, respectively, reduced germination to 50 per cent of maximum germination. Osmotic potential of −0.26MPa inhibited germination to 50 per cent of the maximum for the UP population, whereas the corresponding value for the IR population was −0.37MPa. Seeds placed on or near the soil surface had maximum emergence, and emergence declined with increase in seed burial depth. Seedling emergence of the UP and IR populations was 74 per cent and 13 per cent respectively for seeds placed on the soil surface. For both populations, no seedlings emerged from a soil burial depth of 6cm or more. Germination and emergence responses to light and seed burial depth differed between the two populations of horse purslane. Residues on the soil surface upto 6 mg ha did not influence seedling emergence of either populations.

REFERENCES

Balyan, R. S. and V. M. Bhan 1986a. Emergence, growth and reproduction of horse purslane (*Trianthema protulacastrum*) as influenced by environmental conditions. *Weed Sci.*, 34: 516-519.

Balyan, R. S. and V. M. Bhan 1986b. Germination of horse purslane (*Trianthema portulacastrum*) in relation to temperature, storage conditions and seeding depths. *Weed Sci.*, 34: 513-515.

Cheema, Z. A., A. Khaliq and M. Mubeen 2003. Response of wheat and winter weeds to foliar application of different plant water extracts. *Pakistan J. Weed Res.*, 9: 89–97.

Cheema, Z. A., M. Asim and A. Khaliq 2002. Sorghum allelopathy for weed control in cotton (*Gossypium arboreum* L.). *Int. J. Agric. Biol.*, 2: 37–41.

Duke, S. O., F. E. Dayan, I. G. Romagni and M.A. Rimando 2000. Natural products as a source of herbicides: Current status and future trends. *Weed Res.*, 40: 99–111.

Hussain, F., Yasmin and F. Rehman 1987. *Trianthema portulacastrum* Linn. Exhibits allelopathy: Advances in weed Sci. Proc. Pak Indo Weed Control Workshop, pp. 325-334, Islamabad.

Lee, J., Bhagirath S. Chauhan and D. E. Johnson 2011. Germination of fresh horse purslane (*Trianthema portulacastrum*) seeds in response to different environmental factors. *Weed Sci.* 59 (4): 495-499.

Mahmood, A., Z. A. Cheema, A. Khaliq and A. U. Hassan 2010. Evaluating the potential of allelopathic plant water extracts in suppressing horse purslane growth. *Int. J. Agric. Biol.*, 12: 581–585.

Muhammad Saeed, Muhammad Ashfaq and Bakhtiar Gul 2011. Effect of different allelochemicals on germination and growth of horse purslane. *Pak. J. Bot.*, 43(4): 2113-2114.

Prakongvongs. 1999. Germination and growth of horse purslane (*Trianthema portulacastrum* L.). Proc. the 17th Asian Pacific Weed Sci. Soc. Conf., November, 22-27. Bangkok, Thailand.

Randhawa, M. A., Z. A. Cheema and Muhammad Anjum Ali. 2002. Allelopathic effect of sorghum water extract on the germination and seedling growth of *Trianthema portulacastrum*. *Int. J. Agri. Biol.*, 4 (3): 383-384.

Umarani, R. and A. Selvaraj Jacqueline. 1994. Studies on dormancy and germination of *Trianthema portulacastrum*. *Indian J. Weed Sci.* 26 (3 & 4): 82-88.

2.28. *Tribulus terrestris*

Common names: Ground burnut, Puncture weed, Tack weed, Land caltrops, Puncture vine, Yellow vine, Goat head

Botanical name: *Tribulus terrestris* L., Syn. *T. lanuginosus* L.

Family : Zygophyllaceae

Vernacular names: Gokshura, Gukhurkata (Assamese); Gokhru kanta (Bengali); Be tha gokharu, Nana gokharu, Mithogokharu (Gujarati); Choto gokhru (Hindi); Negalu, Sennaneglu (Kannada); Michirkand, Pakhda (Kashmiri); Nerinjil (Malayalam); Sarate, Gokharu (Marathi); Gukhura, Gokhyura (Oriya); Bhakhra, Gokhru (Punjabi); Goksurah, Svadamstra (Sanskrit); Nerunji, Palleru mullu (Tamil); Palleru, Cinnapalleru (Telugu); Khar-e-Khasak Khurd, Gokhru (Urdu)

Origin and distribution: It is native to southern Europe (Parker, 1972), Africa, temperate and tropical Asia. Plant is widely distributed in different parts of India upto 3000m altitude.

Habitat: A common dry land weed.

Habit : An herbaceous annual with a deep root (Fig. 1).

Root: Tap root deep, slender, branched, often somewhat woody, with a network of numerous fine rootlets.

Stem: The round stems radiate from the crown to a diameter of about 10cm to over 1m, often branching. They are usually prostrate, forming flat patches, though they may grow more upwards in shade or among taller plants.

Leaf: Leaves are opposite, compound with lanceolate stipules, evenly pinnate, shortly petiolate with 10-14 small oblong leaflets, oblique at base, in unequal pairs.

Flower: Peduncled flowers are 4-10mm wide, solitary in leaf axils, pseudoaxilary, white or yellow; sepals five, imbricate; petals five, spreading, imbricate, disc annular, ten lobed; stamens ten or five; ovary sessile 5 to 12 celled, ovules one to five in each cell (Fig. 2).

Flowering, pollination and fruiting

A week after each flower blooms, it is followed by a fruit that easily falls apart into four or five single seeded nutlets. Flowering is rapid, with the first flowers appearing within three weeks and the first fruit within five to six weeks of germination. Flowering may continue for several months until the first frosts in autumn/early winter, when plants die.

Boydston (1990) reported that flowering occurred within three to four weeks of emergence when temperatures were consistently above 20°C, regardless of planting date. It flowers March through October in Arizona, although primarily from July to August (Parker, 1972). Once the plant begins to flower, it is continuous throughout the plant's life (Reddi *et al.*, 1981). The flowers are cross pollinated by insects (foragers include: Coleoptera, Diptera, Hymenoptera, Lepidoptera, Thysanoptera) (California Department of Food and Agriculture, Encyclopedia, 2002, Washington State Noxious Weed Control Board, 2001) along with being self pollinated, which occurs at the end of each flower's receptive period (within one day) (Reddi *et al.*, 1981). Self pollination is accomplished when the petals begin to close and push the stamen inward toward the stigma, the longer anthers making direct contact; the potential of this system is 100 per cent seed set (Reddi *et al.*, 1981). Fruits mature in approximately two weeks, and subsequently split apart into segments soon afterward (Holm *et al.,* 1991). Plants continue to reproduce and produce fruit until the cool season begins. Boydston (1990) reported during trials in Washington, fruit/burr production stopped in October when average temperatures were under 20°C.

Fruit

Schizocarp; woody burrs, gray to yellow-tan, hairy, to approximately 1-1.8cm in diameter, more or less flattened, lobed; separates into five (occasionally four) wedge shaped, indehiscent nutlets (cocci), each with two stout dorsal spines 4-7mm long and spreading and several prickles (Fig. 3). Boydston (1990) reported that seeds planted in May, June, July, and August subsequently produced 5600, 5200, 3600, and 2000 burrs/plant, respectively (Fig. 3).

Seed

Each nutlet can contain 2-5 seeds (Boydston, 1990). The seeds are hard and bear two to three sharp spines, 10mm long and 4 to 6mm broad point to point. It is a trailing and spreading herb, densely covered with minute hair (Fig. 4).

Fig. 1. Plant

Fig. 2. Flower

Fig. 3. Fruit

Fig. 4. Seed

More than 1000 seeds can be produced by a single plant and seed can remain viable for many years.

Propagation : By seeds

Seed propagation : The collection should be done in December. Crush the fruit and sow the seed in polybags. Sow the seeds in the months of May-June. Seed germination takes about 10 to 15 days. Eighty percent seeds germinate.

Seed dispersal

Each fruit section has two sharp divergent spines and several other spines and warty protuberances enabling the fruits to easily attach to animals and humans and to stick onto vehicle tires (cars, farm tractors, airplane), subsequently facilitating long distance dispersion and spread (Ernst and Tolsma, 1988; Ridley, 1930; Squires, 1979; Whitson 1992).

Holm *et al.* (1991) points out that due to the architecture of the fruit, the large and small spines are arranged at different angles with at least one of the spines always pointing upward no matter how the fruits fall from the plant and can easily imbed into feet, hooves, or tires. After getting caught or imbedded into the hooves, feet, and wool of livestock and other animals, the fruits are subsequently broken off as the animals try to rid themselves of the irritation (Ridley, 1930). Furthermore, they can stick to the shoes and clothing of people, and the fur and feathers of animals (California Department of Food and Agriculture, Encyclopedia, 2002).

Life cycle : Caltrop is annual plant which germinates after rainfall in late spring and summer when the maximum air temperature reaches 24°-27°C. Flowers may be produced within three weeks, fruits within six weeks.

Seed longevity: Buried seed can remain viable for several years (California Department of Food and Agriculture, Encyclopedia, 2002), staying dormant in the soil for 4-5 years (Whitson, 1992).

Seed dormancy and germination

Caltrop is annual plant which germinate after rainfall in late spring and summer when the maximum air temperature reaches 24°-27°C. There is a succession of germination throughout the summer following each rainfall event.

Generally, it has a considerable seed dormancy lasting over fall and winter months (Washington State Noxious Weed Control Board, 2001) with some seeds staying dormant for longer periods of time. Its seedlings emerge in the early spring through summer, often in flushes following increased soil moisture (California Department of Food and Agriculture, Encyclopedia, 2002). It germinates after the start of the monsoon rains, on any type of barren soil in southern Arizona (Parker, 1972).

Ernst and Tolsma (1988) reported that dormancy of seeds was very high in both fresh seeds and 3-6 year old seeds. Squires (1979) reported that freshly harvested seeds have a germination rate of ten per cent and dry stored six month old seeds have a germination rate of 84 per cent.

Pathak (1970) reported that the germination is inhibited by low temperatures, low light intensities and wet soil.

Most newly matured seeds are dormant and require an after ripening period of approximately 6-12 months. Imbibition of water differed in dormant versus non-dormant seeds: measurements taken at one hour and periodically through one day of imbibition showed dormant seeds had approximately half of the water content of non-dormant seeds (Ernst and Tolsma, 1988). The largest seed within a nutlet is usually the first to germinate; the remaining seeds may germinate or remain dormant depending on moisture availability (California Department of Food and Agriculture, Encyclopedia, 2002); this large seed is usually positioned near the basal end of the burr (University of California, 1998).

The optimum temperature range for germination to occur is 27-35°C. In Australia, germination occurs when the maximum air temperature is approximately 24-27° (Squires, 1979). During trials, the emergence was initiated when average soil temperatures reached 15°C for at least two weeks and approached 20°C (Boydston, 1990). After initial emergence occurred, emergence was multipeaked with no relationship existing with temperature for the remainder of the summer (Boydston, 1990). In subsequent years, emergence occurred and peaked at similar times regardless of the age of the seed, indicating environmental stimuli influencing emergence (Boydston, 1990).

The seedlings emerge during early spring through summer, often in flushes following increased soil moisture (California Department of Food and Agriculture, Encyclopedia, 2002; Mirsa, 1962 in Holm *et al.,* 1991). During field observations in Botswana, it was observed to germinate and emerge following a rain shower having more than 10mm of precipitation (Ernst and Tolsma, 1988). Maximum germination occurred after a series of heavy rains, facilitating 35 per cent germination rate, with continued germination of seeds lasting for another four months (Ernst and Tolsma, 1988).

On sandy soils, seedlings emerge from depths to approximately 5cm (California Department of Food and Agriculture, Encyclopedia, 2002). Squires (1979) reported that seeds buried more than 10cm deep in sandy soils can successfully emerge.

Germination was irregular, whether seeds remained in the fruit or were isolated (Ernst and Tolsma, 1988). In greenhouse trials, the highest germination rate was achieved by isolated seeds with a maximum of 66.9 per cent and a mean of 37.3±25.1 per cent (ranging from 20/15°C upto 35/30°C, in 5°C increments, alternating day/ night temperatures). Seeds that remained in the fruits rarely germinated synchronously (Ernst and Tolsma, 1988). Ernst and Tolsma (1988) add that because of this pattern, direct competition for water and nutrients is avoided.

Effects of crop production: The roots can develop nitrogen fixing nodules. Athar and Mahmood (1985) observed that plants having root nodules had more lush green healthy growth and greater dry weight versus stunted growth in plants without nodules.

Economic part used : Fruit

Alkaloids: Saponin, diosgenin

Medicinal uses : It has long been used as a tonic and aphrodisiac in Unani system of medicine. It has been used in India and Pakistan as a treatment for impotence and as a stimulant to enhance sexual drive and performance (Brown *et al.,* 2001). It has diuretic and uricosuric effects.

Agricultural uses : The seed extracts effectively inhibited the spore germination, germ tube elongation and mycelial growth of *Sarocladium oryzae* (Selvaraj and Narayanasamy, 1994).

REFERENCES

Athar, M. and A. Mahmood 1985. Observations on nitrogen fixation by *Tribulus terrestris* Linn. under natural habitat. *Geobios,* 12(1): 44-46.

Boydston, R. A. 1990. Time of emergence and seed production of longspine sandbur (*Cenchrus longispinus*) and puncture vine (*Tribulus terrestris*). *Weed Sci.* 38(1):16-21.

Brown, G. A., M. D. Vukovich and E. R Martini. 2001. Effects of andro stenedione herbal supplementation on serum sex hormone concentrations in 30 to 59 years old men. *Int. J. Vitam. Nutr. Res.,* 71(5): 293-301.

California Department of Food and Agriculture, *Encyclopedia.* 2002. Tribulus. Ed by: Healy, E. A., S. Enloe, J. M. DiTomaso, B. Roberson, N. Dechoretz, S. Schoenig, P. Akers, L. Butler, and J. Garvin. Non-Cropland Weed group, UC Extension Service, Weed Science Program, Department of Vegetable Crops. The University of California. Davis, CA. 95616.

Ernst, W. H. and D. J. Tolsma. 1988. Dormancy and germination of semi-arid annual plan species, *Tragus berteronianus* and *Tribulus terrestris. Flora,* 181(3/4):243-251.

Holm, L. G., D. L. Plunknett, J. V. Pancho and J. P. Herberger 1991. The world's worst weeds. Distribution and biology. Krieger Publishing Company, Malabar, Florida, p. 609.

Mirsa, D. 1962. *Tribulus terrestris* weed in arid zone farming. *Indian J. Agron.* 7(2):136-141.

Parker, K. F. 1972. An illustrated guide to Arizona weeds. The University of Arizona Press, Tucson, AZ. 338 pp. http://www.uapress.arizona.edu/online.bks/weeds/

Pathak, P. S. 1970. Contributions to the ecology of *Tribulus terrestris* Linn. II. Habitat studies. *Agra University J. Res. Sci.,* 19(2):149-166.

Reddi, C. S., E. U. Reddi and N. S. Reddi. 1981. Breeding structure and pollination ecology of *Tribulus terrestris. Proc. Indian National Science Academy,* Part B 47(2):185-193.

Ridley, H. N. 1930. The dispersal of plants throughout the world. L. Reeve and Co., Ltd., Lloyds Bank Buildings, Ashford, Kent. p. 744.

Selvaraj, C. and P. Narayanasamy. 1994. Anti-fungal effect of some of the plant extracts against *Helminthosporium oryzae* and *Sarocladium oryzae* in rice. pp. 326-329. In: Crop diseases: Innovative techniques and Management (Eds. K. Sivaprakasam and K. Seetharaman), Kalyani Publishers, New Delhi. 584p.

Squires, V. R. 1979. The biology of Australian weeds. 1. *Tribulus terrestris* L. *J. Australian Instt. Agric. Sci.,*45(2):75-82.

University of California. 1998. The grower's weed identification handbook. Cooperative Extension University of California, Division of Agriculture and Natural Resources, Publication 40: 311.

Washington State Noxious Weed Control Board. 2001. Puncture vine (*Tribulus terrestris* L.). Website: http://www.wa.gov/agr/weedboard/weed_info/puncturevine.html

Whitson, T. D. 1992. Weeds of the West (Eds; Burrill, S. A. Dewey, D. W. Cudney, B. E. Nelson, R. D. Lee and R. Parker). The Western Society of Weed Science in cooperation with the Western United States Land Grant Universities Cooperative Extension Services and the University of Wyoming. p. 630.

Website

http://pi.cdfa.ca.gov/weedinfo/TRIBULUS2.htm

PART - III

HERBACEOUS PERENNIAL MEDICINAL PLANTS

3.1. *Achyranthes aspera*

Common names: Prickly chaff flower, Devil's Horsewhip, Burweed

Botanical name: *Achyranthes aspera* L. Syn. *Achyranthes argentea* Lam.; *Achyranthes aspera obtusifolia* (Lam.) Griseb.; *Achyranthes obtusifolia*

Family: Amaranthaceae

Vernacular names: Apang (Assamese); Apamarga, Apang (Bengali); Chirchita (Dogri); Chirchita, Onga Lathjira (Hindi); Uthraani (Kannada); Kantmogro (Konkani); Kadaladi (Malayalam); Khujumpere (Manipuri); Aghada (Marathi); Opamarga (Oriya); Apamarga, Kharamanjari (Sanskrit); Naayuruvi (Tamil); Antisha (Telugu).

Origin and distribution: The plant is widespread in the world as a weed, in Pakistan, Sri Lanka, Tropical Asia, Africa, Australia and America. In the Northern part of India, it is known as a medicinal plant in different systems of folk medicine

Habitat: Disturbed areas, road sides, waste lands, gardens, crops, grasslands, savanna and forest margins. This species is often found in the moist or shaded areas near trees in savanna or pasture lands where it grows in dense thickets.

Habit: It is an erect herb about 90cm high; sparingly branched, biennial or perennial (Fig. 1).

Stem: Its stems become woody at the base, round, striate and hairy.

Leaf: Its short stalked leaves (dark green above and paler below) are opposite, simple and egg shaped with broad end at base (ovate) upto 10cm long by 8cm wide, densely to sparsely hairy (pubescent) tapering to a point at both ends and shortly stalked.

Flower: Spikes are long, about 25cm long in fruit; bracts membranous, spinescent, persistent; bracteoles spinescent; perianth 4-5, rigid, lanceolate, aristatem strongly ribbed, shiny and hard; stamens 2-5, filaments connate at the base and alternate with staminodes which are toothed lacerate or with a toothed scale at the back; anthers two celled; ovary sub-compressed, oblong, one celled; ovule solitary, pendulous, style filiform, stigma cap-like (Fig. 2).

Fig. 1. Plant Fig. 2. Flower

Fig. 3. Fruits Fig. 4. Seeds

Flowering, pollination and fruiting: Flowering and fruiting is from July to September. Pollination is entomophillus because numerous small flowers are aggregated in dense or congested inflorescences. Pollen dispersal takes place either by wind or animals (Sharma, 1993).

Fruit: The sharp pointed fruits are orange to reddish purple or straw-brown capsules, oblong or ovoid, indehiscent, rounded or areolate at apex; seed inverse, oblong (Fig. 3).

Seed : Sub-cylindrical, truncate at apex, reddish brown (Fig. 4)

Propagation : Propagation is mainly through seeds or sprouting method.

Seed dispersal: The sharp pointed fruits stick to hairs of animals and clothes. They can be dispersed by livestock and wild animals and by people on footwear and clothing.

Seed dormancy and germination: Seeds can be sown in late spring. A seven month period after ripening in dry storage at room temperature increased the germination of utricles from 0 to 50 per cent; however when fresh utricles are washed in running water for 24h, seeds germinated to 100 per cent and the seeds are having physiological dormancy (Pandya and Pathak, 1980).

Economic part used : Whole plant

Medicinal uses : This is pungent and bitter in taste (rasa), pungent in the post digestive (vipaka) and has cold potency (viriya). It alleviates kapha and vata doshas. It possesses light (laghu), dry (ruksha) and sharp (tikshna) attributes. It is diuretic, astringent and a blood purifier. It is useful in the diseases such as obesity, piles, dropsy, hemorrhoids, vomiting, abdominal pain, pruritus and diseases due to ama. The root of the plant is tied to the ear as a remedy for malarial fever in different parts of the country. In the Vaishnodevi hill area of Jammu and Kashmir, a decoction of its roots is given to women after menstruation as an anti fertility agent. It can also be used as a diuretic for goats and for other veterinary purposes.

Other uses: It can be pruned and made into a dome shape. This can be planned and used as a small hedge or as window plants also.

Past work done in Seed Science and Technology

Saraswathi (2004) studied the seed biology of *Datura metal, Achyranthes aspera, Trianthema portulacastrum, Abutilon indicum, Plantago ovata* and *Cassia auriculata* on seed germination, dormancy status of fresh seeds, their response to dormancy breaking treatment, sowing depth, season and storage period. The findings of this study are follows.

(a) Studies on evaluation of dormancy status revealed that the seeds of weed species taken for evaluation were readily germinable by 0-33 per cent, dormant by 22-100 per cent and dead by 0-32 per cent irrespective of species. *D. metal* and *A. aspera* germinated maximum in KNO_3 at 0.5 per cent recording 72 and 96 per cent, respectively for the duration of 16h and 8h. Seed soaking in commercial H_2SO_4 for duration of 30 and 10min respectively enhanced the germination of *A. indicum* (62 per cent) and *C. auriculata* (90 per cent).

(b) Seeds of *A. aspera* alone responded to soaking of seeds in 0.5 per cent thiourea for 8h (96 per cent). The seeds of *P. ovata* recorded the highest germination of 92 per cent by soaking the seeds in water for 16h after removal of fruit / seed coat.

(c) Studies made with *D. metal* and *A. aspera* on different depths of burial *viz.,* top soil, 3 and 6cm in earthern pots indicated that with increase in sowing depth, the germination of seeds decreased irrespective of species.

(d) Studies on season of production on seed germination and their response to identified dormancy breaking treatment with *D. metal* and *A. aspera* expressed that the seeds produced in summer were more dormant than *rabi*. But, the treatment response for removal of dormancy did not vary with season.

(e) The storage studies conducted with *D. metal* and *A. aspera* indicated that the germination of fresh seeds increased with storage period upto 3 months and thereafter, it decreased. But on imposing dormancy breaking treatment, the germination of seeds was in decreasing order with advances in storage period.

REFERENCES

Pandya, S. M and V. S. Pathak. 1980. Seed dormancy imposed by covering structures in *Achyranthes aspera. Geobios (Jodhpur)* 7: 74-76.

Saraswathi, S. 2004. Studies on seed biology in tropical weed species. M.Sc.(Ag.) Thesis, TNAU, Coimbatore.

Sharma, O. P. 1993. Plant taxonomy. Tata McGraw-Hill Publishing Company Limited, New Delhi. pp. 483.

3.2. *Aconitum heterophyllum*

Common name: Indian Atees

Botanical name: *Aconitum heterophyllum* Wall. ex. Royle

Family: Ranunculaceae

Vernacular names: Ataich (Assamese, Bengali); Ativakhani Kali (Gujarati); Atis, Atvika (Hindi); Ati Visha, Athihage (Kannada); Kath (Kashmiri); Ati Vidayam (Malayalam); Ati Vish (Marathi); Atushi (Oriya); Atis (Punjabi); Aruna, Ghuhapriya, Visa (Sanskrit); Athi Vidayam, Athithiranam, Pankurai (Tamil); Ati Vasa (Telugu); Atees (Urdu).

Origin and distribution: It is native of the Western Himalayas and found in Garhwal, Kumaon and Kashmir.

Habitat: This is grown in sub-Alpine and Alpine zone in the temperate zone of about 2500-3900m. It is suitable for sandy, loamy and clay soils. It can grow in semi-shade or no shade. It prefers moist soil.

Habit: It is an erect, 1.5m in height, perennial and tuberous herb, glabrous (Fig. 1).

Root: Root is paired biennials, tuberous, whitish or grey 2.8cm long, and 0.4-1.5cm thick grey brown outside with scattered pointed minute notches. Starchy white inside 4-9 xylem bundles near the periphery embedded in the secondary phloem tissue (Fig. 2).

Tuber: The tubers are upto 3cm long, conical at ends. The mother and daughter tubers occur in pairs. The initial first bud of the daughter tuber is conic in shape.

Stem: Stem erect, simple or branched, from 15-20cm high, glabrous below, finely crispo-pubescent in the upper part.

Leaf: Leaves heteromorphous, glabrous: lowest on long petioles (13cm); blade orbicular- cordate or ovate cordate in outline with a usually narrow sinus (1-1.5cm deep); usually five lobed to the middle, amplexicaul.

Flower: Inflorescence is slender raceme or a lax, leafy panicle, crispo-pubescent; sepals bluish or violet (rarely whitish); navicular obliquely erect, shortly or obscurely beaked, 18-20mm high, 8-9mm wide. Carpels are five, elliptic oblong. Follicles are contagious, linear-oblong, straight, 16-18mm long (Fig. 3).

Fig. 1. Plant

Fig. 2. Root

Fig. 3. Flower

Fig. 4. Seeds

Flowering, pollination and fruiting: Flower from August to September and the seeds ripen from September to October and are pollinated by bees. Anthers dehisced longitudinally between 7.30 and 11.00am, strongly dependent on higher temperature. The pollen grains per anther varied between 2000 and 6000, which means an average of 80,000 pollen grains per flower. Nectar production begins at anther dehiscence and coincides with maximum stigmatic receptivity. Controlled pollination revealed that this species is self-incompatible, although few fruits developed from selfing. Such fruits are smaller than the fruits produced by open pollinated and from hand-crossed flowers and most aborted early in development (Nautiyal *et al.*, 2009).

Fruit: One seeded beaked or plumose, achenes or many seeded follicles or rarely capsules or berry.

Seed: Pyramidal, 3-4mm long, blackish brown, angles acute or more or less winged, faces smooth (Fig. 4).

Propagation: Seeds, tuber segments or young leafy stems can be used as propagules.

Seed collection: The fruits (capsules) that turn light brown (before splitting) are collected during late October to mid-November for better germination of seeds. The seeds should preferably be collected during morning hours.

Seed rate and seed treatment : Approximately 1.5kg seeds are required for obtaining seedlings for one hectare. However, about 111,000 tuber segments are needed for plantation on one hectare. Pre-treatment of seeds with GA_3 (200ppm) results in early germination (9-20 days) and upto 55 per cent germination. Pandey *et al.* (2000) found that KNO_3 (50 and 100mM) treatment significantly enhanced the germination in *Aconitum balfourii* (62%-70%), while in *Aconitum heterophyllum,* thiourea @130mM resulted in higher germination (40%).

Nursery techniques: Seeds have no dormancy period and are sown immediately after collection. They are sown in sand and FYM (farmyard manure)/ compost mixture (1:2) at a depth of 0.5cm in styrofoam trays in a mist house. Nursery beds of 2m × 2m or even smaller size are better for raising seedlings inside the polyhouses. At a depth of 0.5cm, germination is delayed to some extent, but first true leaf initiation occurs earlier as compared to sowing at other depths. Seeds are sown during October–November or March–April in polyhouses at middle altitudes (1800-2200m), during February– March in glass houses at lower altitudes (600-1000m), and during May-June at alpine sites in open beds or in hot houses. Plants raised from seeds have very slow growth and the cotyledonary phase (pseudomonocotyle) persists for at least one growth season (three to four months). Under polyhouse conditions and sandy textured soil, germination takes about two months to complete. However, diurnal temperature variations (25-25°C) and mulching of soil promote germination within 15-25 days.

When tuber segments are used as planting material, they are treated with GA_3 @ 200ppm for maximum rooting percentage and survival. Tuber segments are also planted during the same period as mentioned for seedlings. Propagation using young leafy stems, when inserted in moist soils under scrub canopy or in soil trenches covered with thin layer of moss, induces rooting at every node. Even the top nodal segment roots within 30–35 days under above conditions during July–August. About two to four plantlets can be produced from one leafy stem (Anon., 2008).

Economic part used: Tuberous roots and all other parts.

Alkaloids: Atidine, hetisine, heteratisine, diterpene, heterophylline, heterophyllidine, heterophyllisine, hetidine, atidine and atisenol, a new entatisene diterpenoid lactone from roots.

F-dishydrçatisine, hetisinone, benzyl leteratisine, beta - sitosterol, carotene and 3 - isoatisine from rhizomes.

Medicinal uses

External uses: The seeds crushed in honey are applied locally on throat, in tonsillitis. Nasal insufflations of roots are beneficial in headache.

Internal uses

Respiratory system: The juice of roots along with milk is an expectorant. Root powder is given orally in cervical lymphadenitis.

Digestive system: Seed and root are used in ascites. Seeds are laxative.

Urinary system: The seeds are diuretic, the root decoction reduces burning of urinary tract. It increases volume of urine.

Reproductive system: Root is used in sperrnatorrhoea. The decoction of roots is also used in burning of vagina.

Circulatory system: The juice of leaves along with juice of zinger reduces perspiration.

Toxic effects: Over dosage (More than 5-6g) produces symptoms like dryness of mouth etc.

REFERENCES

Anonymous, 2008. Agro-techniques of selected medicinal plants, Vol. 1, National Medicinal Plants Board, New Delhi.

Nautiyal, B. P., M. C. Nautiyal, V. P. Khanduri and Neelum Rawat 2009. Floral biology of *Aconitum heterophyllum* Wall. a critically endangered alpine medicinal plant of Himalaya, India. *Turkish J. Botany*, 33 (1): 13-20.

Pandey, H., S. K. Nandi, M. Nadeem and L. M. S. Palni 2000. Chemical stimulation of seed germination in *Aconitum heterophyllum* species of medicinal value. *Seed Sci. & Technol.*, 2(1) : 39-48.

3.3. *Adhatoda zeylanica*

Common name: Malabar nut

Botanical name: *Adhatoda zeylanica* Medik. Syn. *Adhatoda vasica* Nees, *Justicia adhatoda* L.

Family: Acanthaceae

Vernacular names: Titabahak, Bahak, Vachaka (Assamese); Baksa, Vasaka, Basak (Bengali); Bangra (Garhwal); Aradusi (Gujarati); Arusa, Bansa (Hindi); Adusoge (Kannada); Bahekar, Baikar, Basuth, Bhenkar (Kashmiri); Adalodakam (Malayalam); Nongmangkha angouba (Manipuri); Vasa, Adulsa (Marathi); Arusa, Basung (Oriya); Bansa, Basuti, Bhekar, Vasaka (Punjabi); Shwetavasa, Vasa, Vasaka (Sanskrit); Adathoda, Kaba kolli (Tamil); Addasaramu (Telugu); Adusa, Basa (Urdu).

Origin and distribution: Distributed throughout India, upto an altitude of 1300m.

Habitat: This plant spread throughout the plains and sub-montane regions of India. It is commonly available in residential areas.

Habit: A tall much branches with dense evergreen shrub, 1.2-6.0m in height (Fig. 1). Branchlets grey pubescent.

Leaf: Leaves are 10-30cm long and 3-10cm broad, lanceolate to ovate-lanceolate, slightly acuminate, base tapering, petiolate, petioles 2-8cm long, exstipulite, glabrescent, 8-10 pairs of lateral vein bearing few hairs, dried leaves dull brown above, light greyish brown below, odour, characteristic, taste, bitter.

Flower: The flowers last for three to five days, they are protandrous and pass through three distinct phases: male, bisexual and female. Dense, short spikes, stalks of the spikes shorter than leaves. Leaf like structures called bracts, present on the spikes; these are conspicuously veined; corolla white, bi-lipped, purple blotched in the throat, stamens two, anthers sub-equal (Fig. 2).

Flowering, pollination and fruiting: Flowering and fruiting occurs during October to March. Flowers are suited for insect pollination because of coloured bilabiate corolla and abundant nectar in hypogynous disc (Sharma, 1993). Two species of carpenter bees, *Xylocopa verticalis* and *Xylocopa* sp are the effective pollinators. Pollen grains are deposited on the dorsal surface of the thorax during *Xylocopa* visit to the flowers in the male phase and the stigma rubs the pollen coated thorax and is pollinated when the bees visit the flowers in bisexual and female phases. There is a high level of geitonogamy. Pollination efficiency under field conditions is high (95%). However, fruit set is poor (6%). The species is self-incompatible. None of the self pollinated flowers sets fruits, but over 50 per cent of the cross pollinated flowers sets fruits. The results indicate that protandry does not prevent self pollination, but reduces interference in export and import of pollen.

Although the flowers have adapted well to achieve a high level of pollination, reproductive success in terms of fruit set is low, largely due to the limitation of compatible pollen (Shivanna, 2009).

Fruit: Capsular, four seeded, capsules clavate, longitudinally channelled, 1.9-2.2cm x 0.8cm (Fig. 3).

Seed: Globular.

Fig. 1. Plant

Fig. 2. Flower

Fig. 3. Fruits

Propagation: Propagation is usually through stem cuttings alone. It prefers loamy soils. Take semi-hardwood and softwood cuttings, 7.5- 10.0cm along with 3-4 nodes. April-June is the best time for planting. Plant the stem cuttings in polythene bags or raised beds. Cuttings will take 15-20 days to sprout and 30 days to root. Ten days after root initiation, they will harden and 80 per cent establishment can be obtained.

Economic part used: Fresh or dried leaves, bark, root.

Alkaloids: Four new quinazoline alkaloids - vasicoline, adhatodine, vasicolinone and anisotine from leaves and one each i.e. vasicinone and vasicol have been isolated from inflorescence and leaves, respectively. A new alkaloid - 9-acetamido-3,4-dihydropyrido[3,4-b] indole and a new glucoside - O-ethyl-a-D-galactoside - along with sitosterol-b -D-glucoside, galactose and deoxyvasicinone are isolated from roots. 2'-hydroxy-4-glucosyloxy chalcone has been identified in flowers. Another alkaloid vasicinolone has been isolated from roots. A new quinazoline alkaloid has been isolated from leaves and characterised as 1,2,3,9-tetrahydro5-methoxy pyrolo[2,1-b]quinazolin-3-ol(I-form). Adhavasinone has been isolated and characterised. (+)vasicinone has been isolated from leaves. Vasicinol, vasicinone, deoxyvasicinone, deoxyvasicine (minor alkaloids) and vasicine isolated from leaves showed seasonal variation in the percentage of minor alkaloids and total alkaloids.

Medicinal uses: It is the source of the drug vasaka, well known in the indigenous systems of medicine for its beneficial effects, particularly in bronchitis. Leaves are rich in vitamin C and carotene. Fresh or dried leaves are reported to be used in bronchial troubles. Leaf juice is reported to be used in diarrhoea, dysentery and glandular tumours. Powdered leaves are supposed to be used for skin affections. The leaves, flowers, fruits and roots are extensively used for treating cold cough, whooping cough and chronic bronchitis and asthma as sedative expectorant, antispasmodic and as anthelmintic.

Agricultural uses: The leaf extracts of this plant can be used as a one of the component in the preparation of insect repellents (Sridhar *et al.*, 2006).

REFERENCES

Sharma, O. P. 1993. Plant taxonomy. Tata McGrow-Hill publishing company Ltd., New Delhi. pp.483.

Shivanna, K. R. 2009. Pollination biology, breeding system and reproductive success of *Adhatoda vasica*, an important medicinal plant. *Current Sci.,* 96(3):408.

Sridhar, S., B. Kriruthika and S. Arumugasamy 2006. Organic farming techniques in bhendi. Centre for Indian knowledge System, Chennai. p. 17.

3.4. *Aloe vera*

Common names: Miracle plant, Healing plant, Plant of immortality, Fountain of youth, Indian aloe, Barbados aloe, Chinese Aloe, True Aloe, Burn Aloe, First Aid Plant

Botanical name: *Aloe vera* (L.) Burm.f. Syn. *A. barbadensis* Mill., *Aloe indica* Royle, *Aloe perfoliata* L. var. *vera* and *A. vulgaris* Lam.

Family : Liliaceae

Vernacular names: Musabhar, Machambar (Assamese); Ghritakalmi (Bengali); Eliyo, Eariyo (Gujarati); Musabhar, Elva, Ghi-kuuar (Hindi); Karibola, Lolesara satva, Lovalsara, Lolesara (Kannada); Musabbar, Siber (Kashmiri);

Chenninayakam, Kattarvazha (Malayalam); Yellia, Khorpad (Marathi); Musabara (Oriya); Kalasohaga, Mussabar, Alua (Punjabi); Kumari (Sanskrit); Sotru Katralai (Tamil); Musambaram, Kalabanda (Telugu); Musabbar, Ailiva, Siber (Urdu).

Origin and distribution: It is originated from Northern Africa, Middle and South America. It is grown in Greece, China, Italy, South Africa, India, Pakistan and Bangladesh. It is naturalised throughout the country, more common along the west coast (Robert and Hentry, 2004). In India, it is largely cultivated in Alwar (Rajasthan), Satnapalli (Andhra Pradesh), Rajapipla (Gujarat), Salem and Tuticorin (Tamil Nadu).

Habitat: The land should be free from water logging problems. It can grow successfully from sandy coastal soils to loamy soils of plains and also over poor marginal soils. The species tolerates higher pH with Na+K salts. However, its growth is faster under medium fertile, heavier soils such as black cotton soils. Higher foliage growth is observed in well drained, loam to coarse sandy loam soils with pH upto 8.5. The plant has wide adaptability and can come up well in localities with warm humid or dry climate. Annual rainfall ranges from as low as 350-400mm to even as high as 1500-2000mm. The plant is found to survive in cold temperatures also but however, protection from severe frost is essential.

Habit: It is a succulent, almost sessile perennial herb growing upto 1m (Fig. 1).

Stem: It is a stem less or very short stemmed succulent plant growing to 80–100cm tall.

Leaf: The leaves are radically arranged in two or three circles and each plant has 12-16 leaves. Each leaf is 30 - 50cm long and 10cm broad at the base having an indented margin. The leaves are fleshy and thick, pea green coloured with occurrence of white spots when young. Leaves are bearing horny prickles on the margins. The mature leaf is 7 -10cm across at the base, weighing 1.5-2.0kg.

Flower: The flowers are produced on a spike upto 90cm tall, hermaphrodite, each flower pendulous, with a yellow tubular corolla 2-3cm long with anthers and stigma protruding. Bright yellow tubular flowers appear in a spike. The ovary is superior, triocular with axile placentation (Fig. 2 and 3).

Flowering and pollination: Usually four years after planting, plant may even bloom; producing a tall stock covered with bright colour coral flowers. The flowers are produced in summer and pollinated by bees.

Fruit: The fruit of the plant are loculicilade capsules.

Propagation : The plant does not produce many viable seeds. It can be propagated with suckers having 4-6 leaves. A healthy mother plant produces about six suckers at a time. Plant the suckers in rows at spacing of 30cm. The rows have to be 60cm apart formed on a well prepared and manure loamy soil having plenty of broken bricks and coarse sand.

It can also be propagated through rhizome cuttings by digging out the rhizomes after the harvest of the crop and making them into 5-6cm length cuttings with a minimum of 2-3 nodes on them. Then, they are rooted in specially prepared sand beds or containers.

Economic part used : Inner gel of the leaves (Fig. 4).

Alkaloids : Aloin A & B (Barbaloin).

Fig. 1. Plant

Fig. 2. Flower

Fig. 3. Fruit

Fig. 4. Gel

Medicinal uses : These are extensively used as active ingredients in laxative and anti-obesity preparation, as moisturizer, emolient or wound healer in various cosmetic and pharmaceutical formulations. Its beneficial properties may be attributed to mucopolysaccharides present in the inner gel of the leaf, especially acemannan (acetylated mannans). It has been used to treat various skin conditions such as cuts, burns and eczema. It is alleged that sap from *Aloe vera* eases pain and reduces inflammation. The gel is also alleged to be useful for dry skin conditions; especially eczema around juice may help some people with ulcerative colitis, an inflammatory bowel disease. If the lower leaf is sliced open, the gel obtained can be applied on the affected area of the skin.

Agricultural uses: Antonysamy (2004) reported that aggergatum onion treated with three per cent of panchagavya along with Notchi, *Calotropis, Nerium, Aloe,* and Pungam in cow's urine for seven days increased the yield of bulbs by 17.5t ha^{-1}.

Other uses: The nectar from the flowers is a loved food for humming birds.

REFERENCES

Antonysamy, 2004. Organic aggregatum onion better packaging and quality. The Hindu News paper. May 20. pp.10.

Robert, B. and T. Hentry 2004. Medicinal plants. Asiatic publishing house, New Delhi. p.308.

3.5. *Alpinia galanga*

Common names: Greater galangal, Thai galangal, Blue ginger, Thai ginger.

Botanical name: *Alpinia galanga* (L.)Willd. Syn. *Amomum galanga* Lour

Family : Zingiberaceae

Vernacular names: Kulinjan, Sugandhbach (Hindi); Kanghu (Manipuri); Kulanjan (Sanskrit); Perarattai (Tamil).

Origin and distribution: It is native to China. This plant naturally found in the Eastern Himalayas and the Western Ghats. It is commonly cultivated in wet lands of Tamil Nadu and Kerala of India, Indochina, Thailand, Malaysia and Indonesia.

Habitat: It is grown well under wetland ecosystems.

Habit: The plant is a rhizomatous, perennial herb, and attains a height of about 1.5–2.5m (Fig. 1).

Rhizome: Slightly aromatic, weighs about 50-60g, dense internodes, spreads upto 1m. Externally, it is reddish brown-white and internally reddish-white (Fig. 2a and 2b).

Leaf: Leaves are leathery, 30-60cm long, glossy on both surfaces, lanceolate and smooth, green above, pale beneath, whitish on margins, median nerve very strong.

Flower: Flowers about 3cm long, greenish white, fragrant, in compound dense bunches; bracts and bracteoles persistent; bracteoles lanceolate, 5-8mm. Calyx tubular, 6-10mm, persistent. Lip of corolla is white, streaked with red. Corolla tube 6-10mm; lobes oblong, 1.6-1.8cm. Lateral staminodes purple, subulate or linear, 2-10mm. Labellum white with red lines, obovate - spatulate, 2cm, apex deeply 2-cleft. Filament 1cm; anther 7mm (Fig. 3).

Fig. 1. Plant

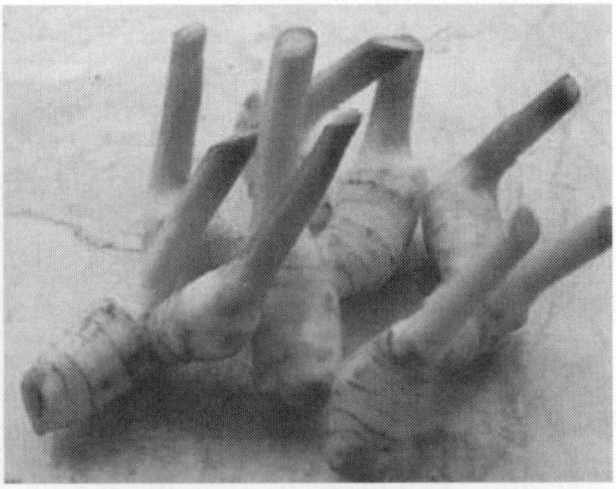

Fig. 2a. Fresh rhizome

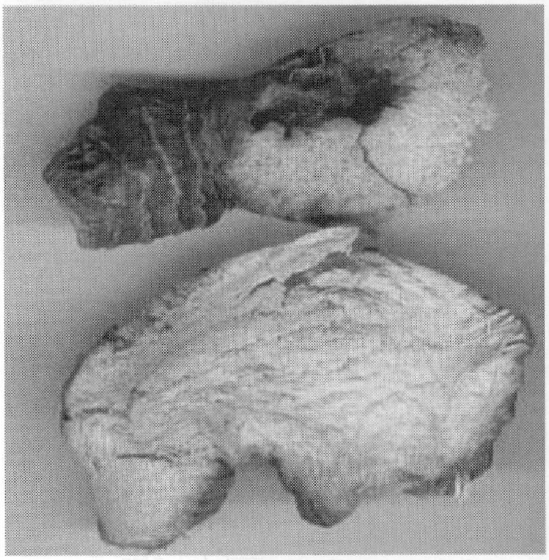

Fig. 2b. Dried rhizome

Fig. 3. Flower

Fig. 4. Fruit

Flowering, pollination and fruiting: Flowering occurs in May and June, while fruiting occurs in August and September (Anon., 2008). It is pollinated by bees.

Fruit: Fruits are capsules, orange to red in colour, thin, glabrous, small cherry size (Fig.4).

Seed : Seeds are 3 – 6 in number.

Propagation: Rhizome splits are used as planting material for commercial cultivation. Seed propagation is not feasible due to low germination rate.

Economic part used: Dried rhizome

Alkaloids: Azarone, Acorine, Calaminol, Alphinin

Medicinal uses: The rhizome is carminative, aphrodisiac, febrifuge and bronchodilator. It expels the phlegm (mucus), improves voice, and is very useful in sore throat and respiratory congestion. It is also used in rheumatism. The seeds have the same uses as the rhizome.

REFERENCE

Anonymous 2008. Agro-techniques of selected medicinal plants, Vol. 1. National Medicinal Plants Board, New Delhi.

3.6. *Alternenthera sessilis*

Common names: Sessile joy weed, Dwarf copper leaf

Botanical name: *Alternenthera sessilis* (L.) R. Br

Family: Amaranthaceae

Vernacular names: Garundi, Guroo (Hindi); Honganne (Kannada); Ponnankannikkira (Malayalam); Phakchet (Manipuri); Kanchari, Koypa (Marathi); Madaranga (Oriya); Matsyaksi (Sanskrit); Ponnankanni, Koduppai, Seethai, Ponni (Tamil); Ponnagantikura (Telugu).

Origin and distribution: Possibly originates from tropical America. Widespread throughout the tropics and subtropics: tropical Africa, southern Asia to Japan, southeast Asia, Australia and the Pacific Islands. In the western hemisphere: the West Indies, Central America, tropical south America including Argentina and the United States.

Habitat: This is often found in and near ponds, canals and reservoirs. It prefers places with constant or periodically high humidity and so may be found in swamps, shallow ditches and fallow rice fields.

Habit: This is a perennial herb with prostrate stems, rarely ascending, often rooting at the nodes (Fig. 1).

Stem: Stem striate, terete below, slightly tetragonous above, solid, sometimes floating in water and fistulose in lower part, stem and branches with narrow lines of whitish hairs and branch and leaf axils with tufts of white hairs.

Leaf: Leaves opposite, simple; petiole upto 5mm long; blade linear-lanceolate, oblong to ovate or obovate, 1-15cm × 0.2-3cm, glabrous to sparsely pilose.

Flower: Inflorescence an axillary, sessile, subglobose head 5mm in diameter, solitary or in clusters of upto five. Flowers bisexual, regular, 5-merous; tepals free, equal, ovate to elliptical, upto 2.5mm long, white to pinkish, 1-veined; stamens united at base into a very short cup, two without anthers; ovary superior, strongly compressed, single celled, style very short (Fig. 2).

Flowering, pollination and fruiting : The flowering and fruiting occurs throughout the year with most vigorous vegetative growth at the onset of the rainy season and the most vigorous reproductive growth at the end of it. The flowers are self-pollinated (Grubben and Denton, 2004).

Fruit: Fruit an obreniform, corky, indehiscent capsule 2mm long, dark brown, single seeded (Fig. 3).

Seed: Seeds lenticular, ovate in outline, 0.9–1.5mm long, 0.8–1mm wide, 0.3–0.6mm thick, with marginal hilar notch at broad end. Hilum small, inconspicuous. Testa glossy, smooth, dull orange or dark brown to black, with faintly visible reticulation. Embryo peripheral, visible on surface under testa, encircling perisperm. The average number of seeds per plant is about 2000 (Fig. 4).

Fig. 1. Plant

Fig. 2. Flower

Fig. 3. Fruit

Fig. 4. Seeds

Propagation: It can easily be propagated by seed and rooted stem parts.

Seed dispersal: The fruits are dispersed by wind and water.

Economic part used: Leaf and stem.

Alkaloids: Triterpenes α-spinasterol, β-spinasterol, stigmasterol, β-sitosterol, oleanotic acid and its derivatives and saturated (aliphatic) esters

Medicinal uses: It is diuretic, tonic and cooling. Juice of this plant, deemed beneficial to eyes, is an ingredient in the making of medicinal hair oils.

Other uses : It is used as an aquarium plant. Young shoots and leaves are eaten as a vegetable.

REFERENCE

Grubben, G. J. H. and O. A. Denton (Editors) 2004. Plant Resources of Tropical Africa 2. Vegetables. PROTA Foundation, Wageningen, Netherlands/ Backhuys Publishers, Leiden, Netherlands/ CTA, Wageningen, Netherlands. Pp. 668.

3.7. *Atropa belladonna*

Common names: Divale, Black Cherry, Devil's Herb, Great Morel, Deadly Nightshade, Devil's Cherries, Naughty Man's Cherries, Dway berry

Botanical name: *Atropa belladonna* L.

Family: Solanaceae

Vernacular names: Yebruj (Bengali); Angur Shefa, Luckmuna, Luckmunee, Sag angur (Hindi); Sagangur (Kashmiri); Suchi (Sanskrit); Bellatona, Pelletonacceti (Tamil); Bikh luffah, Poast bikh luffah (Urdu).

Origin and distribution: It is native to Europe. Widely distributed over central and southern Europe, south west Asia and Algeria; It is found in Kashmir at altitudes of 2000 to 3500m.

Habitat: Found in shade forests lies between 2500-3000m.

Habit: It is a perennial branching herb growing to 1.5m tall. The whole plant is glabrous, though soft, downy hairs may occur on the young stems and the leaves when quite young. Older plants attain a height of 90cm to 150cm, occasionally even 180cm, and the leaves grow about 30cm to 60cm from the ground (Fig.1).

Root: Roots are thick, fleshy and whitish, about 15cm long, or more and branching. The root in three years attains very large dimensions in fact, often so large as to make the lifting a very heavy.

Stem: The stem is purple in colour. It is stout, 60cm to 120cm high, undivided at the base, but dividing a little above the ground into three more rarely two or four branches, each of which again branches freely.

Leaf: Leaves are dull, dark green in colour, oval in shape, acute at the apex, entire and attenuated into short petioles. Leaves are of unequal size, 7.5cm to 25cm long. The lower leaves are solitary and the upper ones are in pairs alternatively from opposite sides of stem, one leaf of each pair much larger than the other. Their leaves are often larger than in fully grown plants and grow on the stem immediately above the ground. The veins of the leaves are prominent on the under surface, especially the midrib, which is depressed on the upper surface of the leaf.

Flower: The plant bears 1-2 flowers in the axils of the lower leaves of flowering branch, in the axil of upper leaves. The flower is bell shaped, about 2.5cm long, drooping, monopetalous, tubular, enlarged below, spreading above, with a short tube. There are five anther stalks, bent below the anthers, two shorter, thicker at the base, hairy, bent inwards at the top, and as long as the tube. The anthers are large and yellow, with slits, and double. The anther stalks lengthen after the anthers are ripe. The pistil is grooved both sides with a honey gland at the base. The stigma ripens first, and projects beyond the anthers. The style is thread like, longer than the stamens, inclined downwards; the stigma is pin headed, two lipped, green (Fig. 2).

Flowering, pollination and fruiting: The flowers appear in June and July and continue blooming until early September. The fruit ripens in July-September. The plant is adapted for cross pollination by medium sized humble bees, bees visiting it and also thrips. Honey is secreted at the base of the ovary and protected by stiff hairs on the stamens.

Fruit: The five cleft calyx spreads round the base of the smooth berry, which ripens, when it acquires a shiny black colour and is in size like a small cherry. It contains several seeds. The berries are globose, about 1.5cm across (Fig. 3).

Seed: Small, flat dull brownish black coloured.

Propagation: Propagation is through seeds, shoot cuttings or root cuttings.

Seed propagation: One gram seed contains about 700 seeds. However, rising of seedlings in nursery or direct broadcasting of seeds in the field is recommended for raising a commercial crop.

Seed germination: The germination of seeds is very poor and takes six weeks if sown in March-April but if sown in last week of June and first week of July, then the seeds germinate within three weeks with good per cent of germination.

Germination of the small seeds is often difficult, due to hard seed coats that cause seed dormancy. Germination takes several weeks under alternating temperature conditions, but can be increase with the use of gibberellic acid. The seedlings need sterile soil to prevent damping off and resent root disturbance during transplanting.

Fig. 1. Plant

Fig. 2. Flower

Fig. 3. Fruit

Seed treatment: Berezenegovskaja and Kuskova (1963) observed that if seeds are soaked for 24h in 0.02 to 0.05 per cent GA_3 solution and their subsequent stratification for six days to two weeks resulted into very high and rapid germination. One year old seeds have 100 per cent germination with above treatments. Seed treatment with sodium hydroxide @ 25 per cent concentration for three minutes significantly enhanced the germination (Gupta and Shah, 1971). Choudhary and Kaul (1973) reported increased germination in the seeds treated with thiourea @ 2 per cent. Menghini and Venanzi (1977) observed that soaking of belladonna seeds for 24h in 100-1000ppm GA_3 stimulates early germination. Formanowiez and Kozlovoski (1980) obtained very high germination by stratifying the seeds on wet sand under 5 to 12° or –2 to 2°C for 10 - 40 days and then putting them for germination on filter paper. The seed treatment with 1000-1400ppm GA_3 increased the seed germination (Suchorska and Ruminska, 1983).

Seed dispersal: The fruit falls around the parent plant or is dispersed by animals, birds, or man and sometimes by water.

Economic part used: Roots and leaves

Alkaloids: Atropine, Hyoscyamine

Medicinal uses: It is used as a tonic, antispasmodic, sedative, curing eye problems, gout, rheumatism and sciatica. Every part of the plant is extremely poisonous and can result in poisoning if not handle properly.

REFERENCES

Bereznegovskaja, L. N. and Z. R. Kuskova 1963. Effect of gibberellic acid on belladonna. *Fiziol. Rast.*, 10: 716-719

Choudhary, D. K. and B. L. Kaul 1973. Note on the effect of thiourea on the germination of *Atropa belladonna. Indian J. Agric. Sci.*, 43(10): 967-968.

Gupta, L. K. and S. C. Shah 1971. Effect of chemicals on germination of *Belladonna* seeds. *Prog. Hort.*, 3 (2): 17-20.

Formanowiez, H. and J. Kozowski 1980. Biology of germination of medicinal plants seeds. Part XII. A. Seeds of species from Solonaceae. *Herba Pol.*, 26 (1): 21-38.

Menghini, A. and G. Venangi 1977. The effect of growth regulators on the germination of seeds of varios medicinal plants. *Annatidella Facotta di Agararia, perugia,* 32(2): 771-783.

Suchorska, K. and A. Ruminska 1983. Changes *Atropa belladonna* L. and *Datura innoxia*, seeds germination process influenced by GA_3. *Acta Hort.,* **132** : 117-124

3.8. *Baliospermum montanum*

Common names: Red physic nut, Wild castor, Wild croton, Wild sultan seed

Botanical name: *Baliospermum montanum* (Willd.) Muell. Arg. Syn. *B. axillare, B. solanifolium, Jatropha montana*

Family : Euphorbiaceae

Vernacular names: Danti, Dantigaacha (Bengali); Danti, Hakum (Hindi); Damti, Kadu haralu, Nagadanti (Kannada); Baktumbo (Konkani); Ceriyadanthi, Nagadanthi (Malayalam); Danti, Hakum, Katari, Sapidi (Marathi); Anukula, Danti, Dantika, Dirgha, Erandhapatrika, Erandhaphala, Makulakah, Nagadanti, Nagavinna, Nikumbha (Sanskrit); Pei amanakku, Neeradi muthu, Nagadanti (Tamil); Adavi amudamu, Kondamudam, Nelajidi, Kanaka pata, nepalamu (Telugu).

Origin and distribution: The species is distributed throughout tropical and subtropical areas receiving rainfall above 1000mm, that is in Himalayan foothills, Kashmir to Khasi hills and particularly in Vindhyas southward. It is very common in North and East Bengal, Chhota Nagpur and peninsular India.

Habitat: Moist deciduous to semi-evergreen forests

Habit: The plant is a stout, monoecious leafy shrub upto 3.5m high, with toothed leaves and stiff branches arising from the root (Fig. 1).

Stem: Branchlets cylindrical, minutely tomentose with vertical lines. Latex watery.

Leaf: Leaves alternate, simple or shallowly three lobed, broadly ovate to elliptic lanceolate, base rounded or sub-cordate, apex acute to acuminate, margin entire or dentate, minutely hairy, with a pair of glands at the base of the leaves; lateral nerves 3-5 starting from the base and with 7-8 other lateral nerves. Leaf stalks upto 6cm long, minutely hairy.

Flower: The flowers of the plant are unisexual. In male flowers, the calyx is globose, 2.5mm long, four to five partite, glabrous or slightly pubescent, membranous, finely mottled with a disc of six glands. Stamens are about 20 in number. Female flowers have ovate-lanceolate and pubescent sepals, and a disc about 2.5mm in diameter (Fig. 2).

Flowering and Fruiting: Flowering and fruiting occurs during November to February.

Fruit: Capsule distinctly three lobed, obovoid, stony, 8-13mm across, minutely densely pubescent (Fig. 3).

Seed : Seeds ellipsoid, smooth, mottled and have oily endosperm (Fig.4).

Fig. 1. Plant

Fig. 2. Flower

Fig. 3. Fruit

Fig. 4. Seeds

Propagation : By seeds and terminal cuttings.

Seed propagation: Sow these fresh seeds on raised beds at a distance of more than 15cm and cover with sand. Water the seedbeds twice daily. Germination commences from the sixth day onwards and will be complete in twelve days. The average germination percentage is around 50. Transplant the seedlings soon after the plant has reached 5-7.5cm in height after about a month.

Seed collection: The best season for seed collection is from February to April. The seeds are collected when the outer tri-lobed covering around fruits starts turning brown and dry. Collect mature fruits in polybags and dry in sun. On drying, the pods open and liberate seeds.

Seed treatment: Seeds require mechanical scarification pre-treatment before sowing to obtain higher seed germination. Soaking of seeds in water for two to three hours and then drying in shade before sowing also promotes germination. It takes about 25 days for complete germination to take place.

Vegetative propagation: Semi-mature, stem cuttings with at least three nodes, measuring 15cm long are used. The cuttings may be treated with rooting hormones such as IBA, NAA or Boric acid and they should be kept in shade houses or mist chambers for better rooting. Plant the cuttings in polybags filled with loamy soil or plant on raised beds. This will also improve the sprouting per cent to nearly 90 to 100 per cent. The stock raised through cuttings takes longer time than seedlings to attain transplantable size.

Economic part used : Roots, seeds, leaves

Alkaloids: Axillarenic acid is present in the seeds, while 12- deoxy-5β-hydroxy-phorbol-13-myristate, 13-palmitate, 12-deoxyphorbol- 13-palmitate, baliospermin, and montanin are reported to be present in the roots (Anon., 2008).

Medicinal uses : Roots, seed, leaves and seed oil are used to treat jaundice, constipation, piles, anemia, conjunctivitis and in wounds it acts as tearing agent for ripe abscess. Root is purgative, anthelmintic, carminative, rubefacient and anodyne. This is used in abdominal pain, constipation, calculus, general anasarca, piles, helminthic infestation, scabies and skin disorders. Root paste is applied to painful swellings and piles. Leaves cure asthma and seeds are used in snakebite.

REFERENCE

Anonymous. 2008. Agro-techniques of selected medicinal plants, Vol. 1. National Medicinal Plants Board, New Delhi.

3.9. *Barleria prionitis*

Common name: Porcupine flower

Botanical name: *Barleria prionitis* L.

Family: Acanthaceae

Vernacular names: Zanti (Bengali); Kantashario, Vajradanti (Gujarati); Kantasaria, Vajradanti (Hindi); Mullu Goranti (Kannada); Kuttivetila (Malayalam); Koranthi (Marathi); Kurantak (Sanskrit); Kundan (Tamil); Goredun (Telugu).

Origin and distribution: Porcupine flower is a native of Tropical East Africa and Asia. Globally distributed in many parts of the world such as United States of America, Australia and parts of Asia in Indonesia, Malaysia, India and Philippines.

Habitat

Porcupine flower grows in a wide variety of well drained soils derived from igneous, metamorphic (including ultramafics) and sedimentary (including limestone) rocks. In the Puerto Rican naturalized range, it grows in areas receiving from about 750 to 900mm of mean annual precipitation. Within the native range, it probably tolerates both higher and lower precipitation. It is found throughout the hotter parts of India and defoliates annually during the dry season. The species is moderately intolerant of shade, growing in both full sunlight and under light forest canopies.

Porcupine flower may grow as single plants or in large, tangled thickets. It is most often found on roadsides, bluffs and bars above streams, overgrazed range, disturbed areas and farmsteads. Widely growing along the hedges and neglected areas near habitations, open scrub forests and rocky habitats throughout the tropical parts of India from south upto foothills of Himalayan ranges.

Habit: It is an erect, prickly shrub, usually single stemmed, growing to about 1.5m tall (Fig. 1).

Root: It has a central tap root with lateral roots branching off in all directions.

Stem: Plants commonly have single stem, but may have multiple stems or branches from near the ground. The stems and branches are stiff, round, and light tan or light gray.

Leaf: Leaves are upto 100mm long and 40mm wide, narrowly to broadly elliptic, lanceolate, entire and appraised, hairy beneath, the entire apex spine 10–20mm long, tipped; tender leaves usually soft hairy beneath.

Flower: Inflorescences are auxiliary and terminal spiked clusters, many flowered, upto 10cm long with long projecting stamens. Flowers are sessile, borne on spine tipped, leaf like bracts, bright orange - yellow in colour (Fig. 2).

Flowering, pollination and fruiting: Flowering and fruiting occurs during August to November. It is pollinated by insects and attracts various species of butterflies (Little and Jones, 1980).

Fruit: Capsules 1.5-2.5cm long, brown, minutely hairy, oval shaped, beaked (Fig. 3).

Fig. 1. Plant

Fig. 2. Flower

Fig. 3. Fruit

Fig. 4. Seeds

Seed : Seeds are two per capsule, orbicular in outline, compressed and hairy. The seeds are 8 by 5mm, flattened and covered with matted hairs (Fig.4).

Seed dispersal: The seeds are transported by water. Also, the seed spread may be quicker along paths or roads. Natural regeneration generally occurs within 1 or 2m of the parent plant in Puerto Rico, perhaps because animal vectors for seed dispersal are lacking. Seedlings can be abundant near fruiting plants.

Propagation : Barleria reproduces by seed, one plant producing upto hundreds of seeds in a season. Seeds require moist conditions to germinate. Sixty fruits collected in Puerto Rico averaged 0.0998 + 0.0044g/fruit. Air dried seeds separated from the same collection averaged 0.0305 + 0.0005g/seed or 33,000seeds/kg. Forty four per cent of the seeds sown on commercial potting mix germinated between 13 and 77 days following sowing.

Seedlings of porcupine flower grow slowly at first. However, established plants add upto 0.5m per year to their height and nursery seedlings reach 0.5m in six months.

Economic part used : Whole plant.

Alkaloids: Iridoid glycosides, barlerin and verbascoside.

Medicinal uses : It has numerous medicinal properties including treating fever, respiratory diseases, toothache, joint pains and a variety of other ailments; and it has several cosmetic uses.

Other uses: Flower furnishes cover for wildlife and protects the soil against erosion. It is widely planted as an ornamental and cultivated in Asia as a hedge plant.

REFERENCE

Little, J. R. and C. E Jones 1980. A dictionary of botany. Van Nostrand Reinhold Company, New York, Cincinnatti, Toronto, Melbourne

3.10. *Boerhavia diffusa*

Common names: Spreading hogweed, Red hogweed, Tar vine, Red spiderling, Horse Purslene

Botanical name: *Boerhavia diffusa* L. Syn. *B. repens* L.; *B. repens* var. *diffusa*; *B. procumbens* Banks ex Roxb.

Family : Nyctaginaceae

Vernacular names: Ranga Punarnabha (Assamese); Punurnava (Bengali); Dholia-saturdo, Moto-satoda (Gujarati); Gadapurna, Lalpunarnava, Snathikari, Survari (Hindi); Sanadika, Kommeberu, Komma, Rakta punarnava, Thopparu (Kannada); Vanjula Punarnava (Kashmiri); Tavilama, Talutama (Malayalam); Ghetuli, Vasuchimuli, Satodimula, Punarnava, Khaparkhuti, Tambadivasu (Marathi); Lalapuiruni, Nalipuruni (Oriya); ltcit, Khattan (Punjabi); Punarnava, Raktakanda, Shothaghni, Varshabhu (Sanskrit); Punarnava, Santhi, Bishkhopra, Sant, Pannai keerai, Mukkarattei, Siru kodi, Saattaranai, Mookkarai saranai (Tamil); Erra galijeru, Punarnava, Atikamamidi (Telugu).

Origin and distribution: It is indigenous to India; it is found throughout the warmer parts of the country upto an altitude of 2000m in the Himalayan region (Chopra, 1969). The plant is also cultivated to some extent in West Bengal (Anon., 1988).

Habitat: A common weed of pastures and waste land.

Habit: A perennial diffuse herb; stem profusely branching, angular, reaching a length of 150cm (Fig. 1).

Root: It has a large root system bearing rootlets. The tap root is tuberous, cylindrical to narrowly fusiform to conical or tapering, light yellow, brown or brownish gray. It is thick, fleshy and very bitter in taste.

Stem: Stem is prostrate, decumbent or ascending, 4-10cm long, rather slender, divaricately branched.

Leaf: Leaves are opposite in unequal pairs, larger ones 25-37mm long and smaller ones 12-18mm long, ovate-oblong or suborbicular, apex rounded or slightly pointed, base sub-cordate or rounded, green and glabrous above, whitish below, margin entire or sub-undulate, dorsal side pinkish in certain cases, thick in texture, petioles nearly as long as the blade, slender.

Flower: Flowers are very small, pink coloured, nearly sessile or shortly stalked, 10-25cm, in small umbells, arranged on slender long stalks, 4-10 corymb, axillary and in terminal panicles, bracteoles, small, acute, perianth tube constricted above the ovary, lower part greenish, ovoid, ribbed, upper part pink, funnel shaped, 3mm long, tube five lobed, stamen 2-3 (Fig. 2).

Flowering, pollination and fruiting: Flowering and fruiting occurs throughout the year, when sufficient water is available. The seeds germinate before the onset of the monsoon. The plant grows profusely in the rainy season and mature seeds are formed in October - November. The flowers are self-pollinated when not cross pollinated. The flowers offer nectar for pollinators, and the flowers are visited by small, medium and large bees, small and medium flies; ants also visited.

Fruit: Fruits are obovoid or sub-ellipsoid, rounded above, slightly cuneate, below, broadly and bluntly five ribbed, very glandular throughout (Fig. 3).

Seed : Seed is obovoid, pale brown (Fig. 4).

Seed dispersal : The mucous coat of the anthocarp shows a distinct sticky swelling when ripe, with which it clings to mammals and birds which helps in its dispersal from one place to another.

Propagation : It is propagated by seeds and rootlets.

Seed propagation: The seeds germinate before the onset of the monsoon. Seeds germinate in seven or eight days. The germination is 70 to 90 per cent. The optimum temperature for seed germination is around 25°C. According to Spellenberg (2000), the seeds have dormancy.

Economic part used : Whole plant

Alkaloids : Punarnavine

Medicinal uses : According to Ayurveda, punarnava is bitter, cooling, astringent to bowels, useful in biliousness, blood impurities, leucorrhoea, anaemia, inflammations, heart diseases, asthma, alternatives etc. The leaves are useful in dyspepsia, tumours, spleen enlargement and abdominal pains. According to Unani system of medicine, the leaves are appetizer, alexiteric, useful in opthalmia, in joint pains. Seeds are tonic expectorant, carminative, useful in lumbago, scabies. The seeds are considered as promising blood purifier. In many parts of India, different parts of punarnava are used as folk medicine.

Fig. 1. Plant

Fig. 2. Flower

Fig. 3. Fruit

Fig. 4. Seeds

Past work done on Seed Science and Technology

Satsangi *et al.* (2010) determine the allelopathic effect of wheat straw on *Boerhavia diffusa* L. Straw treatments linearly decreased the germination and the length of roots and shoots of the test weeds relative to controls. At the higher concentrations of wheat straw leachates, seed germination was affected in *Boerhavia diffusa* under burned and unburned straw treatments, whereas, at lower concentrations (25%) of the burned straw treatment, it remained similar to the water control. Seedling growth was less affected under lower concentrations. Thus, straw management practices should be considered when developing an integrated weed management program (Table 1).

Table 1. Effect of wheat straw leachate on the seed germination (%), root length (cm) and shoot length (cm) of *Boerhavia diffusa* (*in vitro*). Values are the mean ± SD

Leachate concentration (%)	Leachate (unburned straw)			Leachate (burned straw)		
	Germination	Root length	Shoot length	Germination	Root length	Shoot length
0 (Water control)	62.5 ± 2.9	2.0 ± 0.24	4.0 ± 0.30	62.5 ± 2.7	2.0 ± 0.27	4.0 ± 0.19
25	50 ± 2.0	1.8 ± 0.25	3.5 ± 0.26	51.25 ± 2.7	2.0 ± 0.22	4.0 ± 0.23
50	40 ± 2.2	1.5 ± 0.3	3.5 ± 0.26	45 ± 2.3	2.0 ± 0.34	3.8 ± 0.34
100	37.5 ± 2.8	1.5 ± 0.27	2.8 ± 0.19	47.5 ± 2.1	1.5 ± 0.27	2.0 ± 0.27
CD at 5%	3.2	0.22	0.34	4.6	0.76	0.63

CD = critical difference; level of significance ($P<0.05$)

REFERENCES

Anonymous, 1988. The Wealth of India: Raw Materials. CSIR Publication, New Delhi, India, p. 174.

Chopra, G. L. 1969. Angiosperms. Systematics and Life Cycle. S. Nagin & Co., Jalandhar, Punjab, India. pp. 361–365.

Satsangi, G. P., G. Bartariya and A. Saxena 2010. Allelopathic response of wheat (*Triticum aestivum* L.) straw on germination and growth of selected weed species. *Seventeenth Australasian Weeds Conference*. 147-149.

Spellenberg, R. 2000. Blooming behavior in five species of *Boerhavia* (Nyctaginaceae). *SIDA Contributions to Botany* 19(2):311-323.

3.11. *Calotropis gigantea*

Common names: Milkweed, Swallow wort, Crown flower, Crown plant, Giant milkweed, Rubber bush, Bow string hemp

Botanical name: *Calotropis gigantea* (L.) R.Br. Syn. *Calotropis procera*

Family : Asclepidaceae

Vernacular names: Akand, Akan (Assamese); Akanda, Akone (Bengali); Aakado (Gujarati); Aak, Alarkh, Madar (Hindi); Ekka, Ekkadagida, Ekkegida (Kannada); Acka (Kashmiri); Erikku (Malayalam); Hiran Dodi, Shendri, Ghaephul, Dhora, Amri (Marathi); Arakha (Oriya); Ak (Punjabi); Ravi, Bhinu, Tapana (Sanskrit); Vellerukku, Erukku, Erukkan (Tamil); Jilledu (Telugu); Madar, Aak (Urdu).

Origin and distribution: Calotropis is native to Cambodia, Indonesia, Malaysia, Philippines, Thailand, Sri Lanka, India and China (Forster *et al.*, 1996). It has been widely naturalized throughout the New World tropics including the Caribbean and on the continent from Mexico to Brazil.

Habitat: It is common and scattered in dry coastal areas, on beaches and to 180m altitude. Being unpalatable to sheep and cattle, it tends to spread and become common on heavily grazed pastures. It grows on a wide variety of soils, sometimes growing where few plants will grow.

Habit: Large shrub, growing to 4m tall, trunk stout, pale, shoots woolly, perennial (Fig. 1).

Stem: As the plant becomes more tree like it produces a hairy crown of a few twisted branches. There is often a significant amount of dead stems in the canopy of these small trees. The bark is whitish, thick and corky. The wood is whitish and soft. All parts of the plant produce abundant white sap.

Leaf: The leaves are opposite, entire, nearly stalkless or sessile, with broad light yellow mid-vein and main veins. They are broadly elliptic or oblong and heart shaped at the base. The leaves are slightly leathery and pale green or green above. When young, they are whitish green below being covered by a coat of fine soft hair that easily rubs off. Leaf blades are 8 to 20cm long and 5 to 13cm wide.

Flower: The inflorescence is an umbel like cyme at or near the ends of twigs. Many flowers are borne on the inflorescence. The five sepals are 0.5cm long. The corolla has five showy petals that are whitish and tinged with purple at the apex. They are about 2cm long (Fig. 2).

Flowering and pollination: Flowering occurs during winter season. Calotropis is an example of entomophilous pollination and pollination is achieved with the help of bees. The pollen is arranged in a structure named pollinia which is attached to a glandular, adhesive disc at the stigmatic angle. These sticky discs get attached to the legs of visiting bees so that pollinia are pulled out when the bee moves away. When such a bee visits another flower, that flower gets pollinated by the sticky pollinia.

Fruit: Flowers are replaced by a kidney shaped fruit, 7 to 10cm long. It is green and becomes brown when mature. The fruit contains many seeds embedded in a rough fibre. The fruit is a follicle. Follicles recurved, two or one follicles, second more often suppressed, 7.5-10cm long (Fig. 3).

Seed : Seeds are broadly ovate, comose (Fig. 4a and 4b).

Seed dispersal *:* When dry, seed dispersal is by wind. Seeds spread by wind and water over large distances (Smith, 2002).

Seed dormancy and germination: Older seed will not require the cold period to break its dormancy.

Propagation : By seeds and vegetative means *viz.*, rooted cuttings; layers

Fig. 1. Plant

Fig. 2. Flower

Fig. 3. Fruit

Fig. 4a. Seeds

Fig. 4b. Seeds

Vegetative propagation: Cut the branches from a healthy growing established at least 25 to 30cm long. The branch will weep some sap so wait until it stops, place some sprouting hormone or growth stimulating liquid at the tips of the branch. Take the cuttings for propagation and place them at least 2.5cm apart on a flat that has a mixture of vermiculite and black sand. Make sure to put lots of water over all of the cuttings, then cover the entire flat with a thin sheet of plastic, this caused condensation and hold the moisture in the flat. Once notice the cutting start to have little green leaves growing on them, take of the plastic cover, but maintain the moist soil. Then, the roots starting to form and cut the branch and transfer into the pots.

Economic part used : Leaf, root, latex and flower

Alkaloids: The latex contains the cardenolides calotropin, calotoxin and uscharin as well as the proteinase calotropain. The cardiac glycosides calotroposide A and B have been identified in the root bark, together with ß-calotropeol and giganteol, the pregnanone calotropone, several ursane-type triterpenoids and giganticine, a non-protein amino acid. Giganteol, α- and ß-calotropeol were also isolated from the stem, as was ß-calotropeol from the flowers. In addition, the leaves contain the cardenolides calotropin and gigantin, and the roots calotropin, frugoside and 4'-O-ß-D-glucopyranosylfrugoside. In addition to cardiac glycosides, several triterpenes, cyclosadol, cycloart-23-ene-3ß–25-diol, α-lactucerol and lupeol have been isolated from aerial parts.

Medicinal uses : Traditionally it is used alone or with other medicinal plants to treat common disease such as fevers, rheumatism, indigestion, cough, cold, eczema, asthma, elephantiasis, nausea, vomiting and diarrhoea.

Agricultural uses: Different plant part extracts or dried powdered parts, and especially whole plant extracts showed significant anti-feeding, insecticidal, and growth inhibitor activities of the seed storage pests *Callosobruchus maculatus, Sitophilus oryzae* and *Tribolium castaneum* adults, third instar of *Spodoptera exigua* larvae, and eggs of *Helicoverpa armigera.* The latex showed significant nematicidal properties against the root knot nematode, *Meloidogyne incognita* and cowpea cyst nematode, *Heterodera cajani.* Seed treatment with the dry leaf powder of *Calotropis gigantea* resulted in the highest seed germination (98%), and the highest percentage of seedling establishment. Root dip treatment with leaf extracts resulted in a significant reduction of *Meloidogyne incognita* population in the soil at 45 days after transplanting and at harvest, and in increased fruit yield by 24 per cent. Giganticine exhibited a significant antifeedant activity against nymphs of the desert locust *Schistocerca gregaria.* Fresh leaves at an application rate of 200kgha^{-1} showed significant molluscicidal activity against golden apple snail (*Pomacea canaliculata*) in rice. Different extracts of the aerial parts and leaves showed significant antifungal properties against the fungal pathogens *Phyllactinia corylea* (powdery mildew), *Peridiopsora mori* (brown rust), *Pseudocercospora mori* (black leaf spot), *Myrothecium roridum* (brown leaf spot) and different *Fusarium* spp. in mulberry (*Morus* spp). Antonysamy (2004) reported that aggergatum onion treated with three per cent of panchagavya along with Notchi, *Calotropis, Nerium, Aloe,* and Pungam in cow's urine for seven days increased the yield of bulbs by 17.5t.ha^{-1}.

Other uses: It yields a durable fibre useful for ropes, carpets, fishing nets, and sewing thread. Floss obtained from seeds is used for stuffing purposes. Fermented mixture of *Calotropis* and salt is used to remove the hair from goat skins for production of "nari leather" and of sheep skins to make leather which is much used for inexpensive book binding. The flowers last long and in Thailand, they are used in various floral arrangements.

REFERENCES

Antonysamy, 2004. Organic aggregatum onion better packaging and quality. The Hindu News paper. May 20. pp.10.

Forster, P. I., D. J. Liddle and A. Nicholas 1996. Flora of Australia. Asclepidaceae, *Australian Government Publishing Service.* 28: 214 Canberra.

Smith, Nicholas M. 2002. Weeds of the wet/dry tropics of Australia - a field guide. Environment Centre NT, Inc. pp. 28-29.

3.12. *Cassia angustifolia*

Common names : Indian Senna, Tinnevelly Senna

Botanical name: *Cassia angustifolia* Vahl.

Family: Fabaceae

Vernacular names: Sonamukhi (Assamese); Svamamukhi, Sonapata (Bengali); Mindhiaval, Sonamukhi (Gujarati); Hindisana, Sanaya, Sanaka patt (Hindi); Nelavarika, Sonamukhi, Nelaavare, Nelavarike, Nela Aanriake (Kannada); Sena (Kashmiri); Sunnamukhi, Nilavaka, Chinnukki, Adapatiyan (Malayalam); Sonamukhi (Marathi); Sunamukhi (Oriya); Sannamakhi, Sanapati, Sarnapatta (Punjabi); Swarn patri (Sanskrit); Avuri, Nilapponnai, Nilavari, Nilavaikkai, Tirunelveli senna (Tamil); Sunamukhi (Telugu); Sena, Barg-e-Sana (Urdu).

Origin and distribution: It is native to West Asia and distributed in India and Pakistan. In India, it is widely cultivated in Gujarat, Tamil Nadu, Andhra Pradesh, Karnataka, Maharashtra and Rajasthan. The crop is cultivated in Tirunelveli and Ramnathapuram districts of Tamil Nadu, Guntur district of Andhra Pradesh in India and in Sudan.

Habitat: It is a sun loving crop and requires bright sunshine for its successful growth. It is a hardy warm weather crop grown under rainfed and irrigated conditions. It requires dry climate and very low rainfall. This plant can withstand a minimum temperature of 4°C and a maximum of 50°C. It requires very low rainfall. The areas receiving annual rainfall of only 30cm are also suitable for its cultivation. An average annual rainfall of 25-40cm, distributed from June to October, is sufficient for a good harvest. The soil which is prone to water stagnation / logging is not suitable. Though it can be taken up in all types of soils (except saline soils), sandy to loamy soils are best suited. As a pure crop, it is raised on red loams including coarse gravely soils or alluvial loams ranging in pH from 7.0 to 8.5 and even on the rich clayey rice fields.

Habit: It is small perennial, drought resistant under shrub, branched and grown upto 0.7-1.0m height (Fig. 1).

Leaf: Compound pinnate leaves, having 5-8 pairs of shortly stalked leaflet, the leaflets are oval-lanceolate, 2-5x1.5cm, pale green glabrous with smooth margin, acute, apex, oblique base.

Flower: It produces 30-45cm long axillary or sub-terminal raceme, bearing many large brilliant yellow showy flowers. Bracts are membranous, ovate or obovate, caducous. Sepals are obtuse, membranous (Kirtikar and Basu, 2006). Dichogamous with seven stamens and three staminodes, pollen grains are triangular, smooth, 2.8-4.0μ and highly fertile (Fig. 2).

Flowering, pollination and fruiting: The life of a flowering axis is around 35 days and flowering continues till the close of the season. Matured fruits can be harvested at 120 days after sowing. The main flowering season was during August-September, but due to moisture availability, this was also seen in March- May. Fruit set was not evident in apomixis, cross pollinations and natural pollinations during March- May; and apomixis, selfing and natural pollination during July-September. The reproductive capacity was highest (49.8%) during July-September, and the rest of the flowers dropped before fertilization. Low fruit set (0.4-2.05%) was observed under artificial cross pollination and sibbing during July-September. Selfing and sibbing during March- April suggested that the species have both self and cross pollination systems. Of these, cross pollination was dominant (Jindal and Singh, 2003).

Fruit: The pods are slightly curved light green in colour, changing from greenish brown to dark brown or black on maturity. The pods are flat, thin, 3.5 - 6.5cm x 1.5cm in dimension (Fig.3).

Seed: Each pod containing 5-8 obovate yellow to off-cream coloured flat seeds with hard and tough seed coat.

Propagation: By seeds.

Physiological maturity: The seeds attain physiological maturity at 40 days after anthesis associated with colour change of pods from green to brown.

Seed extraction: The pods are harvested at maturity; a second crop pods is obtained in the end of the season when the plants are uprooted during August or April- May. The pods are dried in a thin layer in shade for 4-7 days; the pods are lightly beaten during drying to remove the inert seed material.

Seed grading: Seeds can be processed by using 8/64" round perforated metal sieves.

Fig. 1. Plant Fig. 2. Flower

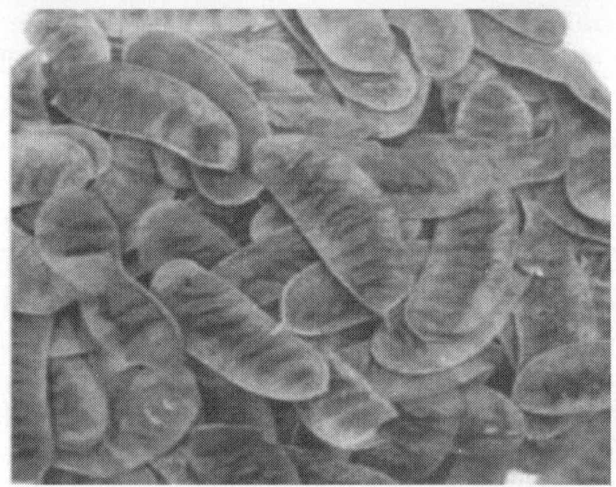

Fig. 3. Fruit

Seed dormancy breaking treatments: The seeds exhibit dormancy for 2 months. Abrasion of seed coat is essential to induce germination. This is obtained by pounding the seeds lightly with coarse sand in a mortar. A traditional method adopted for removing the hard seed coat is mixing the required quantity of seeds with dry and pure sand in the ratio of 1:3 and is gently beaten.

According to Al-Helal *et al.* (1989), seed treatment with H_2SO_4 was more effective for removal of seed coat dormancy than mechanical or boiling water or incision of testa. Hard seedness can be effectively overcome by acid scarification with commercial sulphuric acid @100ml/kg seed for 10min. Bhuse *et al.* (2001) reported that treating the seeds with H_2SO_4 for 12min gave highest germination of 72 per cent.

The seeds should be soaked in water for overnight and then, these will be dried under shade for one hour to enhance the percentage of germination. Seed treatment with thiram or captan (2.5g / kg seed) before sowing is advocated. The seeds start germination in 5-7 days and complete germination in 15 days after sowing.

Seed storage: Seed treatment with carbendazim 50% WP @ 2g/kg of seeds and halogen formulation (Bleaching powder + $CaCO_3$ + Arappu leaf powder @ 5:4:1) @ 3g/kg seed and packed in moisture vapour proof container prolongs viability upto 15 months.

Seed germination: Sow the seeds either during the months of January-February or June - July. Germination starts after 13 to 15 days of sowing and is over in another one week.

Economic part used: Dry leaves and dry pods

Alkaloids: In the leaf, sennosides A and B based on the aglycones sennidin A and B, sennosides C and D which are glycosides of heterodi anthrones of aloeemodin and rhein are present. Others include palmidin A, rhein anthrone and aloeemodin glycosides, some free anthraquinones and some potent, novel compounds of as yet undetermined structure. Senna usually contains more of the sennosides. In the fruit, sennosides A and B and a closely related glycoside sennoside A1. Naphthalene glycosides; tinnevellin glycoside and 6-hydroxymusizin glycoside.

Medicinal uses: India is the largest producer and exporter of senna leaves, pods and sennoside concentrate in the world market. It lowers bowels, increases peristaltic movements of the colon by its local action upon the intestinal wall. It is used as expectorant, wound dresser, antidysentric, carminative, purgative and laxative. Useful in loss of appetite, hepatomegaly, spleenomegaly, indigestion, malaria, bronchitis, skin diseases, jaundice and anaemia. Leaves are made into a paste and applied to various skin diseases.

Other uses: Hussain *et al.*, (2007) reported that senna can be successfully used in management of weeds, especially for *Avena fatua* in wheat. The senna can be used as an allelopathic agent and will be a new but eco-friendly, cheaper and effective mode of weed control.

Past work done in seed science and technology

According to Suryawanshi *et al.* (1998) top of paper media was found to be the best for conducting germination test.

Kalavathy (1996) carried out the experiments on seed technological aspects in senna and periwinkle and the findings are given below.

(a) In senna, higher pod and seed yield plant^{-1} associated with higher shelling outturn and hard seed percentage were obtained from the crop sown in January. The crop sown in April took more number of days for flowering and maturity. The qualities of seeds produced were better from November - December and June sowings.

(b) In senna and periwinkle, the seeds attained physiological maturity at 40 days after anthesis associated with the colour change from green to brown in cassia and from green to transluscent yellow fruits with black seeds in periwinkle.

(c) For processing the bulk seeds of senna, 8/64" (3.2mm dia) sieve is recommended which gave higher recovery of quality seeds. The seed lot of periwinkle can be hand processed using BSS 16 wire mesh sieve (1.4mm dia) with a high recovery of quality seeds.

(d) The hard seeds prevailed upto 67 per cent in senna and to overcome this, treatment with commercial H_2SO_4 @ 100ml kg^{-1} (v/w) for 10min. followed by thorough washing, significantly improved the germination without affecting seed quality.

(e) For germination test, the optimum temperature required is 30°C constant or 25/30°C alternating for all the two kinds of seeds. The seeds of senna and periwinkle germinated well in sand medium. The final count of the germination tests can be taken on 10th day in senna and on 14th day in periwinkle.

(f) The seeds of periwinkle possessed dormancy upto three months with initial germination of 22 per cent. Upon breaking of dormancy, the germination has reached the peak value of 95 per cent within six months.

(g) The seeds of senna and periwinkle can be safely stored at 8-9 per cent moisture content with high vigour and viability under ambient conditions in cloth bag or polyethylene bag with or without thiram treatment (2g kg^{-1}) over 15 months.

Ramasamy (2006) carried out the experiments in periwinkle and senna to evolve a suitable seed production technology and findings of the experiments are given below.

(a) The seeds attained physiological maturity at 40 days after anthesis associated with the colour change from green to yellowish green pods with black seeds in perwinkle and green to brownish green pods with yellow seeds in senna.

(b) Foliar application of panchagavya (4%) significantly increased the yield parameters whereas, resultant seed quality parameters were found to be maximum by the foliar application of 0.2 per cent humic acid in both periwinkle and senna.

(c) Standardization of quick viability test revealed that soaking in 0.1 per cent (or) 0.2. per cent tetrazolium chloride solution for three hours and soaking in 0.1 per cent tetrazolium chloride solution for two hours (or) 0.2 per cent tetrazolium chloride solution for one hour was found to be optimum for assessing the viability with a proper staining intensity for periwinkle and senna, respectively.

(d) The germination of periwinkle could be improved by seed soaking with GA_3 @ 600ppm for 12h (or) KNO_3 (1.0%) for 24h. The periwinkle seeds stored for three months improved the germination to 92 per cent with higher vigour potential.

(e) The periwinkle and senna seeds stored in 700 gauge polythene bag with halogen mixture 3g/kg seed treatment proved to be the best practices for storing seeds with minimum loss in viability and vigour.

REFERENCES

Al-Helal, A. A., M. M. Al-Farraj, R. Ael-Desoki and I. Al-Habashi 1989. Germination response of *Cassia angustifolia* L. seeds to sodium, salt and temperature. *J. Univ. of Kuwait Sci.*, 16(2) : 281-287.

Bhuse, V. H., B. L. Lad and S. T. Ghule 2001. Effect of various seed treatment for enhancement of seed germination and seedling vigour in senna (*Cassia angustifolia*). *Res. Crops,* 2(3) : 359-363.

Hussain, S., S. U. Siddiqui, S. Khalid, A. Jamal, A. Qayyum and Z. Ahmad 2007. Allelopathic potential of senna (*Cassia angustifolia* Vahl.) on germination and seedling characters of some major cereal crops and their associated grassy weeds. *Pak. J. Bot.*, 39(4): 1145-1153.

Jindal, S. K. and M. Singh 2003. Breeding system in senna. In: Proceedings of the National Symposium on arid legumes, for food nutrition security and promotion of trade, Hisar, India, 15-16 May 2002 (Eds.) Henry, A., D. Kumar, N. B. Singh *Advances in Arid Legumes Research*. pp. 144-147

Kalavathy, D. 1996. Seed production, processing, testing and storage studies in medicinal plants of senna (*Cassia angustifolia* Vahl), periwinkle (*Catharanthus roseus* g. Don cv. *roseus*) and roselle (*Hibiscus sabdariffa* Linn.). Ph.D. thesis, TNAU, Coimbatore.

Kirtikar, K. R. and B. D. Basu 2006. Indian medicinal plants, Vol.II. International Book Distributors, Dehradun.

Ramasamy, V. 2006. Seed technological studies in periwinkle (*Catharanthus roseus* (L.) G. Don cv. *Roseus*) and senna (*Cassia angustifolia* Vahl). M.Sc (Ag.) Thesis, TNAU, Coimbatore.

Suryawanshi, Y. B., R. B. Patil, and N. D. Moholkar 1998. Studies on germination procedure in some medicinal species. *Seed Tech. News.,* 28: 57-58.

3.13. *Cassia auriculata*

Common name: Avaram Senna

Botanical name: *Cassia auriculata* L.

Family : Caesalpiniaceae

Vernacular names: Tarvar, Tarval (Hindi); Thangadi (Kannada); Ponnaviram (Malayalam); Tarvad, Aval (Marathi); Avartaki (Sanskrit); Aavaarai (Tamil); Thangedu (Telugu).

Origin and distribution: Plant occurs throughout India and Sri Lanka. Distribution of the plant is pan-tropical. It is a small bush, which grows wild in South India with flowers and pods throughout the year (Matthew, 1983).

Habitat: A gregarious weed on black cotton soil, dry, stony and hilly land. It thrives well along roadside, in degraded forests and on wastelands.

Habit: Tall, branching shrub; branches and underside of leaves grayish (Fig. 1).

Stem: Branchlets finely pubescent.

Leaf: Leaves are compound, even pinnate, arranged in alternate manner, 8-12cm long, with an erect linear gland between the leaflets of each pair; leaflets 12-24, upto 3cm long and 1.5cm broad, oblong in shape, blunt at both ends, apex mucronate, margins slightly overlapping, smooth or very thinly pubescent with short stalk. Leaves are subtended below by prominent, persistent, ear shaped stipules, which is unique character to this species.

Flower: Flowers large, yellow, showy, in dense corymbs; calyx five lobed, tube short; corolla imbricate, petals five, subequal; stamens ten, 6-7 fertile, rest staminodes; ovary superior, one celled, many ovuled, style incurved, stigma truncate (Fig. 2).

Flowering, pollination and fruiting: Flowering and fruiting is almost throughout the year, but in India there are usually two main flowering periods, one in the early monsoon and another in the late monsoon. It is cross pollinated plant.

Fruit: Pods are flat, thin, papery, oblong, obtuse at each end, stalked, pale brown, transversely veined, pubescent, 10-20 seeded, ligulate, straight, glabrous, flexible with a distinct space between the uniseriate seeds (Fig. 3a and 3b).

Seed: Seeds are minute, quadrangular bluntly pointed with 5-6 shallow transverse furrows. The average weight of the seed is 0.13g (Fig. 4).

Propagation : By seeds and stem cuttings.

Seed germination: Collection of seeds can be carried out more or less throughout the year. These seeds are soaked in cold water for 48h and then, sown in polybags or in raised mother beds. It takes 10-15 days for germination. The obtained per cent of germination is 75. In case of raised mother beds, seedlings of 5cm height can be transplanted to polythene bags. Six month old seedlings can be planted in the field.

Seed dispersal: Seeds are dispersed when the legume fruit splits open.

Seed dormancy and germination: For quick germination, seeds are scarified and held in running water. The seedlings are fairly resistant to desiccation. Stem cuttings are planted 5 – 12.5cm apart in rows.

In most leguminosae, the inner ovular integument is completely destroyed during seed development and does not take part in seed coat formation (Prakash, 1987).

Fig. 1. Plant

Fig. 2. Flower

Fig. 3a. Immature pod

Fig. 3b. Mature pod

Fig. 4. Seeds

Vegetative propagation: Hard stem cuttings with 5-6 nodes are most suitable. Cuttings are planted in polybags in a way that two nodes inside the soil and two nodes in aerial. Soil mixture containing sand, soil and manure in ratio of 2:1:1 favours early sprout. Sprouting starts within 15 days, rooting after 3-4 weeks. Observed success rate by this method is 60 per cent.

Economic part used : All parts

Medicinal uses : This plant is reported to contain cardiac glucoside and pacifies vitiated pitta and kapha complaints. A decoction of root is used against fevers, diabetes, diseases of urinary system and constipation. Bark contains tannins. Leaves yield anthraquinone and have laxative properties. The dried flowers and flower buds are sun dried and powdered; this powder is used as tea in case of diabetes. It is also used to improve the complexion in women. The powdered seed is applied to cure eye diseases, gonorrhoea and gout.

Other uses: Its numerous stems and flowering twigs make this shrub as a focal point in garden landscape. It can be planted as single specimen, regular clump or as running hedge.

REFERENCES

Matthew, K. M. 1983. The Flora of the Tamil Nadu Carnatic. The Diocesan Press, Madras, Tamil Nadu, India.

Prakash, N. 1987. Embryology of the leguminosae. **In:** Advances in legume systematic. [Ed.] Stirton C. H. Royal Botanical Gardens, Kew, UK. 241 – 278

3.14. *Cassia tora*

Common names: Chinese senna, Foetid cassia, Java bean, Low senna, Peanut weed, Stinking cassia, Sickle pod, Sickle senna, Wild senna.

Botanical name: *Cassia tora* L. syn. *Cassia obtusifolia* L., *Senna tora*

Family : Fabaceae

Vernacular names: Bon medelua, Dari diga, Medeluwa (Assamese); Panevar, Chakunda (Bengali); Kawaria (Gujarati); Chakunda, Chakvad, Charota, Panwar (Hindi); Sogata (Kannada); Chakramandrakam, Takara (Malayalam); Thaunum namthibi (Manipuri) Takala (Marathi); Kelbe-on (Mizo); Chakunda (Oriya); Chakramarda, Dadmari, Dadrughra, Taga (Sanskrit); Senavu, Tagarai (Tamil); Chinnakasinda (Telugu); Panwar, Panevar, Tarota (Urdu).

Origin and distribution: The plant is native to the tropics. In China, this herb is produced in Anhui, Guangxi, Sichuan, Guangdong, etc., and cultivated everywhere in the south and north of China. Mainly found in the states of Uttar Pradesh and Madhya Pradesh in India.

Habitat: In India, it occurs as wasteland rainy season weed.

Habit: It is a small erect hairlesss shrub, about 1m tall, widely spreading with numerous ascending, hairless branches (Fig. 1).

Leaf: It has pinnate leaves, which are about 10cm long. Each leaf has three pairs of leaflets that are opposite, ovate, oblong and oblique at the base.

Flower: Inflorescence: Axillary racemes; Flower: Usually in sub-sessile pairs in the axils of the leaves, the upper crowded; Calyx: Sepals glabrous, divided to the base, ovate, acute and spreading; Corolla: Petals five, each about 1cm in diameter, pale yellow; Androecium: Stamens ten, the three upper reduced to minute staminodes, the remaining seven perfect, unequal; Gynoecium: Ovary sessile at the bottom of the calyx, ovules many, style short, stigma many (Fig. 2).

Flowering, pollination and fruiting: The plant bears flowers in the rainy season and fruits in the winter. Flowers are pollinated by insects.

Fruit: Pods are 20cm long, obliquely septate, puberulous, sub-tetragonous, much curved when young, not reticulate, the sutures very broad (Fig. 3).

Seed : The seeds are rhombohedral and brown in color, about 30 to 50 in number (Fig.4).

Fig. 1. Plant

Fig. 2. Flower

Fig. 3. Fruit

Fig. 4. Seeds

Seed dispersal: Seeds are dispersed by wind.

Propagation : By seeds

Economic part used : Roots, leaves, seeds and bark.

Alkaloids

Roots: 1, 3, 5-trihydroxy-6-7-dimethoxy-2-methyl anthroquinone and beta-sitosterol.

Seeds: Naptho-alpha-pyrone-toralactune, chrysophanol, physcion, emodin, rubrofusarin, chrysophonic acid-9-anthrone.

Medicinal uses : Their roots have purgative and antihelminthic properties, and their leaves are used to treat skin diseases, dysentery and opthalmia (Quisumbing, 1951).

Other uses: Recommended for reclamation of saline, alkaline and brackish soils. Used as green manure crop in acidic soils. Dried seed contain protein (up to 24%) and is given as a protein rich feed for livestock and birds. Roasted seeds are substituted for coffee like *Tephrosia* seeds. Seeds yield tannins and dyes (yellow, blue and red) fields a gum (7.50%), which is a good agent for suspending and binding. The aqueous extracts of whole plant and leaves produces inhibitory allelopathic effects on common weeds. Recommended to grow in *Parthenium* infested areas as smother crop. Seeds used in preparation of sweet dishes. Leaves are popular potherb. In organic farms of India, it is used as natural pesticide. Fungicidal activity of chrysophanic acid-9-anthrone from it has been reported.

REFERENCE

Quisumbing, E. 1951. Medicinal plants of the Philippines. Department of Agriculture and Commerce. *Philippine Islands Tech. Bull.*, 16:1-1234

3.15. *Catharanthus roseus*

Common name: Periwinkle

Botanical name: *Catharanthus roseus* (L.) G. Don. Syn. *Vinca rosea, Lochnera rosea*

Family: Apocynaceae

Vernacular names: Gul feringi, Nayantara (Bengali); Sadabahar (Hindi); Kasikamgalu (Kannada); Ushamalari (Malayalam); Sadaphal (Marathi); Ainskati (Oriya); Rattanjot (Punjab); Nithyakalyani (Tamil); Pillaganneru (Telugu).

Origin and distribution: It is native to Madagascar and widely distributed in West Indies and Madagascar. In India, it is cultivated in Tamil Nadu, Karnataka, Gujarat, Madhya Pradesh and Assam.

Habitat: It is suited to all types of soil except saline or alkaline soils and tropical climatic conditions. Well distributed annual rainfall of 100cm or more is ideal for raising as a rainfed crop.

Habit: An erect, highly branched perennial herb growing to a height of 1m and spread of 60-70cm. Lactiferous tissues pervades its aerial parts (Fig. 1).

Leaf: The oblong or obovate leaves are opposite, short petiole, smooth with entire margin. The upper laminary surface is deeper green than the lower surface where the veins are prominent. Stomata occur on both surfaces.

Flower: The pink, pink eyed or white coloured flowers are borne on axils in pairs. The calyx lobes are linear - subulate. The corolla tube is cylindrical. The epipetalous anthers are borne on short filaments inside the bulging distal end of corolla tube converging conically above the stigma. The bicarpellate ovary is basically distinct with fused common style and stigma which is ascribed to post genital carpel fusion (Fig. 2).

Flowering, fruiting and pollination: Flowering and fruiting period is throughout year. 6-8 weeks after germination, the first flowers will appear. It is pollinated by butterflies and moths.

Fruit: The dehiscent fruit consists of a pair of follicles each measuring about 25mm in length and 2.3mm in diameter, containing upto 30 linearly arranged seeds with a thin black tegument. On maturity, the follicles split along the length dehiscing seeds (Fig. 3).

Seed: Seeds oblong, 2-3mm long, grooved at one side, black (Fig. 4).

Seed dispersal: Seeds are dispersed by ants, wind and water.

Propagation: The plants can either be propagated by seeds or vegetatively through cuttings. Since plants propagated by cuttings, flowering is earlier than the plants from seeds. It is recommended that for drug production the plants should be grown from seeds and for seed production from cuttings.

Seed propagation: The seeds are sown in nursery beds. Germination is seen in about ten days. The young seedlings may be transplanted to the main field later.

Seed storage: Freshly collected seeds should be used for sowing, as the seeds are viable for a very short period of about eight months. Gogitidze (1983) reported complete loss of germination when seeds were stored in ambient conditions for a period of four years. Seed treatment with carbendazim 50% WP @ 2g/kg of seeds and halogen formulation (Bleaching powder + $CaCO_3$ + Arappu leaf powder @ 5:4:1) @ 3g/kg seed and packed in moisture vapour proof container prolongs viability upto 15 months.

Seed germination improvement: Usually epigeal germination is occurred. The seeds are soaked in water for 12-24h to increase the germination per cent. Seed may remain dormant for several weeks after maturity. The optimum temperature for germination is 20-25°C, and the germination rate is in general over 95 per cent (Mastalerz, 1976). The alternate temperature of 25/30°C and 20/30°C was suggested for better germination (Ellis *et al.,* 1985). Choudhary and Gupta (1995) reported 100 per cent germination of seeds treated with GA_3@ 500ppm either for 12h or 24h prior to sowing.

Fig. 1. Plant

Fig. 2. Flower

Fig. 3. Fruit

Fig. 4. Seeds

Seed development and maturation: Physiological maturity is attained at 40days of anthesis with maximum dry weight, germination and vigour with a change of pod colour from green to yellow with translucence.

Seed grading: Seeds can be processed by using 4/64" round perforated metal sieves with maximum seed recovery.

Seed testing: Bharath (2008) carried out the studies on standardization of seed testing procedures and storage studies and his findings are given below.

- The good seed and seedling quality parameters were recorded in dark with KNO_3 in between paper method at the temperature of 25°C, which recorded highest germination percentage (90.70) and faster rate of germination (0.361). Highest root length (6.70cm), shoot length (9.10cm), seedling vigour index (1442) and seedling dry weight (15.10mg) were also recorded. However, first count was fixed at seventh day and final was at 14th day.

- The lower values of these parameters were found at 15°C in presence of light for 24h in top of paper method.
- The extent of decline in germination was upto 36.00 per cent. Other seed quality parameters were also decreased as the storage period advanced. The KNO_3 treated seeds were the highest in all the observations and control was the lowest upto sixth month of storage among all the treatments.
- The cow urine recorded lower values from seventh month onwards and upto the end of storage compared to other treatments.

Vegetative propagation: To raise plants by this method, soft wood cuttings obtained from the lateral shoots have proved better than either hard or semi-hard wood cuttings. Cuttings of about 10-15cm length with a minimum of 5-6 nodes are ideal and result in about 90 per cent rooting. Soaking the cuttings over night in NAA solution of 25 or 50ppm concentration has been found to further improve rooting to the extent of 96 per cent. This method can be profitably used for multiplying the clones which have high alkaloid content and also where seed alone is to be produced.

Economic part used: Root, stem and leaves

Alkaloids: Vinblastine, vincrestine, raubasine, serpentine, ajmalicine, reserpine

Medicinal uses: Anti-fibrillic, hypotensive, sedative

Agricultural uses: The plant extract of *Catharanthus roseus* showed 58.6 per cent reduction over control for the incidence of chillies fruit rot caused by *Colletotrichum capsici* (Rajavel *et al.*, 2001).

The leaf extracts of *Acalypha indica* and *Catharanthus roseus* effectively inhibited the mycelia growth, sporulation, spore germination and germ tube elongation of *Sarocladium oryzae* (Yesu Raja and Syamala, 2012).

REFERENCES

Bharath, V. B. L. K. 2008. Standardization of seed testing procedures and storage studies in selected medicinal crops. M.Sc. (Agri.) thesis, University of Agricultural Sciences, Dharwad.

Choudhary, S. and K. Gupta 1995. Studies on the germination mechanisms of *Catharanthus roseus* seeds. Effect of temperature and promoters. *Seed Sci. Technol.,* 23 : 831-842.

Ellis, R. H., T. D. Hong, and E. H. Roberts 1985. Hand book of seed technology for gene banks. Vol. II Compendium of specific germination information and test recommendations, International Board for Plant Genetic Resources, UK, pp. 497- 647.

Gogitidze, T. S. R. 1983, Characteristics of *Catharanthus roseus* seed. *Biology sub tropheskie kultury*, 6 : 102-104.

Mastalerz, J. W. 1976. A manual on the culture of bedding plants as a green house crop. *Pennsylvania Flower Growers Assoc.,* Uni. Park.

Rajavel, R., Y. Valluvaparidasan and C. Ushamalini 2001. Management of chillies fruit rot by using botanicals. *Allelopathy J.,* 8 (2):277.

Yesu Raja, I. and M. Syamala 2012. Biopotency of promising plant extracts against rice sheath rot disease. *Madras Agric. J.,* 99 (4-6): 344-348

3.16. *Centratherum anthelminticum*

Common name: Iron weed

Botanical name: *Centratherum anthelminticum* (L.) Kuntze. Syn. *Vernonia anthelmintica* Willd., *Conyza anthelmintica* Roxb.

Family: Asteraceae

Vernacular names: Chingkora (Assamese); Somraj, Kaliziri, Hakuch, Bakshie, Bapchie, Babchi (Bengali); Somraj, Somraji and Banjira (Hindi); Kadu-jirigay (Kannada); Kattujirakam (Malayalam); Kali jiri, Kadu jire (Marathi); Vakuci (Oriya); Atavi jirakaha, Avalguj (Sanskrit); Kattu shiragam (Tamil); Davijilakara (Telugu); Kalyzeery (Urdu).

Origin and distribution: It is native to India or Eastern Africa. This species is globally distributed in Afghanistan, Pakistan, India and Sri Lanka. It is distributed throughout India upto an altitude of about 1650m especially Himalayas and Khasi hills (Yadava and Barsainya, 1996).

Habit: An erect tall herb stems and leaves covered with minute hairs, tall with a single flower head, to those with vigorous, shrubby plants with multiple stems and flower heads that may reach 270cm in height (Fig. 1).

Leaf: Leaves are 6-10cm long, their margins toothed, base tapering into a petiole, leaves are alternate and sessile, and have toothed margins with taper pointed tips and wedge shaped bases. Leaves are 0.5cm to 5cm wide and upto 25cm in length.

Flower: Florets violet or purple, many flowered, homogamous, solitary, axillary or terminal heads, with a linear bract near the top of the peduncle; involucral bracts linear, hairy. Flower head 1.5-2.5cm in diameter and each head with 30-40min. purplish flowers (Fig. 2).

Flowering and fruiting : Flowering and fruiting occurs during September - December.

Fruit: Achenes, 4.5-6mm long, cylindric, hairy with ten narrow ridges; the tuft of hairs on top of achenes reddish.

Fig. 1. Plant

Fig. 2. Flower

Fig. 3. Seeds

Seed : The seed is 4.5-6mm long, dark brown in colour having a characteristic odour and intensely bitter taste. The surface of the seed is comprised of about ten ridges. The ridges are covered with trichomes. The seed is oblong shaped, pointed from one side and hairy tapered from other end (Fig. 3).

The outer most layer of the seed is a single epidermal layer, there are unicellular bulb shape trichomes. Under this layer, there are round structures of collenchymatous cell arranged in bundles giving an appearance of ridge from outside the seed. A sclerenchymatous layer is just above the seed coat giving a wave like appearance. This serves for mechanical support to the embryo, the whole part comes under pericarp region. Beneath the pericarp, seed coat is there. The outer integument of the seed coat is single layered. The cells are beaker shaped. Inner integument is thinner in comparison to outer integument. In the centre embryonic region is there which contains globoid aluerone grains and lipid globules.

Propagation : By seeds

Economic part used : Fresh and dried seeds

Alkaloids: The chemicals present in this plant are glycosides, carbohydrates, phenolic compounds and tannins, flavanoids, proteins, saponins, sterols, lipids and fats.

Medicinal uses : It is having anthelminitic properties especially threadworms. It is also reported to be used in asthma, kidney troubles and cough. It is tonic, stomachic and cure phlegmatic discharge from the nostrils. It also enters in to the prescription for leucoderma, psoriasis and other skin infections.

REFERENCES

Yadava, R. N. and D. Barsainya. 1996. Analysis of carbohydrates from seeds of *Centratherum anthelminticum* Kuntze. *Asian J. Chem.*, 8(4): 813-814.

3.17. *Costus speciosus*

Common names: Malay ginger, Crepe ginger

Botanical name: *Costus speciosus* (J. Konig) Sm.

Family : Zingiberaceae

Vernacular names: Jom Lakhuti (Assamese); Keu, Kemuk, Keumul (Bengali); Pakarmula (Gujarati); Kusht, Keukand (Hindi); Changalakoshta (Kannada); Anakua (Malayalam); Koshi (Marathi); Pushkarmula (Marathi and Sanskrit); Kostam (Tamil); Kevukinna, Chengalvakoshtu, Kushthamu (Telugu).

Origin and distribution: It is native to South East Asia and widely distributed in India. Its distribution is restricted to few localities in North West Himalayas and has witnessed poor response to domestication. In Uttar Pradesh, it is generally found in Dudhi and Lalitpur forest divisions. It is on the verge of extinction in these areas. Harvesting of this plant from its natural habitat has led to depletion of plant resources (Pandey *et al.*, 2011).

Habitat: This plant can be grown from sea level to about 1500m elevation. But, the areas at elevation between 400 to 600m above mean sea level, having a sub tropical climate with an annual rainfall ranging between 1000 to 1500mm produce good quality material. A good alluvial soil with a pH ranging from 5.7-7.5 is considered ideal. A well marked dry period between monsoon precipitation results in higher diosgenin content in naturally occurring plantation. Coastal regions and regions with high annual rainfall and humidity throughout the year bear poor quality material, in terms of diosgenin content.

Habit: Perennial herb, erect, robust, unbranched, to 2m high or more, arising from a rhizome. The root stock is tuberous and the stem is sub-woody at the base (Fig. 1).

Leaf: Its leaves are large, simple, alternate, 15-30 × 5.7-7.5cm in size, sub-sessile, oblong and spirally arranged, attached atop a sheath, blade oblanceolate to elliptic, 8-25 x 3-7cm, lower surface silky pubescent.

Flower: Flowers are many, borne among purple bracts in a dense, ovoid to globose, terminal head 3-10cm long. Corolla is three fused sub-equal petals 4-6cm long and a larger, cup shaped labellum 5-9cm long and to 10cm wide, crepe like, white, with a yellow anther. The single stamen present is perfect, lip large with incurved margins (Fig. 2).

Flowering and fruiting : Flowering and fruiting occurs during July-September.

Fruit: Fruit an ellipsoidal many seeded capsule (Fig. 3).

Seed : The seeds are obovoid or sub-globose, black with a white fleshy aril.

Seed dispersal: Birds disperse seeds when they feed on the fruits.

Propagation : Although the plant can be propagated from seeds, stem cutting and rhizomes; commercially, it is being propagated only through rhizome cuttings.

Seed treatment: Gupta (2003) reported that dormancy breaking treatments like scarification, hot water treatment, cold stratification and use of growth regulators (0.2%) KNO_3 treatment by using top paper method and between paper method in sunlight as the media and obtained 86 per cent of germination from 40 per cent and an increase in 46 per cent of germination was seen.

Vegetative propagation: The rhizomes have a number of buds most of them being concentrated around the stem scars and the tips. The formation of buds on the rhizome is poor during April. The cuttings of rhizome pieces for propagation should have at least two viable buds. Rhizome pieces weighing around 40g should be selected. The rhizome pieces are placed at a depth of 8-10cm taking care to place the eye buds facing upwards, horizontally in rows 50cm apart and covered with soil. The thick sized pieces sprout slowly only after 40 - 45 days of planting. This is due to the eye buds being dormant on these rhizome pieces which take a longer time to develop, especially in the case of the crop planted in April. After 70-75 days, about 90-95 per cent sprouting is obtained. Rhizomes planted during June and July complete germination within 60 days, but yields obtained are low as the crop has a shorter growth period in the first year before dormancy sets in. The active vegetative growth period of the crop is from July to mid September and maximum tuberization occurs during September - November. Comparing the effect of planting on ridges versus flat beds indicates that there is no significant difference in yields although slightly higher yield of fresh rhizomes was obtained from the crop planted on ridges.

Economic part used : Rhizomes

Alkaloids: Tubers and roots contain diosgenin, sitosterol, dioscin, prosapogenins A and B of diosceins, gracillin and quinines. Various saponins, many new aliphatic esters and acids are reported from its rhizomes, seeds and roots. Rhizomes yield diosgenin (2.12%) and tigogenin.

Fig. 1. Plant

Fig. 2. Flower

Fig. 3. Fruit

Medicinal uses : The rhizomes can be used as purgative, depurative, febrifuge, expectorant, and anathematic and as tonic. The tuberous root is bitter, astringent, cooling, digestive, stimulant and good to heart. It cures kapha and pitta disorders, dyspepsia, fever, cough and other respiratory diseases, diabetes, oedema, blood diseases; leprosy and other skin ailments. It also finds its use in treating leprosy, burning sensation, constipation, worm infection, skin disease, fever, asthma, bronchitis, inflammation and anemia and also in the making of sexual hormones and contraceptives.

REFERENCES

Gupta, V. 2003. Seed germination and dormancy breaking techniques for indigenous medicinal and aromatic plants. NBPGR, New Delhi. *J. Medicinal Aromatic Plants,* pp. 402-407.

Pandey, A., S. Gupta and K. R. Yadav. 2011. Agro techniques of *Costus speciosus* : An important endangered medicinal plant. **In:** National Conference on Forest Biodiversity: Earth's Living Treasure. 22 May, 2011. p. 125-129

3.18. *Datura innoxia*

Common names: Devil's trumpet, Downy thorn apple, Angel's trumpet, Indian apple, Moon flower, Horn of plenty

Botanical name: *Datura innoxia* Mill., Syn. *Datura meteloides* DC. ex Dunal

Family: Solanaceae

Vernacular names: Safed dhatura (Hindi); Sagol hidak (Manipuri); Ummata (Malayalam); Pandhra Dhotra (Marathi); Vellummattai (Tamil).

Origin and distribution: It is native to Central and South America, and introduced in Africa, Asia, Australia and Europe.

Habitat: Weed of degraded areas, farmland and roadsides. Drought tolerant and thrives in a wide variety of well drained soils, but is most impressive when grown in rich, well manured loams.

Habit : It is an herbaceous perennial with a sprawling, mounded habit, upto 90cm high and 60cm across (Fig. 1). The closely related *Datura stramonium* differs in having smaller flowers and tooth edged leaves, and *Datura wrightii* is having wider, five toothed (instead of ten toothed) flowers. *Datura inoxia* differs from *D. stramonium*, *D. metel* and *D. fastuosa* is having about seven to ten secondary veins on either side of the midrib of the leaf which anastomose by arches at about 1 to 3mm from the margin.

Leaf : The leaves are coarse and foul smelling, upto 25cm long and 10cm wide. They are dark green, simple, ovate, often some what cardate, alternate, entire and densely covered by short, soft hairs.

Flower : They first grow upright, and later incline downward. The flowers are large and the corolla is trumpet shaped, usually white and in excess of 15cm long. The calyx is long and tubular, but it breaks off near the base at maturity leaving a narrow, reflexed "collar" at the base of the flower. The stigma is well above anthers (Fig. 2).

Flowering, pollination and fruiting : Flowering and fruiting occurs during September to March. These are pollinated by night flying sphinx moths, which look and act like small humming birds. But, honey bees and other insects are attracted to the sweet flowers too and often manage to squeeze into them before they have opened.

Fig. 1. Plant

Fig. 2. Flower

Fig. 3a. Fruit

Fig. 3b. Fruit with seeds

Fig. 4. Seed

Fruit : The fruit is an egg shaped spiny capsule with numerous slender spines, about 5cm in diameter. All spines are nearly the same length (upto 1cm long). The capsule stalk bends sharply downwards. Capsules produce brown seeds, 4-5mm long. When ripe, the capsule splits open, dispersing the seeds (Fig. 3a and 3b).

Seed : Small numerous flattened seeds that look like tomato seeds (Fig. 4).

Seed dispersal: The fruit spines getting caught in the fur of animals, then they carry the fruit far from the mother plant. The seeds have hibernation capabilities and can last for years in the soil.

Propagation : By seeds, root divisions and vegetative cuttings

Seed germination: Germination of seed is low and irregular due to the presence of an inhibitors (Zutshi and Atal, 1970). The germination can be enhanced by soaking the seed overnight in water and washing repeatedly two to three times with fresh water before sowing. The seeds start germinating within a fortnight and in a month time the germination is complete. The seed germination can also be hastened by alternate exposure to freezing and thawing. This treatment makes the seed coat weaker. Singh *et al.* (1974) reported enhancement of germination of seeds with irradiation @ 500Kr of light and ascorbic acid @ one per cent concentration. Toth (2001) found that seeds treated with KNO_3 and GA_3 increased the germination rate significantly.

Economic part used: Fruits

Alkaloids: Atropine, hyoscine and hyoscyamine

Other uses : It has also been planted throughout the world as an ornamental plant for its attractive large leaves, large white flowers and distinctive thorny fruit.

REFERENCES

Singh, C., A. K. Bhan, and B. L. Kaul 1974. The role of some physical and chemical agents in the improvement of seed germination in *Datura innoxia*. *Seed. Sci. & Technol.,* 2:421-425.

Toth, E. 2001. Changes in germination ability during gene bank storage at some medicinal plant seeds belonging to the solanaceae family. *International J. Hort. Sci.,* 7 (2): 51-55.

Zutshi, V. and C. K. Atal 1970. Scopoletin induced inhibition of germination in *Datura* species. *Herb. Hung.,* 9:51-54

3.19. *Datura metel*

Common names: Thorn apple, Angel's trumpet, Devil's trumpet

Botanical name: *Datura metel* L., Syn. *Datura alba, Datura fastuosa*

Family: Solanaceae

Vernacular names: Kala-dhotra, Dhatura (Hindi); Dhotra (Marathi); Oomathai, Karu oomathai (Tamil).

Origin and distribution*:* It comes from India originally; it has been naturalised and become a cosmopolitan: Spain, France, Italy and tropical Africa. Tunisia, Algeria and Morocco. Rarely grows wild in Tunisia; central and north-eastern Tunisia.

Habitat : It grows in the wild in all the warmer parts of the world and occurs throughout India in waste places.

Habit : It is an annual, ash-green, hairy species 0.40-1m high (Fig. 1).

Leaf : Leaves are simple, 15-20cm long, alternate, dark green, ovate- lanceolate or broadly ovate, acute or acuminate, unequal at the base and often cordate, entire or repand-dentate, densely tomentose on both surfaces and generally glandular, petioles 6.3-9cm long.

Flower : The flowers are strongly scented, with short peduncles, erect and big (15-20cm). Calyx about 7.5cm long, inflated towards the middle, persistent and reflexed in fruit; teeth lanceolate, acuminate, unequal, corolla about twice as long as the calyx, white tinged with green blow, pubescent outside, limb ten toothed. They come in single and double forms, and a variety of colors, from white to yellow and light to dark purple. The flowers are hermaphrodite (Fig. 2).

Flowering, pollination and fruiting: Flowering and fruiting occurs from July to December. It is pollinated by insects.

Fruit : The fruit is a globular pendulous capsule, with little thorns that are not dilated at the base (Fig. 3a and 3b).

Fig. 1. Plant

Fig. 2. Flower

Fig. 3a. Fruit

Fig. 3b. Fruit cross section

Seed : Seeds are many with membranous wings, arranged on central woody septum.

Seed dispersal: The seeds disperse freely over pastures, fields and even wasteland locations.

Propagation: By seeds

Economic part used : Bark, leaves, flowers, fruit and seed.

Alkaloids : The leaves contain about 0.5 per cent of alkaloids, the main one being scopolamine, accompanied with a little norscopolamine, hyoscyamine and meteloidine. A new tropanic alkaloid, datumetine was isolated on the leaves. These are also rich in withanolides (steroidal elements: datumeteline, daturibine, etc.). The seeds contain 0.2-0.5 per cent of alkaloids and the roots 0.1-0.2 per cent. The flowers particularly contain scopolamine (0.26%) and hyoscyamine.

Medicinal uses : The plant finds application in the treatment of diarrhoea and skin diseases. It is used in the treatment of catarrh, epilepsy, insanity, hysteria, rheumatic pains, hemorrhoids, painful menstruation, skin ulcers and wounds. It is also used in the treatment of burns. It is used to calm cough and to treat laryngitis and treacheries.

Other uses: It can be used in the treatment of infectious diseases caused by resistant pathogenic microorganisms (Vadlapudi and Kaladhar, 2012).

REFERENCE

Vadlapudi, V. and D. S. V. G. K. Kaladhar. 2012. Antimicrobial study of plant extracts of *Datura metel* L. against some important disease causing pathogens. *Asian Pacific J. Tropical Disease*. 894-897.

3.20. *Datura stramonium*

Common names: Jimson weed, Thorn apple, Stramonium, Locoweed

Botanical name: *Datura stramonium* L. Syn. *D. tatula* L.

Family: Solanaceae

Vernacular names: Dhatura (Bengali, Gujarati, Hindi and Marathi); Uummatta (Kanarese, Malayalam, Tamil, Telugu); Dhattura, Unmatta, Kanaka, Shivpriya (Sanskrit); Oomathai, Karu oomathai (Tamil).

Origin and distribution : It is native to North America.

Habitat : It grows wild in warm and moderate regions, where it is found along roadsides and in dung heaps. In Europe, it is found as a weed on wastelands and in garbage dumps.

Habit *:* It is a foul smelling, erect annual, freely branching herb that forms a bush upto 60 to 150cm tall (Fig. 1).

Root: The root is long, thick, fibrous and white.

Stem: The stem is stout, erect, leafy, smooth, and pale yellow green. The stem forks off repeatedly into branches and at each fork forms a leaf and a single, erect flower.

Leaf: The leaves are approximately 8-20cm long, smooth, toothed, soft, irregularly undulate. The upper surface of the leaves is a darker green and the bottom is a light green. The leaves have a bitter and nauseating taste, which is imparted to extracts of the herb, and remains even after the leaves have been dried.

Flower: The fragrant flowers are trumpet shaped, white to creamy or violet, 6-9cm long, and grow on short stems from either the axils of the leaves or the places where the branches fork. The calyx is long and tubular, swollen at the bottom, and sharply angled, surmounted by five sharp teeth. The corolla, which is folded and only partially open, is white; funnel shaped, and has six prominent ribs. Stamens five, inserted in the corolla tube and included in it, equal, anthers small, dehiscing longitudinally. Ovary of two carpels, pyramidal, four lobed, covered with close erect processes, imperfectly four celled in the lower part by the outgrowth of a false dissepiment from the dorsal surface of the carpels to the axile placenta. Style about as long as stamens, stigma bilateral, blunt (Fig. 2).

Flowering, pollination and fruiting: The flowers open at night, emitting a pleasant fragrance and providing food for nocturnal moths. Flowering and fruiting throughout the summer. It is an entomophilous.

Fruit: Fruit about 5cm long capsule, erect, ovoid, rounded-quadrangular in section, surrounded at the base by the enlarged and reflexed persistent base of the calyx, thickly set with unequal sharp rigid spines, nearly completely four celled, pericarp leathers, dehiscing by four valves about half way down (Fig. 3).

Seed: The seeds are very numerous, closely packed and flattened horizontally, somewhat kidney shaped, about 0.5cm long; testa very dark brown, with large shallow pits and also a minute reticulation; embryo much curved and surrounded by endosperm (Fig. 4).

Seed dispersal: The seed is thought to be carried by birds and spread in their droppings. It can dormant underground for years and germinate when the soil is disturbed.

Propagation: By seeds

Seed germination: Kevseroglue (1993) found that seed treatment with KNO_3 for 15min. increased the germination percentage.

Economic part used: Leaf, flower, seeds

Alkaloids: Atropine, Scopolamine

Fig. 1. Plant

Fig. 2. Flower

Fig. 3. Fruit

Fig. 4. Seeds (Seedpod, opening up to release seeds inside)

Medicinal uses: This plant is used for treatment of asthma, lumbago, swelling of joints, ear ache, dandruff, loss of hair (Rani and Reddy, 2009).

Other uses: The leaf extracts inhibited the spore germination of *Sarocladium oryzae* (Komala *et al.*, 1988). The efficacy of plants extracts was tested on the fruit rot pathogen of sponge gourd using water extraction and acetone extraction. Twenty per cent acetone extracts showed maximum per cent inhibition of 91.13 (Rawal and Thakore, 2001).

REFERENCES

Kevseroglue, K. 1993. The effects of some physical and chemical treatments on germination of Datura (*Datura stramonium*) seeds collected from natural vegetation. *Turk Tarim Ve Oxmacilile Dergisi*, 17(3): 727-735.

Komala, D., R. Ramabadran and R. Velazhahan. 1988. Effect of some plant extracts on conidial germination of *Acrocylindrium oryzae*. Paper presented in the TNAU-UGC sponsored National seminar on basic research for crop disease management. May 18-20, TNAU, Coimbatore, India. pp.62.

Rani, A. S. and K. J. Reddy 2009. Folklore medicinal uses of some indigenous plants among the tribes of Telangana region, Andhra Pradesh, India. *Ad. Plant Sci.*, 22(1):199-204.

Rawal, P. and B. B. L. Thakore 2001. Anitfungal activity of botanicals on fruit rot pathogen of sponge gourd. Abstracts of the papers presented at the 23 rd annual conference of Indian society of Mycology and Plant Pathology and National Symposium on Bioinoculants for sustainable Agriculture and Forestry held at Kakatiya University. p 128.

3.21. *Digitalis lanata*

Common names: Foxglove, Woolly Foxglove, Grecian Foxglove

Botanical name: *Digitalis lanata* Ehrh.; Syn. *D. purpurea*

Family: Scrophularaceae

Vernacular names: Tilpushpi (Hindi); Naripukayilai (Tamil).

Origin and distribution: It is native to Western Europe and widely distributed in Switzerland, Holland, France, Hungary, Germany, U.S.A and England. In India, it is grown in Kashmir, Himachal Pradesh and higher elevations of south India.

Habitat: The plant prefers part shade and humus rich soil. The plant also prefers sandy, loamy and clay soils. It can grow under dry or moist conditions. It prefers temperate climate (Gupta, 1997).

Habit: It is a biennial or perennial herb upto about 1.2m high. Second year growth produces flowering stems, 90-180cm in height (Fig.1).

Leaf: Lower basal leaves long stalked, hairy, ovate, 15-30cm long, the upper leaves almost without stalks, becoming smaller in size upwards.

Flower: The flowers are tubular and bell shaped with a creamy white color and purplish brown netting as well as a long broad lip. Both flowers and stems are also woolly or hairy as well as pretty. Flowers are hermaphrodite, 5-8cm long, white or purple, with ciliate lobes, borne in 30 -60cm long, one sided bunches (Fig. 2).

Flowering, pollination and fruiting: The flowers usually bloom in the second year. Flowering is in June and July. The flowers are pollinated primarily by bees and then, oval pods containing seeds are produced.

Fruit: Fruit is a pod covered with hooks that attach to anything that passes by allowing it to spread far from the mother plant (Fig. 3).

Propagation: It is commercially propagated by seeds. The seeds are sown in the surface of the soil during early spring in a cold frame. The seed usually germinates in 2 - 4 weeks at 20°C. When they are large enough to handle, prick the seedlings out into individual pots and plant them out in the summer.

Economic part used: Dried leaves

Alkaloids: Gitoxin, digitoxin.

Fig. 1. Plant Fig. 2. Flower

Fig. 3. Mature pod

Medicinal uses: This plant is a widely used herbal medicine with a recognised stimulatory effect upon the heart. It is also used in allopathic medicine as the main source of the cardiac glycosides that are used in the treatment of heart complaints. It has a profound tonic effect upon a diseased heart, enabling the heart to beat more slowly, powerfully and regularly without requiring more oxygen. At the same time, it stimulates the flow of urine which lowers the volume of the blood and lessens the load on the heart.

REFERENCE

Gupta, R. 1997. Medicinal and Aromatic Plants. In: Handbook of Agriculture, ICAR, New Delhi

3.22. *Glycyrrihiza glabra*

Common names: Licorice root, Licorice, Liquorice, Sweet root, Chinese licorice

Botanical name: *Glycyrrihiza glabra* L. Syn. *Glycyrrhiza glandulifera* Waldst. et Kit., *G. glabra* L. subsp. *glandulifera* (Waldst. & Kit.) Ponert, *G. glabra* L. var. *glandulifera* (Waldst. & Kit.) Regel & Herder, *G. glabra* L. var. *typica* L., *G. glabra* L. var. *violacea* (Boiss. & Noe) Boiss., *G. hirsuta* Pall., *G. pallida* Boiss. & Noel, *G. violacea* Boiss. & Noel

Family: Papilionaceae

Vernacular names: Jesthimadhu, Yeshtmadhu (Assamese); Jashtimadhu, Jaishbomodhu (Bengali); Jethimadhu (Gujarati); Jothi-madh, Kubas susa, Mithilakdi, Mulethi, Mulahti (Hindi); Atimaddhura, Jestamaddu, Yastimadhuka (Kannada); Multhi (Kashmiri); Atimadhuram, Iratimadhuram (Malayalam); Jashtimadh, Jeshtamadha (Marathi); Jatimadhu (Oriya); Jethimadh, Mulathi (Punjabi); Jalayashti, Klitaka, Madhu, Madhuka, Madhu yashtikam, Yashti madhuh (Sanskrit); Athimadhuram (Tamil); Atimadhuranu, Yashtimadhukam (Telugu); Mulhatti, Mulathi (Urdu).

Origin and distribution: Native to central and south western Asia and the Mediterranean region. It is cultivated in the Mediterranean basin of Africa, in south Europe and in India. In India, it is cultivated in Kashmir, Punjab, Himachal Pradesh, Uttar Pradesh and Tamil Nadu.

Habitat: They grow in deep, humus rich, moist soils, in sunny sites. It grows best in deep valleys, well drained soils, with full sun, and is harvested in the autumn, two to three years after planting.

Habit: It is a perennial herb/sub-shrub for subtropical and temperate zone. The plant attains a maximum height upto 2m. The underground stem grow horizontally upto 2m length, shortly pubescent, covered with small brownish glandules, sometimes with small admixture of glandular spinules, highly branched consisting of short taproot with large number of rhizomes (Fig. 1).

Root: Root stock throws numerous additional roots or in some forms, there are many branched stems near the root stock remaining underground.

Leaf: Leaves alternate, pinnate, yellow green leaflets 4-7 pairs are covered with soft hairs on underside, 2-3.5cm long and 1-2.5cm wide.

Flower: Flowers appear in axil of terminal and axillary leaves in raceme (4-8cm long), pea like, lavender to purple in color. Flowers are 8-11mm long, numerous. Calyx 5-6mm long and its teeth are lanceolate, shorter than tube. Corolla is whitish violet. Standard is oblong-ovate. Ovary is with 6-7 ovules (Fig. 2).

Flowering, pollination and fruiting: Flowering and fruiting is from August to February (Ashok, 2005). It is cross pollinated entomophilous plant.

Fruit: Pods 2-3.5cm long, linear - oblong, laterally compressed, straight or slightly falcately bent, slightly convex above seeds, studded with short glandular spinules and 2-3 seeded (Fig. 3 and Fig. 4).

Fig. 1. Plant

Fig. 2. Flower

Fig. 3. Pod

Fig. 4. Seeds

Propagation : The crop is propagated through seeds, rooted cuttings, runners or underground stem pieces (10cm length).

Seed germination: The germination rate of the seeds is to be around 20 per cent. To improve the germination rate, the seeds are stratified for several weeks and then, soaked in water for a long period of time. Treated seeds have a much higher germination rate of 80 per cent. When planted, germination will take around two weeks. Cent per cent seed germination could be achieved when 24h water soaked seed were treated with concentrated H_2SO_4 for 20min. (Sheelaverma, 1996). Gupta *et al.* (1997) also reported that the seeds treated with concentrated H_2SO_4 for five minutes improved the germination.

Vegetative propagation: The roots can grow in upwards of 180cm in length and are capable of growing additional plants of them. This allows for the plant to spread without having to reproduce sexually. It also has the ability to re-sprout from harvested roots.

Economic part used : Dry roots

Alkaloids: The major bio-active constituent of rhizomes is a triterpenoids saponin, glycyrrhizin, glycyrrhizinic acid, glabrin A and B, glycyrrhetol, glabrolide, isoglabrolide, isoflavones, coumarins, triterpene sterols etc.

Medicinal uses : Refrigerant, cough, sore throat, bronchitis, abdominal pains, epilepsy.

REFERENCES

Ashok, S. 2005. The Herbs of India. 1st Edition, Vol.2. Gujarat: Hi Scan Pvt. Ltd. p. 566.

Gupta, V., Anjali Kak and B. B. Singh 1997. Studies on seed germination and seedling vigour in liquorices (*Glycyrrihiza glabra*). *J. Medicinal and Aromatic Pl. Sci.*, 19(2) : 412-413.

Sheelaverma, S. 1996. Seed germination of *Glycirrhiza glabra. Crop Res.*, 12 (1) : 114-115.

3.23. *Hibiscus rosasinensis*

Common name: Chinese Hibiscus, Rose of China, Hawaiian Hibiscus, Hawaiian Rose, China Rose, Shoe back plant, Shoe flower.

Botanical name: *Hibiscus rosasinensis* L.

Family: Malvaceae

Vernacular names: Joba (Assamese); Jaba (Bengali); Gurhak, Jasund (Hindi); Dasavala (Kannada); Ayamparutti (Malayalam); Jassvandi (Marathi); Mondaro (Oriya); Japapuspa (Sanskrit); Semparuthi, Mandaram, Sappathu (Tamil); Mandara (Telugu).

Origin and distribution: It is native to China or south east Asia.

Habitat: It is widely grown as an ornamental plant throughout the tropics and sub-tropics.

Habit: Hibiscus is a large ever green shrub or small tree growing 4.5m high in favourable climates. The habit is upright, broad and spreading (Fig. 1).

Stem: It is often many stemmed. Branchlets terete, sparsely stellate pilose. Bark of the plant has a coarse texture. Star like hairs, mucilage cells in the bark and pith and tough fibrous bark.

Leaf: Stipules filiform, 5-12mm, hairy; petiole 5-20mm, villous; leaf blade broadly or narrowly ovate, not lobed, 4-9 × 2-5cm, papery, pilose on veins abaxially only, base rounded or cuneate, margin dentate or lobed, apex acuminate.

Flower: Flowers solitary, axillary on upper branches, usually pendulous, simple or double. Pedicel 3-7cm, sparsely stellate pilose or nearly glabrous, articulate near apex. Epicalyx lobes 6-7, filiform, connate at base, 8-15mm, sparsely stellate, apex obtuse or acute. Calyx campanulate, 2cm, stellate puberulent, lobes five, ovate to lanceolate. Corolla rosy red, reddish, or orange-yellow, funnel-shaped, 6-10cm in diameter, often double; petals obovate, pilose abaxially, apex rounded. Staminal column 4-8cm, glabrous. Style branches five. Capsule ovoid, 2.5cm, glabrous, apex beaked (Fig. 2).

Flowering, pollination and fruiting: Flowering and fruiting occurs year round, but some flower less during the winter months or in hot weather. It is pollinated by insects and humming birds.

Fruit: Simultaneously with the development of seeds from fertilized ovules, the ovary also undergoes ripening changes and forms a fruit. The mature fruit is many seeded, simple, dry, capsule.

Seed: A mature seed is endospermic and hairy. The seeds are liberated as the dry capsule dehisces (Fig.3).

Propagation: It is propagated by seeds, rooted cuttings, air layering, ground layering, budding, cuttings (Randhawa and Mukhopadhyay, 1986).

Propagation through cuttings: Hard wood cutting of 30cm length possessing one or two leaves is cut (Fig.4). A slant cut is given at the bottom and this is planted at an angle in the field. Alternatively, these cuttings can also be rooted in a raised nursery bed and the rooted cuttings can be planted (Fig. 5). The cuttings can also be rooted in a plastic nursery bag filled with sand : red earth : soil in the ratio of 1:2:1 placed under shade and watered (Fig. 6). The cuttings root easily within a month. This is indicated by the fresh sprout of leaves at the tip of the cutting. Cuttings dipped in seradix B1 powder and placed in auto humid chamber lead to 80-85 per cent rooting while all those placed in the open (Maithi, 1977) failed to root.

Economic part used: Dry flowers, dry leaves and roots.

Alkaloids: Taraxeryl, acetate, beta-sitosterol, campesterol, stigmasterol, cholesterol, erogosterol, lipids, citric, tartaric and oxalic acids, fructose, glucose, sucrose, flavonoids and flavonoid glycosides, hibiscetin, alkanes, cyaniding and cyanin glucosides.

Medicinal uses: It is used in the treatment of stomatitis and cough, the labourers use its roots. The roots are boiled in water and decoction is prepared. The decoction is used in form of gargle. In case of chronic cough, they keep the freshly collected root pieces inside the mouth (Pankaj Oudhia, 2002).

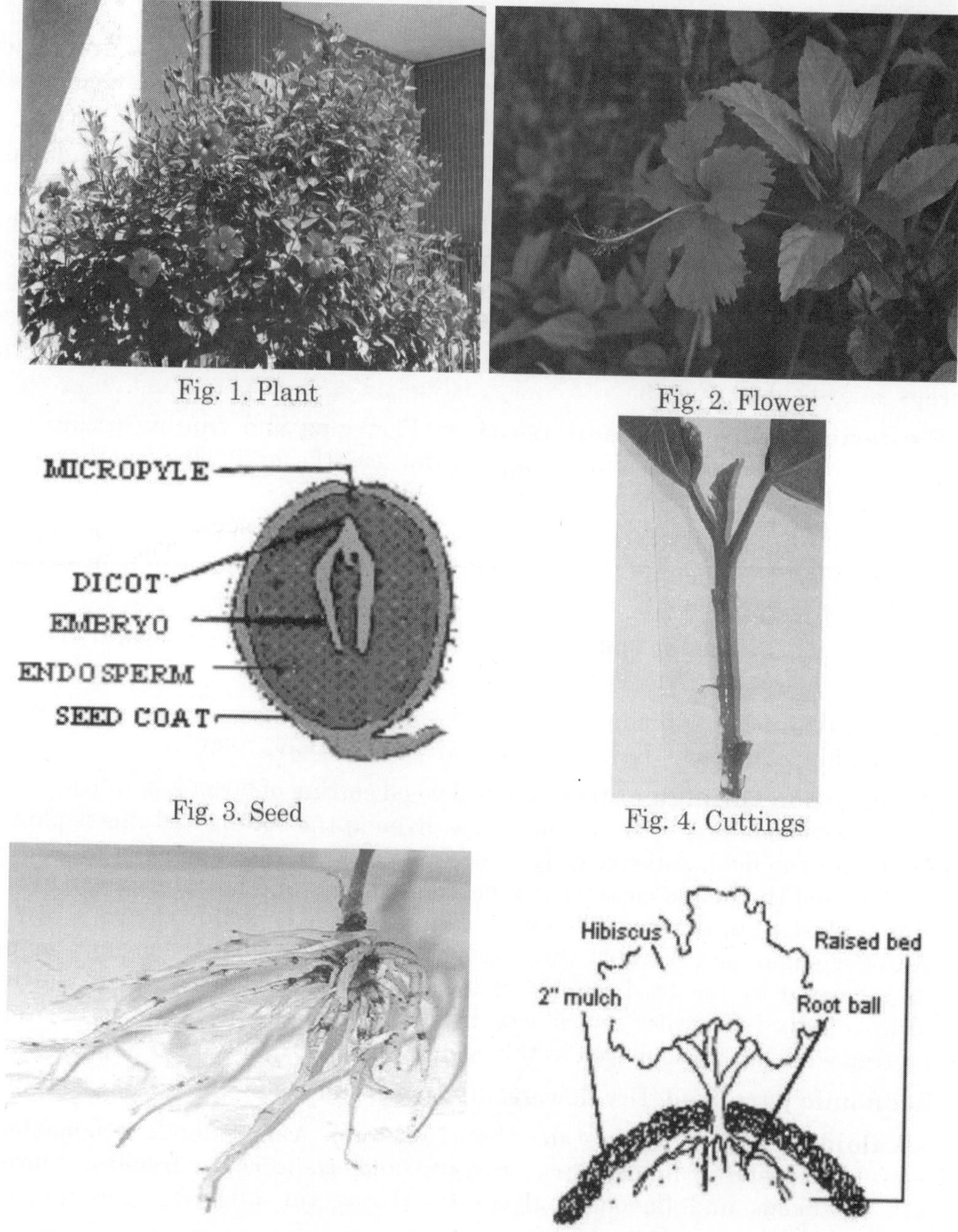

Fig. 1. Plant

Fig. 2. Flower

Fig. 3. Seed

Fig. 4. Cuttings

Fig. 5. Rooted cuttings

Fig. 6. Planting of *Hibiscus*

A handful of crushed leaves kept overnight in water, filtered and drunk to check dysentery. Fresh leaves are applied on forehead to relieve headache. Fresh leaf paste is also applied for abscess. Dried flowers pounded with cooked rice along with three black peppers and made as a cake is eaten three times daily for 2-3 days to treat piles (Girach *et al.*, 2003).

The crude aqueous ethanolic extract of the aerial parts contain spasmogenic and spasmolytic constituents mediating their effect through cholinergic, which explains the traditional use of plant in the gastro-intestinal disorders like constipation and diarrhoea (Gilani *et al.*, 2004).

Other uses: Plants are often used for hedges and screens.

REFERENCES

Gilani A. H., S. Bashir, K. H. Janbaz and A. J. Shah. 2004. Studies on the presence of spasmogenic and spasmolytic constituents in *Hibiscus rosasinensis* aerial parts. *Iranian J. Pharmaceutical Research Supplement,* 2: 60-60

Girach, R. D., Aminuddin and Ismail. 2003. Malvaceae in ethono-medicine of Orissa. *J. Medicinal and Aromatic Plant Sci.,* 25: 350-353.

Maithi, R. G. 1977. Regeneration of terminal greenwood cuttings of some ornamental shrubs under auto-humid chamber. *Plant Sci.,* 6: 104-105.

Pankaj Oudhia. 2002. Jason (*Hibiscus rosasinensis*). www. http.// Botanical. com.

Randhawa, G. S. and A. Mukhopadhyay. 1986. Floriculture in India. Allied Publishers, New Delhi. pp : 147-149.

3.24. *Hybanthus enneaspermus*

Common names: Spade flower, Pink ladies slipper

Botanical name: *Hybanthus enneaspermus* (L.) F. Muell., Syn: *Ionidium suffruticosum*, *Viola enneasperma* L.

Family: Violaceae

Vernacular names: Munbora (Bengali); Ratan purush (Hindi); Purushara-thna (Kannada); Orilathamarai (Malayalam); Rathanparas (Marathi); Amburuha, Athichara, Avyatha, Charati, Chariti, Padma, Padmacharini, Sthala padmini, Sthala kamala, Rathnapurusha (Sanskrit); Orilaitamarai, Orithal tamarai (Tamil); Nilakobari, Purusharatham, Suriyakanti, Ratnapurusha (Telugu).

Origin and distribution: It is native to Indonesia. Abundant from Bundelkund and Agra to Bengal and Sri Lanka, Tropical Asia, Africa and Australia.

Habitat: Distributed in plains from the coast to 900m, in exposed places, common in arable, pasture and wastelands.

Habit: Perennial herb or small shrub to 60cm high, smooth or hairy (Fig.1).

Root: Roots 10-12cm. Long wiry slightly woody, but easily breaks up, pale creamy yellow in colour with small lateral roots, pleasantly aromatic with slight bitter taste.

Stem: Erect stems longitudinally ribbed and clothed in scabrid hairs.

Leaf: Leaves are linear to lance like, 1-5cm long, margins recurved to revolute, occasionally flat. Leaf blades are 16-60 x 2 x 3-9mm, petioles absent or upto 1mm long. Leaf blade clothed in short erect hairs, side veins obscure, margin strongly inrolled (revolute). Stipules are 0.75-2.5mm long, with a few hairs at the base.

Flower: Peduncles are about 4mm long. Pedicels are 2-7mm long, thin. Sepals are lanceolate, 2.5-3mm long, keeled, glabrous with only a few hairs at base or keel densely hairy. Petals variable; two petals cream to green, 2.5 x 1mm; two petals 4 x 1.25mm, cream to green with a pink to orange apex; one petal larger, spathulate, 8-10 x 4-5mm, orange - pink, with a short (0.5-0.7mm long) saccate spur. Anthers are 0.75mm long, filaments about 1mm long, anterior filaments with penicellate nectaries. Pollen is cream. Some staminal filaments have dorsal, plumed appendages. Ovary is about 1mm diameter. Style is about 1-1.25mm long (Fig. 2).

Fig. 1. Plant

Fig. 2. Flower

Fig. 3. Fruit

Flowering, pollination and fruiting: Flowering and fruiting occurs during July - November. It is pollinated by butterflies.

Fruit: Capsule 4-9mm long and sub-globose (Fig. 3)

Seed: Seeds 5-12, ovoid, acute, longitudinally striate, yellowish white, about 1.5mm long.

Propagation: By seeds and cuttings.

Micro-propagation: Prakash *et al.* (2002) developed a procedure for plant regeneration through seed derived callus. Seeds demonstrated a high induction frequency and a high yield of light-yellow friable callus on Murashige and Skoog's (MS) medium containing NAA and BA within four weeks of incubation. Shoot differentiation resulted when callus was transferred to MS medium supplemented with BA (8.8ppm) and NAA (2.6ppm).

Economic part used: Leaves

Alkaloids: Plant extract contains an alkaloid, aurantiamide acetate, β-sitosterol and isoarborinol. Nectar contains high amount of amino acids including valine, leucine and glutamic acid.

Medicinal uses: Plant is tonic, demulcent and diuretic; removes urinary calculi; used in strangury, pain, dysentery, vomiting, asthma and cough. Root is diuretic; prescribed in bowel complaints of children; infusion is used in gonorrhoea and urinary affections. Dried powdered leaves are used in asthma.

REFERENCE

Prakash, E., S. Reddy and K. R. Rao. 2002. Regeneration of plants from seed derived callus of *Hybanthus enneaspermus*, a rare ethnobotanical herb. *Plant Cell Reports* 18: 873-878.

3.25. *Jatropha curcas*

Common names: Barbados nut, Purging nut, Physic nut, Black vomit nut

Botanical name: *Jatropha curcas* L.

Family: Euphorbiaceae

Vernacular names: Mogali erand (Hindi); Chandar-jyoti (Marathi); Danti, Pratyanshrani (Sanskrit); Kattaamanakku (Tamil).

Origin and distribution: It is originated from South America but introduced to India by the Portuguese. It is distributed throughout India especially in the dry tropical regions of the country.

Habitat: It grows on gravely, sandy or saline soils and also on the poorest stony soils and rock crevices. It is widely distributed in the tropics and sub-tropics.

Habit: Semi-evergreen exotic shrub or small tree, reaching a height of 6m (Fig.1).

Stem: Bark greenish white, smooth, peeling off in thin flakes, latex acrid.

Leaf: In general, the leaves are green to pale green, alternate to sub-opposite, broadly cordate, glabrous and three to five lobed with a spiral phyllotaxis. The petiole length ranges from 6.1mm to 23.1mm.

Flower: The inflorescence can be formed in the leaf axils. Flowers are yellowish green in terminal or axillary panicled cymes. Plants are monoecious and also have hermaphroditic flowers occasionally. Male and female flowers are produced on the same inflorescence, averaging 20 male flowers to each female flower or ten male flowers to each female flower (Fig. 2).

Flowering, pollination and fruiting: It is a quick maturing plant species that starts bearing fruits within a year of its planting. Flowering occurs during the wet season and two flowering peaks are often seen. In permanently humid regions, flowering occurs throughout the year. The seeds mature about three months after flowering. Early growth is fast and with good rainfall conditions nursery plants may bear fruits after the first rainy season, direct sown plants after the second rainy season. The flowers are pollinated by insects especially honey bees.

Fruits are produced in winter, or there may be several crops during the year if soil moisture is good and temperatures are sufficiently high. Most fruit production is concentrated from mid summer to late fall with variations in production peaks where some plants have two or three harvests and some produce continuously through the season.

Fruit: This plant bearing trilocular fruits. Fruits ovoid, three lobed, rugose on drying. Each fruit bears three black seeds, which are the source of oil (Fig. 3a and 3b).

Seed: The seeds are mature when the capsule changes from green to yellow. Seeds dull brownish black (Fig. 4).

Propagation: This can be propagated both by seeds as well as vegetative means.

Seed propagation: Seeds are collected in May - June. The seeds are soaked overnight in water and the large heavy seeds are sown at the onset of monsoon. Seeds are sown in polybags or in raised beds at a depth of 5cm. Seeds can also be dibbled at the rate of two seeds for each spot at a depth of half a finger. Seed germination is 95 per cent and takes 10 to 15 days.

Propagation through seed leads to a lot of genetic variability in terms of growth, biomass, seed yield and oil content.

Fig. 1. Plant

Fig. 2. Flower

Fig. 3a. Mature fruit

Fig. 3b. Dry fruit

Fig. 4. Seed

Fig. 5. Cutting

Fig. 6. Grafting

Fig. 7. Air layering

Vegetative propagation: Vegetative propagation has been achieved by stem cuttings, grafting, budding as well as by air layering techniques (Fig. 5, 6 and 7). Semi-hardwood cuttings with 30-40cm length and 5cm thickness from the base of the stem with short internodes are planted in polybags. The root initiation takes place in 15 to 20 days and about 80 per cent of the cuttings will root. The cuttings should be taken preferably from juvenile plants and treated with 200ppm of IBA to ensure the highest level of rooting in stem cuttings.

Economic part used: Leaves, seeds and roots

Alkaloids: The seeds contain around 20 per cent saturated fatty acids and 80 per cent unsaturated fatty acids, and they yield 25-40 per cent oil by weight. In addition, the seeds contain other chemical compounds, such as saccharose, raffinose, stachyose, glucose, fructose, galactose and protein. The oil is largely made up of oleic and linoleic acids. Furthermore, the plant also contains curcasin, arachidic, linoleic, myristic, oleic, palmitic, and stearic acids and curcin.

Medicinal uses: The leaves are reported to be used to treat foul ulcers, tumours and scabies. The latex is effective against wounds and ulcers. Seeds are reportedly used to treat haemorrhoids, wounds and skin diseases. The yellow oil from the roots has a strong anthelmintic action. Oil from the seed is used externally for rheumatism and paralytic affections. Seeds are used as a contraceptive in south Sudan.

Agricultural uses: Sprays of plant extracts of *Clerodendrum inerme, Jatropha curcas* and *Ocimum sanctum* had 100 per cent inhibition of cowpea aphid borne mosaic virus upto 20, 22 and 30 days, respectively (Patel *et al.*, 2001).

Work done in Seed Science and Technology

Kathiravan (2004) carried out the investigations to standardize seed technological practices in *Jatropha curcas* for the production of quality seed. The salient findings of this study are given below.

(a) In germination test, the first and final count could be taken at 7th and 14th day after sowing, respectively and sand was identified as the suitable medium.

(b) Soaking of seeds in cold water for 12h and in hot water for 45min improved the germination by 10 to 12 per cent compared to control (74%).

(c) The seeds fortified with each 1.0 per cent solution of $FeSO_4$, $ZnSO_4$ and $MnSO_4$ recorded the highest survival percentage (81, 82 and 80) than control (66%) with higher seedling vigour.

(d) Seed hardening with each one per cent solution of either $CaCl_2$ or KCl for 24h, soaking and drying back to original moisture content recorded 14 and 11 per cent higher seedling survival, respectively than untreated seeds with increased the seedling vigour.

(e) The seeds germinated well in clay alluvial soil enhanced establishment, survival and seedling growth rate measured upto 90th day after sowing compared to black, laterite and red soils.

(f) The stem cuttings with 40cm length and 2.5 to 3.0cm thickness with a circumference of 9.5-11.1cm was found to be very suitable for quicker regeneration seedling quality characters.

(g) The plants propagated by seeds produced maximum fresh and dry weight, germination and seedling quality characters than propagated through cuttings.

(h) The fresh and dry weight of seed, germination, seedling vigour, oil content and protein content reached maximum at 75 days after anthesis which was associated with the visual symptoms in change of fruit colour from green to yellowish orange.

(i) Perforated sieve with 12mm round was more suitable for getting maximum good seed recovery and seedling production.

(j) The seeds treated with bavistin + halogen mixture @ 1 + 1.5gkg^{-1} (1:1) and packed in high density polyethylene inter woven bag (HDPE) or polylined cloth bag could be stored safely for 12 months with minimum loss of viability and vigour.

Gurunathan (2006) carried out the investigations on seed handling techniques in *Jatropha curcas* for maximization of quality characteristics of seed at nursery and on storage. The findings of this experiment are as follows.

(a) Fresh fruits of yellowish brown colour, dried fruits of black colour, fruits of bigger size and fruits from the seed source Odukkampalayam (Coimbatore) are recorded higher seed and seedling quality characters as fresh (initially), on storage (six months) and at nursery.

(b) Use of hand operated thresher followed by beating with pliable bamboo sticks were feasible for extraction of quality seeds that performed best both at nursery and after storage. Seed extraction using power operated thresher recorded higher extraction efficiency, but with higher percentage of kernel damage.

(c) The bigger sized, heaviest / heavier and black coloured seeds recorded higher seed germination and vigour both at nursery and after storage expressing a positive association between the seed quality characters and seed size, seed weight and seed colour individually.

(d) The seeds lost their total germination and viability within nine days of accelerated aging at 40°C and 100 per cent RH. However, the duration of viability varied with seed source.

(e) Seed treatment with bavistin @ 2g.kg^{-1} of seed as slurry and the seeds packed in 700 gauge polyethylene bag maintained the seed viability at higher order after six months of ambient storage.

(f) The seeds should be positioned as caruncle end downwards at 2cm depth for maximization of nursery emergence and production of quality seedlings.

(g) Seed fortification with two per cent Na_2HPO_4 or two per cent *Vitex negundo* leaf extract for 6h or riboflavin 200ppm for 8h or IBA 200ppm at ambient temperature for 12h or 24h or seed pelleting with *Azospirillum* (100g.kg^{-1} of seed) or pungam leaf powder (300g.kg^{-1} of seed) maximized the nursery emergence and also favoured the production of quality seedlings.

Soumiya *et al.* (2012) reported that Jatropha seeds extracted from fruits harvested at four stages of maturity *i.e.*, green [approaching to maturity, 45 days after anthesis (DAA)], yellow (mature, 60 DAA), brown (ripened, 65DAA) and black (ripened and dried) were used for estimation of seed quality parameters. Fruits of Jatropha harvested at yellow colour stage recorded higher seed quality parameters *viz.*, germination (91%), mean seedling length (41.90cm), mean seedling dry weight (363mg), seedling vigor index-I (3838) seedling vigor index-II (33284), total dehydrogenase activity (1.72), protein content (24.64%) and with low electrical conductivity (162dS^{-1}m). Therefore, harvesting of fruits at yellow stage (Table 1 and 2) would be better for seed purpose. However, fruits of black stage yielded higher seed oil per cent (43.75%).

Table 1. Effect of maturity stages on seed moisture content, germination, mean seedling length, mean seedling dry weight and seedling vigour indices of Jatropha

Fruit maturity stages	Pod moisture content (%)	Germin-ation (First Count)(%)	Germi-nation (Final Count) (%)	Mean seedling length (cm)	Mean seedling dry weight (mg)	SVI-I	SVI-II
Green (45DAA)	52.49	15	35	32.60	186	1149	6703
Yellow (60DAA)	53.90	75	91	41.90	363	3838	33284
Brown (65DAA)	37.12	44	85	40.62	342	3463	29203
Black (70DAA)	11.44	64	84	38.77	319	3265	26948
S Em ±	0.12	1.33	1.74	0.58	10.11	92	944
C.D (0.05 P)	0.35	3.97	5.22	1.72	30.31	276	2840
C.V (%)	0.69	5.94	5.27	3.34	7.46	7.02	8.80

Note: DAA –Days After Anthesis, SVI- Seedling Vigour Index

Post harvest handling techniques for seed production in *Jatropha curcas:* Fruits at yellow stage gave 95 per cent germination indicating the attainment of physiological maturity. The experiment revealed that soaking the seeds for 24h in water followed by two hours drying helps in recovery of good seeds. The infection in the seeds was confirmed through the X-ray images. The most suitable material to store the seeds as revealed by the study would be cotton cloth bag or jute bag (Anon., 2008).

Table 2. Effect of maturity stages on total dehydrogenase activity, electrical conductivity, pH, oil and protein in seeds of Jatropha

Fruit maturity stages	TDH (Absorbance at 480nm)	EC (dSm^{-1})	pH	Oil (%)	Protein (%)
Green (45DAA)	1.473	245	8.50	12.74	20.53
Yellow (60DAA)	1.718	162	7.96	30.27	24.64
Brown (65DAA)	1.690	185	7.80	36.83	23.44
Black (70DAA)	1.624	226	7.24	43.75	22.60
S Em ±	0.013	8.20	0.123	0.729	0.367
C.D (0.05 P)	0.039	24.59	0.370	2.187	1.099
C.V (%)	1.77	8.95	3.50	5.28	3.59

Note: DAA –Days After Anthesis

REFERENCES

Anonymous 2008. Annual Report (2007-08). Institute of Forest Genetics and Tree Breeding, Coimbatore.

Gurunathan, N. 2006. Studies on seed management techniques in *Jatropha curcas*. M.Sc. (For.) thesis, TNAU, Coimbatore.

Kathiravan, M. 2004. Seed production, processing, testing and storage techniques in Jatropha (*Jatropha curcas* Linn.). Ph.D. Thesis, TNAU, Coimbatore.

Patel, D. J., A. B. Patel, S. M. Pandey and A. Mishra 2001. Evaluation of plant extracts against cowpea aphid borne mosaic virus. Abstracts of the papers presented at the 23rd annual conference of Indian society of Mycology and Plant Pathology and National Symposium on Bioinoculants for sustainable Agri and forestry held at Kakatiya University: 122.

Sowmya K. J., R. Gowda, P. Balakrishna and M. R. G. Rao 2012. Effect of fruit maturity stages on seed quality parameters in Jatropha (*Jatropha curcas* L). *Indian J. Plant Sci.*, 1(1):85-90.

3.26. *Kaempferia galangal*

Common names: Aromatic ginger, Sand ginger, Resurrection lily

Botanical name: *Kaempferia galangal* L.

Family: Zingiberaceae

Vernacular names: Chandramula (Assamese); Ekangi, Bhui champa (Bengali); Chandramula, Sidhoul (Hindi); Kachchura, Kachhoora (Kannada); Kacholam, Katjulam (Malayalam); Kachri, Kapur Kachri (Marathi); Chandra moolika, Corakah, Karcurah, Sathi, Sati, Gandhamula (Sanskrit) Kacholum, Pulankilanku (Tamil).

Origin and distribution: It is native to south India and widely distributed in China, Taiwan, Cambodia and India.

Habitat: It thrives best in slightly shaded places such as open forest, forest edges, and bamboo forest, on various soils, upto 1000m altitude.

Habit: It is a perennial rhizomatous aromatic herb with very fragrant underground parts (Fig. 1). The plants become dormant in winter. The leaves die down in late autumn and rhizomes remain underground through winter.

Root: Spreading horizontally, rhizome, the subunits are flat elliptical cross section, it has a dark reddish brown skin and the soft interior is nearly white (Fig. 2).

Leaf : The plant has thick rounded leaves with short petioles that lay flat on the ground. New leaves start growing in mid-spring from the small dormant rhizomes. The leaves are usually 2-3 (-5), with 1.5-5cm long sheaths. The blade is often horizontal and lying flat to the soil, broadly elliptical to slightly flat with circular outline, the top surface is smooth while the bottom surface is cobweb hairy.

Flower: The inflorescence is sessile, emerging from between the leaves. It is 4-12 (-15) flowered. The sepal is 2-3cm long. The petal is white, with tube 2.5-5cm long and lobes 1.5-3cm long. Their lip is broadly reversed egg shaped, divided to about halfway or more, white or pale purple with violet to purple spots at the base. Each lateral lobe is about 2-2.5cm x 1.5-2cm. Other abortive stamen has an imperfect anther that is oblong reversed egg shaped to oblong lance shaped, 1.5-3cm long and white. Their fertile stamen is 10-13mm long, with two lobes deeply connective with abruptly bent lobes (Fig. 3).

Flowering and fruiting: Flowering occurs during August to September. A fruit occurs in winter.

Fruit: The fruit is three celled capsule.

Propagation: The plant is vegetatively propagated through rhizomes and 500-700kg rhizome is required for one hectare.

Micro-propagation: *In vitro* plantlet production has been achieved on 0.75 x Murashige and Skoog (MS) medium supplemented with 12μM BA, 3μM α-naphthaleneacetic acid (NAA) and 3% sucrose. The procedure ensures 13-fold rate of plantlet production every four weeks. Hardened plantlets produced normal storage roots as the parent plants (Shirin *et al.*, 2000).

Economic part used: Rhizomes

Alkaloids: Essential oils, Methoxy cinnamate

Fig. 1. Plant

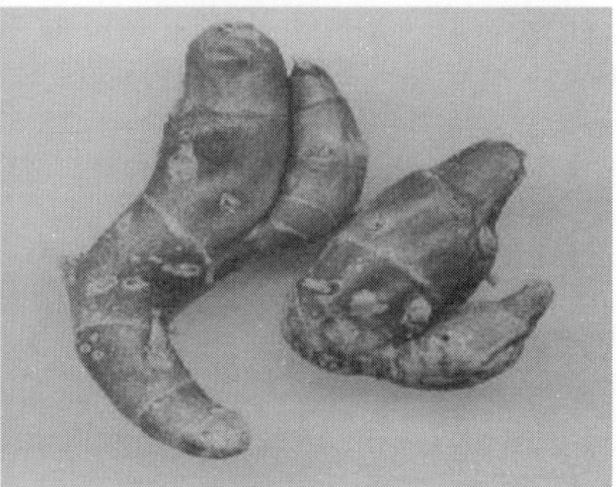

Fig. 2. Rhizome

Fig. 3. Flower

Medicinal uses: The rhizomes are good for skin diseases, rheumatism, asthma, bronchitis, etc.

Reference

Shirin, F., S. Kumar and Y. Mishra 2000. *In vitro* plantlet production system for *Kaempferia galanga*, a rare Indian medicinal herb. *Plant Cell, Tissue Organ Culture* 63: 193–197.

3.27. *Mentha* spp.

Common name: Mint

Botanical name: *Mentha* spp.

Family: Lamiaceae

Vernacular names: Pudina (Tamil, Hindi).

Origin and distribution: It is native to northern Hemisphere. The genus has distribution across Europe, Africa, Asia, Australia, and North America. In India, it is largely confined to North India in the States of Uttar Pradesh, Punjab and Haryana. It is widely cultivated in Indo- Gangetic plains, Punjab, Kashmir and Tamil Nadu (Kodaikanal, Ooty, Hosur, Yercaud and Kolli hills).

Habitat: It can be found in many environments, mostly grow best in wet environments and moist soils. Well drained loamy and acidic soil high in organic matter content is suitable. Sub-tropical areas receiving an annual rainfall of 100 - 150cm are good.

Habit: A perennial herb rises to a height of 40-80cm (Fig. 1).

Root: Creeping root stalk.

Stem: Erect stem, 1-2 quadrangulate branched with short dense hair.

Leaf: Leaves are 2.5-5cm long. The leaves are arranged in opposite pairs, from oblong to lanceolate, often downy, and with a serrate margin. Leaf colours range from dark green and gray green to purple, blue, and sometimes pale yellow.

Fig. 1. Plant

Fig. 2. Flower

Fig. 3. Fruit

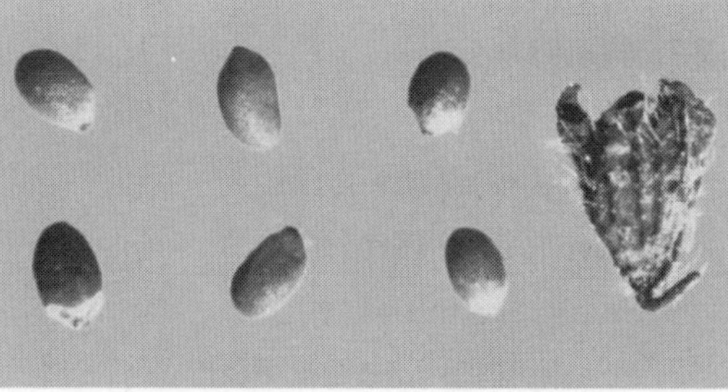

Fig. 4. Seeds

Flower: Flowers are in auxiliary whorls, none at the top. The flowers are white to purple and produced in false whorls called verticillasters. The corolla is two lipped with four sub-equal lobes, the upper lobe usually the largest (Fig. 2).

Flowering and pollination : Flowers appearing in May-June and again in September-November under cultivation. Male sterility has been found and both genetic and cytoplasmic in nature. It is responsible for frequent cross pollination (Sobti, 1969) by honeybees (Abrol, 2012).

Fruit: The fruit is small, dry capsule containing one to four seeds (Fig. 3).

Seed : It does not produce viable seeds (Fig. 4) under tropical conditions even under controlled crossing (Gupta, 1995).

Propagation: Rooted stem cuttings, root suckers.

Economic part used: Leaves

Alkaloids: Thymol, carvacrol, p-cunene-2, 3- diol, biphehyls, flavonoids.

Medicinal uses: In medicine, it is used against stomach disorders, rheumatism, in ointments for headaches, in cough drops, inhalations etc. The oil and dried plants are antiseptic, carminative, refrigerant, stimulant and diuretic.

REFERENCES

Abrol, D. P. 2012. Pollination biology, biodiversity conservation and agricultural production. Springer London, p.792.

Gupta, R. 1995. Japanese Mint. **In:** Advances in Horticulture. Vol. 11 (Eds.) Chadha, K. L. and R. Gupta. Malhotra publishing house, New Delhi. pp.689-716.

Sobti, S. N. 1969. Cytological studies in genus *Mentha*. CSIR News, 19 (10): 3-4

3.28. *Mimosa pudica*

Common name: Touch me not (Sensible plant), Sensitive plant, Shy plant, Humble plant, Sleeping grass, Tickle plant and Modest Princess.

Botanical name: *Mimosa pudica* L.

Family : Marsileaceae

Vernacular names: Lajjabati (Bengali); Lajwanti, Chui-mui (Hindi); Lajja, nachike, Mudugu davare; Tintarmani (Malayalam); Lajalu (Marathi); Tottalvadi, Thottal chinungi (Tamil); Attapatti, Peddanidrakanni (Telugu).

Origin and distribution: Native to south and central America, where it was first documented in Brazil. It's now considered a pan-tropical weed with distribution throughout Africa, Asia, Australia and the Pacific Islands.

Habitat: Small woody herbs or low spreading under shrub with hairy and prickly branches, hairs glandular. This herb grows in full sun or light shade in warm climates, because it cannot tolerate frost. It thrives and becomes a ground cover in poor soil because of its ability to convert nitrogen for its own use.

Habit: The plants live one to two years. Seedlings grow slowly for two or three months and then accelerate, reaching 0.5 to 2m of extension at the end of the first year. Growth of plants that survive into the second year is much slower. Potted and field grown individuals are sensitive to overwatering (Bui, 2001) (Fig. 1).

Stem: The stem is erect in young plants, but becomes creeping or trailing with age. The stem is slender, branching, and sparsely to densely prickly, growing to a length of 1.5m.

Leaf: Bipinnate, sensitive to touch, pinnae 1-2 pairs, leaflets 10-20 pairs, linear, glabrous.

Flower: Heads small, peduncled, globose, auxilalry, pink-purple, calyx campanulate, petals crenate towards base (Fig. 2). The flowers are globe shaped, about 2.5cm across.

Flowering, pollination and fruiting: The plant flowers in mid to late summer. The flowers are pollinated by the wind and insects. In China, the flowering season from March to October, with fruiting season from May to November.

Fruit: The fruit consists of clusters of 2-8 pods from 1-2cm long each, these prickly on the margins. The pods break into 2-5 segments and contain pale brown seeds some 2.5mm long (Fig. 3).

Seed : The seeds have hard seed coats which restrict germination. 2.4-2.8 (-3.3) mm x 1.7-2mm; plumply lentiform, fawn brown, with an open pleurogram. 1000 seed weight is 5.3g (Fig. 4).

Propagation : By seed.

Seed dispersal: Pods can be spread by attaching to fur of animals, attaching to clothing, by mud on vehicles and by water (Smith, 2002).

Seed dormancy and germination: Seed will germinate in 14 to 21 days at 70°F. The dormancy is induced by a hard seed coat that restricts germination. Scarification released the seeds from dormancy and stimulated germination, though the germination of the scarified seeds was not influenced by light. The germination increased markedly with the exposure to high temperature pretreatment (e.g. 150°C), which was achieved by placing non-scarified seeds in an oven for 5min. followed by incubation at 35/25°C day/night temperatures for 14 days. The germination of the scarified seeds was tolerant of salt and osmotic stress, as some seeds germinated even at 250mmol l^{-1} NaCl (23%) and at an osmotic potential of −0.8 MPa (5%). The germination of the scarified seeds was >74% over a pH range of 5-10. The seedling emergence of the scarified seeds was 73-88% at depths of 0-2cm and it gradually decreased with an increasing depth, with no seedling emergence at the 8cm depth. The rice residue applied to the soil surface at rates of ≤6t.ha^{-1} did not influence the seedling emergence and dry weight (Chauhan and Johnson, 2009).

Fig. 1. Plant

Fig. 2. Flower

Fig. 3. Fruit

Fig. 4. Seeds

The germination of *M. pudica* is independent of light exposure. Optimum germination (percentage and germination speed index) temperature for the species was continuous, at 30°C or alternate temperature of 25-35°C, even though the seeds germinated satisfactorily in continuous temperature above and under 30°C, and in all combinations of alternate temperatures. Comparatively, the germination speed index was more sensitive to the effects of the temperature than the germination percentage.

Economic part used : Whole plant

Alkaloid: Mimosine

Medicinal uses : According to Ayurveda, root is bitter, acrid, cooling, vulnerary, alexipharmic and used in treatment of biliousness, leprosy, dysentery, vaginal and uterine complaints, inflammations, burning sensation, fatigue, asthma, leucoderma, blood diseases etc. According to the Unani system of medicine, root is resolvent, alternative, useful in diseases arising from blood impurities and bile, bilious fevers, piles, jaundice, leprosy etc.

It is also grown as garden herb, green manuring, fixes nitrogen, used as fodder.

REFERENCES

Bui, L. D. 2001. *Mimosa pudica.* http://bio.maimi.edu/mimosa/mimosa.html. 7 p.

Chauhan, B. S. and D. E. Johnson 2009. Germination, emergence and dormancy of *Mimosa pudica. Weed Biology Management,* 9 (1): 38–45.

Smith, N. M. 2002. Weeds of the wet/dry tropics of Australia - a field guide. Environment Centre NT, Inc. 112 pp.

3.29. *Nardostachys jatamansi*

Common name: Jatamansi, Nard, Nardin, Indian Valerian, Tagara, Indian Spikenard, Bhuutajatta, Tapaswini and Muskroot

Botanical name: *Nardostachys jatamansi* (D. Don) DC. Syn. *Nardostachys grandiflora, Valeriana jatamansi*

Family : Valerianaceae

Vernacular names: Jatamansi, Jatamangshi (Assamese, Bengali, Marathi, Oriya); Baalchad, Kalichad (Gujarati); Balchara (Hindi); Bhootajata, Ganagila maste (Kannada); Bhutijata (Kashmiri); Manchi, Jatamanchi (Malayalam); Billilotan, Balchhar, Chharguddi (Punjabi); Mansi, Balchara, Sumbul-ut-teeb (Sanskrit); Jatamansi (Tamil); Jatamams (Telugu); Sumbul-ut-teeb (Urdu).

Origin and distribution: It is found to occur between 3,300 and 6,000m in the Himalayas: Kashmir, Himachal Pradesh, Uttarakhand, Nepal, Sikkim, Bhutan, South West China including Tibet, Yunnan and South West Sichuan.

Habitat: In nature, the plant is generally found clinging to steep rocky cliffs, stony and grassy slopes, and is rarely found as lithophytes. It is also reported from ravines of the rocks and crevices, small depressions, somewhat found in wet meadows and by the banks of the rivulets in the high valleys and peaks ranging from 3,000 - 4,000m.

Sandy loam and acidic soil rich in organic carbon and nitrogen is found best for better survival of seedlings and productivity.

Habit: An erect perennial herb, 10-60cm in height, with woody stout (Fig. 1).

Stem: Rootstock covered with reddish brown fibres of the petioles of radical leaves.

Leaf: Leaves radical, longitudinally nerved; flower pale pink or blue.

Flower: Flowering shoot is 30-40cm high and bears 40-50 flowers at summit in three to seven condensed cymes. Flowers are campanulate, pinkish red to bluish white. Corolla tube is 6mm long and hairy (Fig. 2).

Flowering, pollination and fruiting: Flowering time is July to August, while fruiting occurs in September to November. The anthesis takes between 7:00–11:00h. Anthers disperse pollen grains after 24h of anthesis. Stigma is of protogynous type. This plant is self-pollinated, but dependent on pollinators for optimum pollination. Cross pollination may have adaptive value for species as it compensate failure of autogamy and also maintains genetic variability in the population.

Fruit: Fruit 4mm long, covered with ascending white hairs, around by the ovate, acute, often denate calyx teeth (Kirtikar and Basu, 2006).

Rhizome: It has a stout unbranched or sparsely branched woody aromatic rhizome, covered with reddish brown thick fibers of remnants, a petiole of withered radical leaves with a single long tap root, typically with 2-7 rhizomes, but sometimes upto 15. The rhizome may be 6-9cm long with about 20-30cm long yellowish taproot.

Propagation : By seeds (Fig. 3) and vegetative propagation.

Seed propagation : Seeds are sown in small plots of 1m × 1m size or in styrofoam trays during November - December in polyhouse at lower altitude, during March - April in open beds at middle altitude and during May in alpine area. Seed germination varies between 74 per cent and 80 per cent and occurs within 12–30 days when planted at a depth of 0.5cm in sandy soil inside the polyhouse (Anon 2008). Approximately 600g of seeds are required for raising seedlings for transplanting in one hectare of land. Seedlings are transplanted after six to eight weeks in the field. At lower altitude, root growth as well as number and length of leaves increase, rapidly as compared to higher elevation. However, fibrous root formation takes place only after third year of growth when, plants are raised by seedlings. About 1,10,000 plants are planted in one hectare of land.

Seed treatment: Both seeds and rhizomes are treated with GA_3 at 100ppm and 200ppm for 48h for rapid germination/ sprouting.

Seed collection: Seeds are collected in October when they turn pale brown in colour.

Vegetative propagation: Propagation through splitting of roots is found most successful and observed better for multiplication as well as for higher production within a short period than cultivation through seedlings.

Economic part used : Rhizomes and roots

Chemical constituents: Rhizomes, rich in sesquiterpenoides, yield jatamanshic acid, jatamansone, patchouli alcohol, nor-seychelanone, seychellen, and alpha and beta patchoulene. The dried rhizomes are steam distilled to yield 1.5 – 1.9 per cent of a pale yellow essential oil, commercially known as spikenard oil, emitting a pleasant odour. The essential oil yields several hydrocarbons, a new oxide, beta-cudesmol, elemol, angelicin and jatamansinol.

Fig. 1. Plant

Fig. 2. Flower

Fig. 3. Seeds

Medicinal uses : In Indian medicine, rhizomes and rootlets of the plant are used in prescriptions for their sedative, antispasmodic, antiepileptic, antihysteric, hypotensive, anti-inflammatory and cardio-tonic properties.

REFERENCES

Anonymous. 2008. Agro-techniques of selected medicinal plants Vol. I. National Medicinal Plants Board, Government of India, New Delhi.

Kirtikar, K. R. and B. D. Basu 2006. Indian Medicinal Plants Vol. II. International Book distributors, Dehradun.

3.30. *Ocimum basilicum*

Common name: Sweet basil, French basil, Common Roman basil

Botanical name: *Ocimum basilicum* L.

Family : Lamiaceae

Vernacular names: Damaro, Nasabo, Sabza (Gujarati); Bubui tulsi, Gulal tulsi, Kali tulsi, Marua, Falooda (Hindi); Kamakasturi, Sajjagida (Kannada); Niazbo (Kashmiri); Marwa, Sabza (Marathi); Dhalatulasi, Kapur kanti (Odiya); Furrunj-mushk, Baburi, Niyazbo, Panr (Punjabi); Munjariki, Surasa, Varvara (Sanskrit); Karpoora Thulasi, Thiruneetrupachilai, Uruthira sadai (Tamil); Bhutulasi, Rudrajada, Vepudupachha (Telugu).

Origin and distribution: It is native to Africa, Central Asia and North West India and cultivated throughout India, Sri Lanka and Myanmar.

Habitat: Cultivated, but occasionally found in wastelands.

Habit: Erect, shiny herb and can grow upto 90cm high. Entire plant is very intoxicating some times. Sweet basil possesses a clove like scent with an aromatic, somewhat saline taste.

Stem: Four angled, glabrous or sparsely hairy and branching.

Leaf: Both the sides of the leaves are glossy. Opposite, petiolate, lanceolate, lanceolate-ovate or ovate, oblong narrowed at both the ends and have numerous dot like oil glands.

Flower: Terminal clusters of whorled flowers (vertcilillasters). Protandrous, small, inconspicuous and greenish in colour. Corolla white, bilabiate, 1cm long. Stamens exerted beyond corolla. Calyx five lobed, upper lobe expanded into a lid or cap over others.

Flowering, pollination and fruiting: Flowering and fruiting occurs during July - October. It is pollinated by honey bees (Abrol, 2012).

Fruit: The fruit consist of four one seeded nutlets enclosed in the enlarged membraneous, strongly veined recurved calyx.

Seed : Ellipsoid, black and become mucilaginous when wetted. When in contact with water, it bloats and become sticky. The seeds are non-endospermic or have a very little endosperm. The embryo is straight and the cotyledons are usually flat. The seeds are odourless with an oily, slightly pungent taste. When steeped in water, they liberate mucilage which is semi-transparent and nearly tasteless.

Propagation : It can be propagated through fresh seeds and terminal cuttings. Best season for sowing is October-November in plains and March-April in hills. The seeds germinate with in 10-15 days under direct sowing and 8-12 days under nursery.

Pre-sowing seed treatment : Gupta (2002) studied on overcoming seed dormancy in seven species of *Ocimum,* where he used GA_3 at 5 to 30ppm level. But, most effective was 20ppm. Also he used salts like KCl, KNO_3 and H_2O_2 at one to ten per cent concentration and in that most effective was five and ten per cent doses, which increased seed germination capacity, germination velocity index and seedling vigour index (SVI) significantly in seven species of *Ocimum*.

Pre-soaking of *Ocimum basilicum* seeds in 2% KNO_3 for 3h or 150ppm GA_3 for 6h or osmoprimed with PEG @ 0.2MPa for four days improves its germination (Swapna, 2003) (Table 1).

Table 1. Effect of pre-soaking seed treatments on germination of *Ocimum basilicum*

Treatment	Germination (%)	Drymatter production (mg / 10 seedlings)	Vigour Index
GA_3			
Control	51	12.5	389
50ppm for 6h	65	13.5	501
100ppm for 6h	67	14.5	529
150ppm for 6h	82	15.1	662
200ppm for 6h	65	13.6	510
CD (P = 0.05)	2.6	0.17	33.2
KNO_3			
0.5% for 3h	76	13.4	546
1.0% for 3h	81	15.4	656
2.0% for 3h	89	16.4	748
3.0% for 3h	81	13.0	626
CD (P = 0.05)	2.7	0.66	22.2
PEG			
- 0.2MPa for four days	81	16.2	706
- 0.4MPa for four days	79	15.9	655
- 0.6MPa for four days	71	14.5	564
- 0.8MPa for four days	70	13.9	545
- 0.1MPa for four days	65	13.4	490
CD (P = 0.05)	1.4	0.27	22.9

Economic part used : Green leaves, seeds

Alkaloids: Eugenol, Linalool, Methyl chavicol

Medicinal uses : It can be used as a carminative, anthelmintic, antipyretic, expectorant, stimulant, antioxidant, antispasmodic, digestive, galactogogue, stomachic and tonic. An infusion of the plant is given for joint pain and used as a gargle for foul breath. The juice of the leaves is used for ear ache and also for ringworm. Seeds possess cooling properties. Popularly used in sherbats, falooda (Ice creams). Seeds are also used in poultices for sores and sinuses.

Other uses: The volatile oil is used as a flavouring agent and also as perfume. The oil can be used for flavouring confectionery, baked goods, sauces, ketchups, tomato pastes, pickles, vinegar, spiced meats, sausages and beverages. It is also used for scenting dental and oral preparations and in certain perfume compounds, notably jasmine blends, to impart strength and smoothness. Basil oil possesses insecticidal and insect repellent properties; it is effective against houseflies and mosquitoes. It is also bactericidal.

Varieties: European, Reunion

Past work done on Seed Science and Technology

Seed testing : For conducting germination test, between paper is found to be best for *O. basilicum* and *O. sanctum*, while top of paper is the best for *O. gratissimum* (Table 2) (Swapna, 2003). The optimum temperature is an alternate temperature of 25 / 30°C or 20 / 25°C with alternate light (16h) and dark (8h) (Table 3) (Swapna, 2003). The first and final count of the germination could be taken on 4th and 14th day for all the *Ocimum* spp (Table 4).

Table 2. Determination of a suitable germination medium for *Ocimum* species

Media	*O. basilicum*			*O. sanctum*			*O. gratissimum*		
	Germination (%)	DMP (mg Sl^{-10})	Vigour index	Germination (%)	DMP (mg Sl^{-10})	Vigour index	Germination (%)	DMP (mg Sl^{-10})	Vigour index
Top of paper (TP)	48	13.4	212	28	7.4	77	86	4.1	220
Between paper (BP)	66	15.3	508	36	8.5	152	74	4.0	208
Sand	27	14.6	173	18	8.2	59	42	4.0	112
CD (P = 0.05)	1.6	0.31	17.4	1.4	0.1	6.8	1.6	0.06	6.8

Table 3. Fixing a suitable temperature and light for germination test of *Ocimum* species – Germination

Temperature °C	*O. basilicum*			*O. sanctum*			*O. gratissimum*		
	Light (24h)	Dark (24h)	Alternate light (16h) and dark (8h)	Light (24h)	Dark (24h)	Alternate light (16h) and dark (8h)	Light (24h)	Dark (24h)	Alternate light (16h) and dark (8h)
25	70	75	78	50	40	62	63	44	83
30	55	26	72	42	13	54	75	-	76
20/25	72	78	81	51	40	68	85	72	89
25/30	78	79	86	68	56	76	82	73	92
Mean	69	64	79	53	37	65	76	47	85
CD (P = 0.05)	T	L	T x L	T	L	T x L	T	L	T x L
	1.2	1.0	2.0	1.3	1.1	2.3	1.4	1.2	2.5

For conducting quick viability test, the seeds of all the *Ocimum* spp. should be preconditioned with water soaking for 18-24 hours. The best suited preparation method is longitudinal cutting (Table 5) along the side of the seed coat (Swapna, 2003). For proper staining as well as to facilitate the precise evaluation of viable and non-viable seeds in *Ocimum* spp., the following procedure can be followed. For

O. basilicum and *O. sanctum,* the seeds are to be placed in pertiplate containing 0.5 per cent tetrazolium chloride solution and incubate in dark for 1h at 40°C, whereas for *O. gratissimum,* the seeds are to be placed in pertiplate containing 0.2% tetrazolium chloride solution and incubate for 2h at 40°C in dark.

Table 4. Determination of days to first and final count for *Ocimum* species

Days after sowing	*O. basilicum*			*O. sanctum*			*O. gratissimum*		
	Germination (%)	Drymatter Production (mg sl^{-10})	Vigour index	Germination (%)	Drymatter Production (mg sl^{-10})	Vigour index	Germination (%)	Drymatter Production (mg sl^{-10})	Vigour index
3	14	8.2	54	5	4.3	11	-	-	-
4	37	8.6	155	17	4.9	40	48	1.9	92
5	45	9.1	206	24	5.4	59	79	2.2	158
6	49	9.7	243	30	5.7	78	82	2.4	178
7	55	10.1	299	33	6.1	91	87	2.7	194
8	58	10.5	337	34	6.5	98	87	2.9	201
9	60	10.9	362	34	6.8	101	87	3.3	205
10	60	11.2	391	34	7.1	105	87	3.6	208
11	60	11.5	403	34	7.4	109	87	3.9	211
12	60	11.9	414	34	7.8	114	87	4.0	213
13	60	12.2	429	34	8.5	121	87	4.1	215
14	60	12.5	447	34	8.5	138	87	4.1	216
15	60	12.5	447	34	8.5	138	87	4.1	214
16	60	12.3	443	34	8.5	137	87	4.1	213

Table 5. Standardization of quick viability test for *O. sanctum, O. basilicum* and *O. gratissimum* : Viability (%)

Tetrazolium treatments (T)	*O. basilicum*			*O. sanctum*			*O. gratissimum*		
	LC	P	Mean	LC	P	Mean	LC	P	Mean
0.2% TZ 40°C for 1h	25	10	17	0	0	0	0	0	0
0.3% TZ 40°C for 1h	40	15	27	0	0	0	0	0	0
0.5% TZ 40°C for 1h	75	54	64	0	0	0	0	0	0
0.75% TZ 40°C for 1h	73	51	62	16	7	11	16	7	11
1.0% TZ 40°C for 1h	71	49	60	65	42	53	65	42	53
0.2% TZ 40°C for 2h	22	7	14	87	66	76	87	66	76
0.3% TZ 40°C for 2h	34	12	23	0	0	0	0	0	0
0.5% TZ 40°C for 2h	73	46	59	0	0	0	0	0	0
0.75% TZ 40°C for 2h	71	43	57	17	8	12	17	8	12
1.0% TZ 40°C for 2h	69	41	55	45	31	38	45	31	38
1.5% TZ 40°C for 2h							83	54	68
2.0% TZ 40°C for 2h							95	73	84

LC : Longitudinal cut; P : Water soaking for 18-24 h

The weight of working sample is 4.0, 1.0 and 2.0g for *O. basilicum, O. sanctum* and *O. gratissimum*, respectively. Therefore, the weight of submitted sample is 40, 10 and 20g for *O. basilicum, O. sanctum* and *O. gratissimum* respectively (Swapna, 2003).

Hamada *et al.* (1993) reported that the alternate temperature of 30/20°C gave better germination.

REFERENCES

Abrol, D. P. 2012. Pollination biology, biodiversity conservation and agricultural production. Springer London, p.792.

Gupta, S. C. 2002. Seed dormancy studies in some *Ocimum* species and its control through chemical treatment. *J. Med. Arom. Pl. Sci.,* 24 : 957-960.

Hamada, A. A., W. Koch, A. Hamdoun, M. Kunisch and J. Saverbon 1993. Effect of temperature, light and simulated drought on the germination of some weed species from the Sudan. *Angewandte Botanik.,* 67 (1/2): 52-55.

Swapna, M. 2003. Tracing seed maturation pattern and development of seed testing procedure for *Ocimum spp.* M.Sc.(Ag.) Thesis, TNAU, Coimbatore.

3.31. *Ocimum gratissimum*

Common name: Shrubby basil, Spice basil, Clove basil

Botanical name: *Ocimum gratissimum* L.

Family: Lamiaceae

Vernacular names: Jangli tulsi (Hindi); Ram tulsi, Elumichai thulasi (Tamil).

Origin and distribution: It is widely distributed in India, Sri Lanka, Java and Tropical Africa.

Habitat: In its native area, *O. gratissimum* occurs from sea level upto 1500m altitude in coastal scrub, along lake shores, in savanna vegetation, in sub-montane forest, and disturbed land. In South East Asia, it is not frequently found in open locations like roadsides and clearings, but more often cultivated as a hedge plant, upto about 300m altitude. *O. gratissimum* is found throughout the tropics and subtropics, both wild and cultivated. Its greatest variability occurs in tropical Africa (from where it possibly originates) and India. In South East Asia, it is cultivated mainly as a home garden crop, only in Vietnam it is grown on a commercial scale as well.

Habit: Aromatic, perennial herb, 1-3m tall (Fig. 1).

Stem: Stem is erect, round quadrangular, much branched, glabrous or pubescent, woody at the base, often with epidermis peeling in strips.

Leaf: Leaves opposite; petiole 2-4.5cm long, slender, pubescent; blade elliptical to ovate, 1.5-16cm x 1-8.5cm, membranaous, sometimes glandular punctate, base cuneate, entire, margin elsewhere coarsely crenate-serrate, apex acute, puberulent or pubescent.

Flower: Inflorescence a verticillaster, arranged in a terminal, simple or branched raceme 5-30cm long; rachis lax, softly pubescent; bracts sessile, ovate, 3-12mm x 1-7mm, acuminate, caducous; pedicel 1-4mm long, spreading or ascending, slightly curved; flowers in 6-10-flowered verticillasters, small, hermaphrodite; calyx two lipped, 2-3mm long, fruit 5-6mm, pubescent, upper lip rounded and recurved, reflexed in fruit, lower lip with four, narrow, pointed teeth, central pair of teeth minute and much shorter than the upper lip; corolla campanulate, 3.5-5mm long, two lipped, greenish- white, pubescent outside, upper lip truncate, four fid, lower lip longer, declinate, flat, entire; stamens four, declinate, in two pairs, inserted on the corolla tube, filaments distinctly exserted, upper pair with a bearded tooth at the base; ovary superior, consisting of two carpels, each two celled, style two fid (Fig. 2).

Fig. 1. Plant

Fig. 2. Flower

Fig. 3. Fruit

Flowering, pollination and fruiting: Flowering and fruiting occurs during October- November. Flowering started after 136 days and continued until 195 days. Seed matured after 259 days. In South East Asia, flowers can be found year- round. In northern India, oil content of young plants was low (2.3%) until the seed setting stage, then remained constant at 2.8 per cent until the seed maturation stage. It is highly cross pollinated crop.

Fruit: Fruit consisting of four, dry, one seeded nutlets enclosed in the persistent calyx (the lower lip closing the mouth of the fruiting calyx); nutlet sub-globose, 1.5mm long, rugose, brown; outer pericarp not becoming mucilaginous in water (Fig. 3)

Seed : Brown, globose and do not become mucilaginous when wetted.

Propagation : Seeds, terminal cuttings.

Pre-sowing seed treatment: Obaremi *et al.* (2002) reported that light and temperature of 25°C is the ideal for germination and leaching of seeds in running water for 12h gave cent per cent germination. In *Ocimum gratissimum,* germination has been to be phytochrome dependent, which had a typical high irradiance response (Roche *et al.,* 2002). For improving the germination and vigour, the seeds are to be soaked in 1% KNO_3 for 3h (Swapna, 2003) (Table 1).

Table 1. Effect of KNO_3 seed treatment on germination of *Ocimum gratissimum*

Treatment	Germination (%)	Drymatter production (mg / 10 seedlings)	Vigour index
Control	29	4.1	58
KNO_3			
0.5% for 3h	47	4.8	116
1.0% for 3h	53	4.5	116
2.0% for 3h	49	4.1	101
3.0% for 3h	49	4.0	102
CD (P = 0.05)	1.5	0.06	4.5

Economic part used: Leaves.

Alkaloids: Eugenol, Linalool, Camphor

Medicinal uses: It cures stomach ache, cough and cold, rheumatism and paralysis.

REFERENCES

Obaremi, O., O. O. Obenb and D. A. Agboola 2002. Effect of pre-treatments on the germination of seeds of *Ocimum gratissimum*. *Seed Res.,* 30 (2) : 315-317.

Roche, S. F. R., F. C. M. Chaves, L. C. Ming and F. M. Scarda 2002. Phytochrome and light influence on germination and vigour of *Ocimum gratissimum* seeds under high irradiance conditions. *Acta Hort.,* 569:33-39.

Swapna, M. 2003. Tracing seed maturation pattern and development of seed testing procedure for *Ocimum* spp. M.Sc.(Ag.) Thesis, TNAU, Coimbatore.

3.32. *Ocimum sanctum*

Common name: Sacred basil, Holy basil

Botanical name: *Ocimum sanctum* L. Syn. *O. tenuiflorum*

Family: Lamiaceae

Vernacular names: Tulsi (Hindi); Tulasi (Kannada and Malayalam); Tulas (Marathi); Tulasi, Vishnu priya, Divya (Sanskrit); Thulasi (Tamil).

Origin and distribution: It is native to south and western Asia, as well as Australia. This herb is widely distributed throughout the entire Indian subcontinent upto 1800m in the Himalayas as well as throughout India and Pakistan.

Habitat: It is now cultivated under tropical conditions.

Habit: Much branches, erect annual herb, evergreen, 75cm height, hairy all over, quadrangular. Branches are often purple (Fig. 1).

Leaf: The leaves are opposite, ovate to lanceolate 5cm long, margins entire or toothed, hairy on upper as well as lower surface, dotted with minute glands, aromatic, light green in colour. The petioles are very slender and usually slightly hairy.

Flower: The flowers are 0.75 to 1.25cm long, borne on long terminal racemose inflorescence, simple or much branched. The calyx is five toothed, greenish in colour, the upper tooth is rounded and shorter than the others and two lower teeth are ovate and lanceolate with a brittle point, two laterals shorter than the lower. The corolla is 0.72 to 1.25cm long white, pink or pale purplish two lipped tube. There are four protruding stamens. Filament length is 1mm and colour is white. The ovary is bicarpellary, syncarpous and bicolour, the stigma is bifid. It has single style and the colour is white (Fig. 2)

Flowering, pollination and fruiting: Flowering and fruiting occurs during September-December. It is highly cross pollinated crop.

Fruit: The fruits are small, ellipsoid nutlets and are mucilaginous when wet.

Seed : Seeds are dark brown, sub-globose and do not become mucilaginous when wetted.

Propagation : By seeds and stem cuttings

Seed propagation: Generally, the seeds are collected during January / February months and sown in raised mother beds within a week from the day of collection. Sowing should not be deeper than 1.25cm. Irrigation is done by sprinkling every day morning. Germination starts after ten days and continues for another seven days. Range of germination observed is between 70-90 per cent. The seedlings can be transplanted in the field after they reach a height of 15-20cm or seedlings having 10-15 leaves.

Fig. 1. Plant Fig. 2. Flower

Fig. 3. Fruit

Pre-sowing seed treatment: Limbale *et al.* (2001) reported that treatment with GA_3 (400ppm) (32h) resulted better germination. Gupta (2003) reported that seeds soaked in (0.2%) KNO_3 and sown in top paper method and between paper method in sunlight obtained 86 per cent germination from 40 per cent.

Soaking of seeds in one per cent KNO_3 for three hours (Swapna, 2003) or 100ppm GA_3 or 75ppm kinetin or 50ppm ethrel or 25ppm boric acid (Dey and Choudhuri, 1982) is found to increase the germination.

Seed testing: Bharath (2008) carried out the studies on standardization of seed testing procedures and storage studies and his findings are given below.

- Top of paper method was found to be better and it has registered highest germination (84.80%) at an alternate temperature of 20/30°C in presence of light for 24h. The higher rate of speed of germination was recorded (0.400).
- The first count was taken on third and final count was on ninth day.
- Seed germination and other quality attributes declined significantly over storage. The reduction in germination was to the extent of 34.00 per cent at

the end of ten months of storage. However, the seeds treated with KNO_3 maintained better quality among all the treatments.

- The GA_3 treated seeds maintained higher quality upto three months and gradually decreased thereafter.
- The control recorded lower seed quality parameters upto six months of storage and seventh month onwards cow is urine showed very lowest values of seed quality parameters.

Vegetative propagation: Vegetatively it can be propagated by terminal cuttings with about 90-100 per cent success during October-December months. For this purpose, cuttings with 8-10 nodes and 10-15cm length are used. They are so prepared that except for the first 2-3 pairs of leaves, the rest are trimmed off. Later, they are planted in the well prepared nursery beds or polythene bags. In about 4-6 weeks time, the rooting is complete and they are ready for transplanting into the main field.

Economic part used : Fresh herbage.

Alkaloids: Rosemarinic acid eugenol.

Medicinal uses : Tulsi is an incomparable, omnipotent herb possessing preventive and curative ability against various health complaints such as fever, cough, cold, vomiting, diarrhoea, tooth ache, eye diseases, stomach ache, head ache, nasal discomfort, rheumatism, worms, boils, skin diseases, muscular pains, loss of appetite, indigestion, sore throat, cuts and wounds and so on. There is also a practice of planting tulsi to repel mosquitoes.

Agricultural uses: Mahapatra and Tewari (1994) reported that application of *Ocimum sanctum* ethanolic extracts for the control of collar rot and yellow root diseases in groundnut increased seed germination due to fungitoxicant.

Kadadiya and Dhruj (2001) found that the extract of *Ocimum sanctum* has served as a good inhibitor of growth as well as spore germination of *Cercospora canescens* in mungbean.

Sprays (more than one) of pyhtoextracts of *Clerodendrum inerme, Jatropha curcas* and *Ocimum sanctum* had 100 per cent inhibition of cowpea aphid borne mosaic virus upto 20, 22 and 30 days, respectively (Patel *et al.*, 2001).

Prakasam *et al.* (2001) suggested that combination of different extracts of *Azadirachta indica, Ocimum sanctum* and V*itex negundo* var. *purpurescens* could be suitably applied to carrot for effective and safe management of post harvest rot disease of carrot.

The efficacy of five plants extracts *viz., Azadirachta indica, Ocimum sanctum, Datura stramonium, Lantana camara* and *Diosphros cordifolia* were tested on the fruit rot pathogen of sponge gourd using water extraction and acetone extraction. Twenty per cent acetone extracts of *Datura stramonium* (91.13%) and *Azadirachta indica* (90.02%) showed maximum per cent inhibition followed by *Ocimum sanctum*

(88.90%), *Lantana camara* (88.90%) and *D. cordifolia* (81.14%) (Rawal and Thakore, 2001).

The combined application of *Albizia lebbeck* (Green leaf manure @10 t.ha^{-1} incorporated during field preparation and allowed 45 days for decomposition) + *Ocimum sanctum* (seed soaked in five per cent solution for 30min.) recorded the highest percentage (85.73) of seed germination in bhendi *cv.* Arka anamika (Anbukkarasi *et al.*, 2011).

REFERENCES

Anbukkarasi, V., A. Sadasakthi and S. Mariappan 2011. Studies on performance of bhendi under "Vrkshayurvedic farming". Lambert Academic Publishing GmbH & Co. KG, Saarbruckeen, Germany. p.148.

Bharath, V. B. L. K. 2008. Standardization of seed testing procedures and storage studies in selected medicinal crops. M.Sc. (Agri.) thesis submitted to the University of Agricultural Sciences, Dharwad.

Dey, B. B. and M. A. Choudhuri 1982. Seed germination as affected by plant age, growth and development stages of *Ocimum sanctum. Seed Sci. & Tech., 10:* 243-255.

Gupta, V. 2003. Seed germination and dormancy breaking techniques for indigenous medicinal and aromatic plants. NBPGR, New Delhi. *J. Medicinal Aromatic Plants,* pp. 402-407.

Kadadiya, H. J. and I. U. Dhruj 2001. Antifungal property of some plant extracts against *Cercospora canescens* mungbean leaf spot instant. *J. Mycol. Pl. Pathol.*, 31(1):103.

Limbale, S. D., S. T. Ghule, V. H. Bhuse and R. D. Wash 2001. Stimulation of germination and dormancy studies in some important medicinal plants. National Seminar on Production, Processing and Marketing of Medicinal, Aromatic and Dye Yielding Crops, p.188.

Mahaptra, T. K. and S. N. Tewari 1994. *Ocimum sanctum* L. leaf extracts toxicity against collar rot (*Aspergillus niger*) and yellow root (*Aspergillus flavus*) diseases of groundnut. *Allelopathy J.,* 1(2):114-117.

Patel, D. J., Patel, A. B., Pandey, S. M. and Mishra, A. 2001. Evaluation of plant extracts against cowpea Aphid borne mosaic virus. Abstracts of the papers presented at the 23 rd annual conference of Indian Society of Mycology and Plant Pathology and National Symposium on Bioinoculants for sustainable Agri and forestry held at Kakatiya University: 122.

Prakasam, V., S. Abraham and C. Kannan. 2001. Management of post harvest fungal diseases of carrot using botanicals. *South Indian Hort.* 49 (Special): 271-274.

Rawal, P. and B. B. L. Thakore 2001. Anitfungal activity of botanicals on fruit rot pathogen of sponge gourd. Abstracts of the papers presented at the 23rd annual conference of Indian Society of Mycology and Plant Pathology and National Symposium on Bioinoculants for sustainable Agri and forestry held at Kakatiya University: 128.

Swapna, M. 2003. Tracing seed maturation pattern and development of seed testing procedure for *Ocimum spp.* M.Sc.(Ag.) Thesis, TNAU, Coimbatore.

3.33. *Plumbago zeylanica*

Common names: Ceylon leadwort, Doctor bush, White leadwort

Botanical name: *Plumbago zeylanica* L.

Family : Plumbaginaceae

Vernacular names: Agiyachit, Agnachit, Boga agechita (Assamese); Sitarak (Bengali); Chitrakmula (Gujarati); Chira, Chitra, Chitrak (Hindi); Chitramula, Vahni, Bilichitramoola, Chitramulika (Kannada); Chitra, Shatranja (Kashmiri); Thumpokkoduveli, Vellakoduveli (Malayalam); Telhidak angouba (Manipuri); Chitrak (Marathi); Chitamula, Chitoparu, Ogni (Oriya); Chitra (Punjabi); Agni, Vahni, Chitraka, Dahana, Hutabhuk (Sanskrit); Kodi mooli, Sengodi veli, Sivappu chithira moolam, Karimai, Sengodi mooli (Tamil); Chitramulamu (Telugu); Cheetah (Urdu).

Origin and distribution: It is native to South East Asia (Aditi *et al.*, 1999). In the recent decades, the plant is widely spread in tropical and sub tropical regions of Australia, Asia and Africa (Vijver and Looter, 1971), including Ethiopia (Abera, 2003). The species is largely cultivated in gardens throughout India. It grows wild in South India, West Bengal, and some parts of Madhya Pradesh and Chhattisgarh.

Habitat: It is often occurring as a road side weed, sometimes cultivated in gardens. Occasional in waste lands, hedges and forests. In natural habitats, the plant prefers moist soil with high organic content. It needs full sun to partial shade with intermediate to warm temperatures.

Habit: It is an evergreen herbaceous plant with a height of 1.5–2.0m. During summer, the plant remains almost leafless under natural conditions, but under irrigated conditions, the active growth starts in April (Fig. 1).

Root: Roots are light coloured inside, when fresh and reddish brown when dry. Outer surface of the root is brown and striated.

Stem: The stems are glabrous, angled, much branched, branchlets striate, climbing prostrate or erect.

Leaf: It has dark green leaves that are ovate to 30cm long by half as wide. The leaves are alternate, petiolate or sessile and have lance-elliptic, or spatulate to oblanceolate blades that measure 5-9 × 2.5-4cm in length. Bases are attenuate while apexes are acute, acuminate, or obtuse.

Flower: Inflorescences are dense racemes, 3-15cm in length and have glandular, viscid rachises. Bracts are lanceolate and 3-7 × 1-2mm long. Flowers are bisexual, scented and white in colour. Calyx is persistent and tubular, with conspicuous viscid glands. The heterostylous flowers have white corollas 17-33mm in diameter and tubes 12.5-28mm in length (Fig. 2).

Flowering, pollination and fruiting: Flowering occurs from September to November, while fruiting occurs from January to February. It is cross pollinated crop. The final flowering percentage (95.3 ± 1.71%) and/or seed set (89.4 ± 1.41%) was obtained under glass house condition compared to the nursery, which dropped as low as 50 per cent in flowering and seedset (Abera *et al.*, 2008).

Fruit: Fruits are green coloured, with sticky hairs when young and become dark brown when mature. Capsules are 7.5-8mm long, oblong, pointed, pericarp thin below, thick and hardened above (Kirtikar and Basu, 2006) and contain reddish brown to dark brown seeds (Fig. 3).

Seed : Seed oblong, 5-6mm long, reddish brown to dark brown.

Propagation : It is propagated by semihardwood cuttings, division of older plants or by seeds.

Seed propagation: Freshly collected seeds are soaked in water for 24h. The seeds are dibbled in polythene bags filled with the rooting medium prepared by mixing equal quantities of sand, red earth and manure. About 70 per cent of seeds germinate within 7-12 days.

Seed germination: Seeds rapidly germinated on a mixture of nursery soil and cattle dung in a ratio of 3:1 filled in pots or on cultivated soil under nursery. Germination was cent per cent when seeds were cut at the micropylar and chalazal ends and incubated in moist filter paper (Menon *et al.,* 2001). The study found that rainy season, plant size, leaf number, low temperature, cross pollination and glass house conditions were found to be the most determining factors for the phenology (Abera *et al.*, 2008).

Vegetative propagation: The plant is also propagated through semihardwood cuttings. The stem cuttings should be obtained from base till the third node at the apex. The cuttings with a length of 10-15cm (3-4 nodes) are planted closely on the raised beds during March-April three to four months before planting in the field in July. These cuttings should be treated with 500ppm NAA to promote quick rooting. The maximum success rate is obtained from the basal cuttings and it reduces gradually towards stem apex. Plant propagation can be done throughout the year in a mist chamber, and 80-100 per cent success rate is obtained from basal cuttings extending upto seventh to ninth nodes. Under open conditions, the mortality rate increases upto 70-90 per cent during the summer season. The prepared cuttings should be planted within 24h in raised nursery beds (15cm) during rainy season and flat nursery beds during winter and summer. The bed size 10m × 1m should be made under partial tree shades. In mist chamber, the cuttings should be planted in trays filled with sand. Out of three nodes, one node must be buried in the sand, as the roots would sprout from this node. The cuttings should be planted in rows with plant-to-plant distance of 5cm and row-to-row distance of 15cm in nursery. The cuttings start rooting within one month of planting in nursery. The sprouting percentage and growth are better in mist chamber than in the open conditions. These rooted cuttings are planted in main field during July.

Economic part used : Roots and milky juice.

Alkaloids: The root contains an acrid crystalline, principle called 'Plumbagin'.

Fig. 1. Plant

Fig. 2. Flower

Fig. 3. Fruit

Medicinal uses : It is used in treating intestinal troubles, dysentery, leucoderma, inflammation, piles, bronchitis, itching, diseases of the liver, and alcohol consumption. The leaves of this herb work well for treating laryngitis, rheumatism, diseases of the spleen, ring worm, scabies, and it acts as an aphrodisiac. A tincture of the root bark is used as an anti-periodic. Its root helps to improve digestion and it stimulates the appetite. It is also an acro-narcotic poison that can cause an abortion.

REFERENCES

Abera, B. 2003. Medicinal plants used in traditional medicine in Jimma Zone, Romia, Southwest Ethiopia. *Ethiop. J. Health Sci.,* 13: 85-94.

Abera, B., L. Negash and J. Kumlehn. 2008. Reproductive biology in the medicinal plant, *Plumbago zeylanica* L. *African J. Biotechnology,* 7(19): 3447-3454.

Aditi, G., G. Anjali and J. Singh 1999. New naphtoquinones from *Plumbago zeylanica. Pharm. Biol.,* 37: 321-323.

Kirtikar, K. R. and B. D. Basu. 2006. Indian Medicinal Plants Vol.II. International Book distributors, Dehradun.

Menon, T. S., S. T. Amma and E. V. Nybe. 2001. Stimulation of seed germination in white flowered ledwort (*Plumbago zeylanica*). *J. Tropical Agric.*, 39(2) : 180-181.

Vijver, L. M. and A. P. Looter. 1971. The constituents of the roots of *Plumbago auriculata* and *P. zeylanica* responsible for antimicrobial activity. *Plant. Med.*, 20: 8-13

3.34. *Podophyllum hexandrum*

Common names: Himalayan May apple, Indian May apple

Botanical name: *Podophyllum hexandrum* Wall. Ex Hook.f. & Thoms.

Family: Podophyllaceae

Vernacular names: Bankakri (Hindi); Bankakkri (Dogri); Vanatrapusi (Sanskrit).

Origin and distribution: Global; Afghanistan, Pakistan, India, Nepal, Bhutan, South Western China and Myanmar National; Jammu and Kashmir, Himachal Pradesh, Uttarakhand, Sikkim and Arunachal Pradesh.

Habitat: It can be grown in the altitude of 2000 - 4500m. It is distributed in restricted pockets throughout the alpine Himalayan region (Airi *et al.,* 1997). It is suitable for light (sandy) and medium (loamy) soils and prefers well drained soil. It can grow in full shade (deep woodland) or semi-shade (light woodland). It prefers moist soil.

Habit: It is an erect perennial, blabrous, fleshy rhizomatous herb growing 15 - 40cm tall (Fig. 1). Leaf emergence is at the onset of snow melt and dispersal of seed at the onset of winter. During winter, the plant is in dormant condition and lives in the rhizomatous form. In spring, the plant has to produce flowers, successfully pollinate and reproduce/ disperse seeds within a short span of four months.

Stem: Stems are erect, unbranched bearing two large terminal lobed leaves encircling the single large flower.

Leaf: Leaves usually two; blade rounded in outline 10-25cm, deeply cut into three ovate toothed lobes, sometimes further lobed.

Flower: Flowers are hermaphrodite, white or pale pink, fragrant, cup shaped 2-4cm across, with six petals, six stamens and a single, unilocular pistil with 50-150 ovules (Fig. 2).

Flowering, pollination and fruiting: Flowering and fruiting occurs during July to September. One of the major problems for the cultivation of this plant is its long juvenile phase and poor fruit setting ability. During flowering, the plant attracts pollinators with its bright, white colored flowers and a background of pigmented dark colored leaves. The flower lacks nectar, but has abundant pollen.

The flowers are protrandrous; the anthers often dehisce before the flower has opened, stigmas remain receptive even as the stamens and petals abscise from the flower, approximately one to two weeks after anthesis (Rajkumar and Ahuja, 2010).

Fruit: Fruits are large scarlet or reddish berry 2.5-5cm, with many seeds imbedded in pulp (Fig. 3).

Propagation: By seeds and rhizomes

Seed dormancy and germination: In the natural habitat, seed (Fig. 4) germination and seedling establishment are very poor (Alam *et al.*, 2009). Seed germination is blocked by numerous problems such as coat and endosperm dormancy, poor and non-uniform germination and long interval for plant establishment due to hypocotyl dormancy. Seed dormancy has been considered to be a major constraint in these seeds and most of the earlier reports recommended dormancy breaking pretreatments such as chilling, GA_3, etc. Seeds remain dormant for about ten months under natural conditions. Badhwar and Sharma (1963) reported 44 per cent germination upon sowing seeds with fruit pulp immediately after collection. Nadeem *et al.* (2007) reported five fold improvements in seed germination following pretreatment with sodium hypochlorite and also improved vegetative multiplication using rhizome segments treated with IBA.

Kharkwal *et al.* (2008) reported that seeds of the 14 accessions exhibited no dormancy. Besides genetic make up of accession, collection of seeds with high moisture content could be one of the reasons for lack of dormancy. The germination of seeds in sand at 25°C for high and reproducible results within a shorter period of time compared with earlier reports. Treatment of cotyledon leaves of one week old seedlings with 200ppm GA_3 resulted in true or functional leaf emergence within seven days and the resultant plants were also more vigorous than the ones obtained from manual removal of cotyledon leaves. This study helped to advance the establishment of seedlings by one growing season.

The uptake of water by seeds during germination showed a triphasic pattern with a reduced rate of water uptake during Phase III, due to the presence of thick walled endosperm cells which presented a physical barrier to the protrusion of the radicle. During germination, loss of cellular integrity of the endosperm cells in the micropylar region facilitated radicle protrusion. The existence of the multilayered endosperm and the thick testa appears to provide protection of the embryo during extreme environmental conditions by preventing germination until conditions are more favorable (Sreenivasulu *et al.*, 2009).

Seed storage: Seed storage studies indicated that viability could be maintained for about six months at 4°C. After ten months, it diminished with 50 per cent loss in germination. However, when stored for one year, the seedling vigour was affected as evident from no or reduced cotyledon leaves of the seedlings.

Economic part used: All parts of the plant

Alkaloids: The roots contain podophyllin and berberine.

Fig. 1. Plant

Fig. 2. Flower

Fig. 3. Fruit

Fig. 4. Seeds

Medicinal uses: The fruit is used as remedy for cough. The root is used as a purgative. The rhizome contains podophyllin which is used in treatment of tumors and cancer.

REFERENCES

Airi, S., R. S. Rawal, U. Dhar and A. N. Purohit 1997. Population studies on *Podophyllum hexandrum* Royle - a dwindling medicinal plant of the Himalaya. *Plant Genetic Resources Newsletter,* 110: 29-34.

Alam, M. A., P. Gulati, A. K. Gulati, G. P. Mishra and P. K. Naik 2009. Assessment of genetic diversity among *Podophyllum hexandrum* genotypes of the North Western Himalayan region for podophyllotoxin production. *Indian J. Biotechnology,* 8: 391-399.

Badhwar R. L. and B. K. Sharma 1963. A note on germination of *Podophyllum* seeds. *Indian Forester,* 89:445–447.

Kharkwal, A. Bhattacharya, P. K. Nagar and P. S. Ahuja 2008. An efficient method of propagation of *Podophyllum hexandrum*: an endangered medicinal plant of the Western Himalayas under *ex situ* conditions. *J. Nat. Med.* 62(2): 211-216.

Nadeem M, L. M. S. Palni, A. Kumar and S. K. Nandi 2007. Podophyllotoxin content, above and below ground biomass in relation to altitude in *Podophyllum hexandrum* populations from Kumaun region of the Indian Central Himalaya. *Planta Medica*, 73: 388–391.

Rajkumar, S. and P. S. Ahuja 2010. Developmental adaptation of leaves in *Podophyllum hexandrum* for effective pollination and dispersal. *Curr. Sci.*, 99 (11): 1518-1519.

Sreenivasulu, Y., S. K. Chanda and P. S. Ahuja 2009. Endosperm delays seed germination in *Podophyllum hexandrum* Royle - an important medicinal herb. *Seed Sci. & Technol.*, 37(1): 10-16.

3.35. *Rauvolfia serpentina*

Common name: Serpentina root, Rauvolfia root, Serpentine, Indian snakeroot

Botanical name: *Rauvolfia serpentina* Benth., Syn. *Ophioxylon sepentinum* L.

Family: Apocynaceae

Vernacular names: Bhungmaraja (Arunachal Pradesh); Arachoritita (Assamese); Chandra (Bengali); Amelpodee (Gujarati); Chandrabhaga, Chota chand, Sarpagandha (Hindi); Sarpagandha (Kannada); Churannavilpori, Suvapavalporiyam (Malayalam); Hadki, Sarpagandha (Marathi); Patalagarur, Sanochado (Oriya); Sarpagandha, Chandrika, Patalguruda (Sanskrit); Chavandaavalpori, Sivan Amalpodi, Sarpagandha (Tamil); Patalaguni, Patalagaruda, Sarpagandha (Telugu).

Origin and distribution: It is native to India and distributed in all parts of India. It is cultivated in India, Sri Lanka, Thailand, Bangladesh, Indonesia and Malaysia. It can be grown upto 1000m altitude, common in sub-mountain regions of Himalayas and in lower ranges of eastern and Western Ghats. It has been categorized as globally endangered.

Habitat: It prefers tropical to sub-tropical climate, receiving heavy monsoon rains between June and August, and the temperatures between 10°C and 38°C are considered ideal for its growth (Gupta, 1997). It flourishes well under hot humid tropical climates in open or partial shade. It requires slightly acidic to neutral soils for good growth. However, for commercial cultivation soils ranging from sandy loam to medium black cotton soils rich in organic matter with soil pH of 6-8 and good drainage facility are suitable.

Habit: It is an erect evergreen glabrous perennial shrub, 30-75cm height (Fig.1).

Root: Prominent, tuberous, soft tap root, reaching a length of 30-50cm, root bark 40-60 per cent of the whole root, rich in alkaloids, fresh roots emit a characteristic acrid aroma and very bitter in taste.

Leaf: Leaves whorled, usually three, ovate lanceolate, elliptic lanceolate, 7-16 x 2.5-5cm, acute or acuminate, entire, dark green above and pale green beneath, membranous, with 8-12 pairs of lateral nerves, petiole 5-15mm long, with glands in the leaf axils.

Flower: Inflorescence is terminal and consists of small flowers in compact cymes forming a hemispherical head at the end of a long peduncle. Flowers are small, pedicillate, complete and hermaphroditic with five deep red and glabrous sepals. Five petals in gamopetalous condition form a tubular corolla which is swollen in the middle and white to pink in colour. The flowers have highly narrow and long tubular corolla. Five stamens in epipetalous condition are enclosed within the dilated portion of the corolla tube. Two connate carpels have a filiform style and large bifid stigma; a bilocular ovary has two ovules in each locule (Fig. 2).

Flowering, pollination and fruiting

Sihag and Wadhwa (2011) studied the floral and reproductive biology of Sarpagandha and their findings are presented below.

1. Flowering occurred during peak summer, last week of May, when maximum and minimum temperatures ranged between 35.2-43°C and 20-29°C, and within eight days. Flowering remained mediocre (quantified in terms of numbers/m^2) for about two weeks from early June to mid June when maximum and minimum temperatures ranged between 33.4-42.7°C and 20.5-28.5°C respectively. The peak flowering, continued for 43 days, when ambient temperature fluctuated between 31-40.9°C maximum and 20.9-30.7°C minimum. Thereafter, decline started till cessation in the first week of August when maximum and minimum temperatures fluctuated between 32.6-36.3°C and 25.0-26.5°C. Thus, the plant remained in blooming stage from last week of May to first week of August over wide range of temperature.
2. Flowers started opening in the early morning between 05.00-05.30h when ambient temperature fluctuated between 24-29°C. However, anthers did not dehisce on the first day of flowering; it took place on the second day of flower opening between 07.00-07.30h at relatively higher temperature range of 28-31°C. Nectar secretion started on the first day of flower opening between 08.00 to 08.30h in the morning and continued upto 13.00 to 13.30h in the evening; it again started between 15.00 to 15.30h to continue till dusk at 17.30h on both the days. Nectar production was for a longer time and at wider range of temperature, 27 to 44°C. The diurnally opening tubular flower with liberation of floral reward during day time further makes it an example of psychophilous pollination syndrome. The flower longevity was very short, most probably due to very high ambient temperature regime.
3. As pollen was not available on the first day of flower opening, self-pollination was not possible on day one. Whether stigma was receptive or not on the day of flower anthesis, could not be ascertained. On the second

day, after anthesis self-pollen was available, but self-pollination resulted in no seed set. Either stigma was not receptive in second day flowers or due to self-incompatibility in the flowers. These possibilities were tested through cross pollination experiments. Crossing also was not possible between any two first day flowers as well as between a first day flower as pollen donor and a second day flower as recipient due to non-availability of pollen in the donor flower. Cross pollination between second day flower as pollen donor and second day flower as pollen recipient also resulted in no seed set. This again indicates self-incompatibility or non-receptivity of stigma on the second day. However, cross pollination between first day flower as pollen recipient (stigma) and second day flower as pollen donor (pollen available) produced seeds in 100 per cent recipient flowers indicating protogyny. However, fresh self-pollen (from second day flower) could not fertilize the ovary of second day flower confirming that stigma was not receptive on the second day after anthesis. These experiments revealed that in the two days' age of flower, its stigma was receptive only on the first day as anthers dehisced on the second day when stigma had become non-receptive. Therefore, flowers need to be cross pollinated.

4. The pollen remained almost fully viable and stigma fully receptive only for four hours. Thereafter, viability of pollen and/or receptivity of stigma declined. These were low after eight hours and very low after ten hours. On the next day, flower completely lost stigmatic receptivity, even fresh pollen did not produce seeds in the pollen recipient flowers. Therefore, pollination has to be completed on the first day itself, that too within a short period (<6h).
5. The number of inflorescences per plant ranged from 17 to 63 and flowers in the inflorescences ranged from 36 to 54. Each flower produced four seeds (n = 100) on an average and the normal plant indicates a potential to produce 7355 seeds which is the absolute / maximum reproductive potential.

Fruit: The fruits are single or didymous drupe and connate more or less deeply (Kirtikar and Basu, 2006), 0.5cm in diameter, small, round, dark purple or shiny blackish when ripe.

Seed: Seeds often flat with straight embryo and a crown of hairs.

Seed dispersal: Dispersal is affected mainly by wind because of winged seeds and presence of crown of hairs (Sharma, 1993).

Propagation: It can be propagated by seeds and also by vegetative means like root cuttings, root stumps, stem cuttings and leaf cuttings. But, the growth of plant and root yield is better in those raised from seeds (Badhwar *et al.*, 1956).

Seed propagation: Commercially, the plant is usually propagated by seeds. Seeds are sown in the middle of May.

Fig. 1. Plant

Fig. 2. Flower

Seed collection: The ripe seeds collected from the beginning of June to the end of October or even November. Only a few fruits ripen at a time and, if they are not collected immediately, they shed and are lost. Therefore, the collection of ripe fruits twice a week is necessary.

Seed extraction: The fruits, after collection are free from their pulpy covering by rubbing them against old gunny bags or on rough flooring. The cleaned seeds are thoroughly dried in the sun.

Seed yield: Seed yield of 100-120kg.ha^{-1} is obtained from three years old plantation.

Seed storage: The seeds stored in air tight bins, retained their viability for about six months. The viability of the seeds drops markedly with the increase in the interval of time between collection and sowing.

Seed germination: Irregular and low percentage of germination of seeds is the main difficulty in the propagation due to presence of cinnamic acid and its derivatives in the seed (Mitra, 1976). The percentage of germination of seeds is quite variable, ranging from 10-75 (Illahi *et al.*, 19988). This is partly attributed to the adverse influence of the stony endocarp. Another serious factor is the absence of embryos, though the fruits may appear perfectly normal externally. This may possibly be an effect of parthenocarpy. The rate of germination, however, depends on the percentage of fully matured, heavy seeds in a particular lot. Fresh seeds, collected from ripe fruits and immediately sown, show a higher percentage of germination (58-74%) as their viability losts for only six months. Germination starts after 15-20 days and continues upto 40-50 days after sowing. The nursery should be kept moist throughout the germination period. Santapan (1956) observed poor germination which was due to the immature embryo of freshly collected seed and it requires post harvest ripening period for its full development. Freshly harvested seeds are heavier and germinated better (Choudhary, 1963). These seeds did not require any rest period and they germinated soon after harvest (Dutta, 1963). Treating the seeds of *Rauvolfia serpentina* with hot water (80°C for 5min.) and then, cooling down to room temperature produced highest germination percentage and better seedling vigour index (Bhuyar *et al.*, 2000).

Seed treatment: The seeds should be soaked in water overnight and the light seeds, which float can be discarded. The seeds can be treated with thiram @ 3g.kg^{-1} of seeds. About 5.5kg of seeds sown in 500m^2 area will yield seedlings sufficient to plant one hectare.

Vegetative propagation: As collection of seeds from wild sources is both laborious and costly, vegetative propagation by root or shoot cuttings has been advocated for raising plantations to collect the seeds, as well as to multiply quickly the genetically superior clones.

By root cuttings: Large tap roots with a few lateral secondary rootlets are used. Cuttings of 2.5-5.0cm length are planted in holes, at the beginning of the monsoon and are almost completely covered with earth, leaving only 1cm above the surface. Nearly 50 per cent of the root cuttings sprout in about a month. Trials have shown that under irrigated conditions, root cuttings of about 0.25cm diameter planted during March-June give a 50-80 per cent success rate. About 100kg of root cuttings are required to plant 1ha. The highest percentage of success obtained by using root cuttings makes it more preferable than propagation by seeds. However, the recovery of roots from them has been found to be not as high as in plants raised from seeds.

By root stumps: Propagation by using about 5cm of root with a portion of the stem above the collar has also been attempted. This method gives about 90-95 per cent success, sometimes even 100 per cent. Such plants were transplanted in the month of May- July into irrigated fields and become well established by the end of September. This method has its limitations, as only one plant can be raised from a single stump.

By stem cuttings: Stem cuttings taken from woody twigs have also been tried as a source of propagation. Hardwood cuttings have been found better than soft wood cutting. Cuttings of 15-22cm length, with three internodes are the most suitable. Nearly 60-100 per cent of rooting is obtained by treating hard wood cuttings with β indole acetic acid solution of 30ppm for 12h and the treated cuttings root within 15 days. Stem cuttings, though they start sprouting on 3-4 days after planting, naturally strike roots mostly after about 75 days. Stem cuttings have been found less satisfactory than root cuttings, since many of them do not root easily.

Economic part used: Dry roots

Alkaloids: The major alkaloid present in root, stem and leaves of the plant is reserpine varies from 1.7 to 3.0 per cent. The root barks has more than 90 per cent of the total alkaloids in roots. The minor alkaloids present in the plant are ajmalicine, ajmaline, isoajmaline, ajmalinine, chandrine, rauvolfinine, renoxidine, rescinnamine, reserpiline, reserpin, reserpinine, sarpagine, serpentine, serpentinine, tetraphyllicine, yohimbine, 3-epi-a-yohimbine. The root contains ophioxylin and resin.

Medicinal uses: It cures snake bite, insect stings, hypertension, eye problems, cholera and fever.

Past work done in Seed Science and Technology

Seed scarification with sand paper, increases the germination percentage to some extent. Seed treatment with concentrated sulphuric acid upto 30min. increased germination percentage to some extent. Pre-sowing seed treatment with potassium nitrate and thiourea improved germination to some extent Table 1. As mechanical and chemical scarification of seeds could not improve germination of seeds, it appears that germination inhibitors may be located inside the seeds. It appears that seed germination is very complex and further work is necessary to understand the mechanism of germination (Paul *et al.*, 2008).

Table 1. Effect of pre-sowing seed treatment on germination percentage of *R. serpentina*

Treatment	Germination (%)	Treatment	Germination (%)
Control	26	**Heat treatment 70°C**	
Scarification		16h	28
Sand paper	48	24h	28
Grind stone	32	48h	32
Nicking	30	72h	26
Hot water		96h	28
3min.	25	**Heat treatment 80°C**	
5min.	25	16h	28
10min.	26	24h	30
15min.	30	48h	32
30min.	38	72h	30
45min.	30	96h	28
90min.	25	**Heat treatment 90°C**	
Hydrochloric acid		16h	26
3min.	24	24h	28
5min.	48	48h	30

10min.	72	72h	30
15min.	96	96h	28
30min.	32	1% Boric acid	32
45min.	28	1% Calcium hydroxide	30
90min.	27	1% Sodium dihydrogen phosphate	30
		1% Potassium dihydrogen phosphate	30
		1% Potassium nitrate	40
		0.5% Thiourea	30
		100ppm GA_3	32
		100ppm NAA	30

REFERENCES

Badhwar R. L., G. V. Karira and S. Ramaswami. 1956. Methods of propagation and their effect on root production in *Rauvolfia serpentina. Indian J. Pharm.* 18: 170-175.

Bhuyar, S. A., S. G. Wankhade, J. T. Paturde and P. T. Khode. 2000. Seed germination studies in sarpagandha (*Rauvolfia serpentina* Benth). *Res. Crops*, 1(2) : 189-191.

Choudhary, T. K. 1963. On the germination of seeds of the species of *Rauvolfia. Sci. Cult.*, 29 (3): 156-157.

Dutta, P. K. 1963. Viability of seed of *Rauvolfia serpentina* on storage. *Sci. Cult.,* 29 (3): 148-149.

Gupta, R. 1997. Medicinal and Aromatic Plants. **In:** Handbook of Agriculture, ICAR, New Delhi.

Illahi, I., M. Akram and L. Kraus. 1988. *In vitro* studies for mass propagation and alkaloid synthesis. *Pakisthan J. Sci. Industrial Res.*, 31: 114-117.

Kirtikar, K. R. and B. D. Basu. 2006. Indian Medicinal Plants Vol. II. International Book Distributors, Dehradun.

Mitra, G. C. 1976. Studies on the formation of viable and nonviable seeds in *R. serpentina. Indian J. Exp. Biol.,* 14: 54-54.

Paul, D., N. K. Paul and P. K. Basu. 2008. Seed germination response of *Rauvolfia serpentina* Benth to certain physical and chemical treatments. *J. Bio. Sci.* 16: 129-131.

Santapan, H. 1956. Studies on *Rauvolfia* seed germination. *Indian J. Pharm.*, 18: 117-118.

Sharma, O. P. 1993. Plant taxonomy. Tata McGraw-Hill publishing company Limited, New Delhi. pp.483.

Sihag, R. C. and N. Wadhwa. 2011. Floral and reproductive biology of sarpagandha *Rauvolfia serpentina* (Gentianales: Apocynaceae) in semi-arid environment of India. *J. Threatened Taxa* 3(1): 1432-1436

3.36. *Salvia officinalis*

Common names: Garden sage, Kitchen sage, Common sage, Golden sage, True sage, Culinary sage, Broadleaf sage and Dalmatian sage

Botanical name: *Salvia officinalis* L.

Family : Lamiaceae

Vernacular names: Bui tulasi (Bengali); Salvia, Sefakus (Hindi); Salvi tulasi (Malayalam); Sathi (Punjabi); Salvitulasi (Tamil).

Origin and distribution: Sage is a native of Mediterranean area. It grows wild in the Dalmatian region of Yugoslavia. It is cultivated in Yugoslavia, Italy, Albania, Turkey, Portugal, Spain, Cyprus, England, Canada and USA. In India, it is sparingly cultivated in Jammu.

Habitat: Dry banks and stony places, usually in limestone areas and often where there is very little soil.

Habit: It is a perennial, evergreen, hardy sub-shrub (Fig. 1).

Stem: Stems are white, angled, woody, 32-60cm tall.

Leaf: The leaves are oblong, aromatic, petiolate, ranging in size upto 6.4cm long by 2.5cm wide. Leaves are grey green, rugose on the upper side and nearly white underneath due to the many short soft hairs. Sometimes the hairs are glandular and secrete volatile oils that typically give a distinct aroma to the plant. When the hairs are rubbed or brushed, some of the oil bearing cells are ruptured, releasing the oil. This often results in the plant being unattractive to grazing animals and some insects. On drying, leaves turn silvery grey with soft velvetty texture.

Flower: The flowers are produced in racemes or panicles and generally produce a showy display with flower colors ranging from blue to red, with white and yellow less common. The calyx is normally tubular or bell shaped, without bearded throats, and divided into two parts or lips, the upper lip entire or three toothed, the lower two cleft. The corollas are often claw shaped and are two lipped. The upper lip is usually entire or three toothed. The lower lip typically has two lobes. The stamens are reduced to two short structures with anthers two celled, the upper cell fertile and the lower imperfect. The flower styles are two cleft (Fig. 2).

Flowering, pollination and fruiting: The flowering and fruiting occurs in late spring or summer. *Anthophora dufourii* male pollinates *Salvia hierosolymitana*. The stamens deposit pollen on the bee's back. It is pollinated by honey bees (Abrol, 2012).

Fruit: The fruits are smooth ovoid or oblong nutlets and in many species, they have a mucilaginous coating (Fig. 3).

Seed : The seeds are blackish brown, contained in long, toothed husks, as an old writer describes the calyx (Fig. 4).

Propagation : It can be propagated by seeds and cuttings

Fig. 1. Plant

Fig. 2. Flower

Fig. 3. Fruit

Fig. 4. Seeds

Seed propagation: One thousand weight of the seeds is 8.03g. The rate of water absorption is 49 per cent within 24h. The highest rate of germination capacity reached 85.33 per cent and the germination index is 45.56 (Lieu *et al.,* 2006).

The seeds are sown during March / April or September in a greenhouse. Germination usually takes place within two weeks. Prick out the seedlings into individual pots when they are large enough to handle and plant them out in early summer. In areas where the plant is towards the limits of its hardiness, it is best to grow the plants on in a greenhouse for their first winter and plant them out in late spring of the following year.

Economic part used : Leaves, whole herb.

Alkaloids: Carnosol, camosic acid, rosmanol, rosmarinic acid.

Medicinal uses : Sage is used as a mild tonic, astringent and carminative. It is diaphoretic and anti-pyretic. Sage oil is used in perfumes as a deodorant. Sage and sage oil exhibit anti-oxidant properties.

REFERENCES

Abrol, D. P. 2012. Pollination biology, biodiversity conservation and agricultural production. Springer London, p.792.

Lieu, L., Q. S. Guo, Y. P. Wang and R. M. Zhao. 2006. Study on characteristics of seed germination of *Salvia officinalis, Zhongguo Zhong Yao Za Zhi.* 31(19):1587-1589

3.37. *Solanum khasianum*

Common names: Medicinal Solanum, Tropical soda apple

Botanical name: *Solanum khasianum* Clarke., Syn. *Solanum viarum.*

Family: Solanaceae

Vernacular names: Marunthu kathiri, Sundakai (Tamil)

Origin and distribution: It is native to India, Brazil and Argentina. It is naturalized introduction, growing wild in the hills of North Eastern India and South India (Babu and Hepper, 1979). It is widely distributed in Indian states like Assam, Manipur, Nagaland, West Bengal, Sikkim, Odisha and Gangetic plains, Tamil Nadu and states in United States *viz.,* Alabama, Florida, Georgia, Mississippi and South Carolina.

Habitat: It has been observed as a weed in agricultural land and in natural areas. Well drained loamy soils are suitable. It grows well under moderate climatic conditions.

Habit: Erect annual and much branched, height 0.75-1.40m, more of straight prickles often mixed with a few curved spines on the stem. Gadwal (1977) reported that the presence of spines on the leaves and stems (Fig. 1).

Leaf: Both surfaces of the leaves are armed with much larger prickles (3-25mm long) that are found along the veins. These leaves (6-20cm long) are borne on thick stalks (3-7cm long) and their margins are wavy with several rounded lobes. Both leaf surfaces are also densely hairy.

Flower: The inconspicuous star shaped flowers (about 1.5-2.0cm across) are produced in small clusters below the newer leaves, with 1-6 flowers in each cluster. They are borne on stalks 3-6mm long and have five white or greenish white petals (about 10mm long and 2.5mm wide) that are usually bent backwards. The flowers are bisexual and also have five white stamens that are held erect and have slender anthers 6-7mm long (Fig. 2).

Flowering, pollination and fruiting: In parts of India where it is hot, tropical soda apple can be an annual that produces flowers three months after germination and mature berries in another two months. Saini and Biswas (1967) observed that in the Khasi and Jantia hills of eastern India, flowering and fruiting extended round the year except for a short spell during autumn. They recorded highest fruit set in June sown plants and lowest in March sown. It is a self pollinated crop and artificial pollination is feasible at the time of anthesis in long styled flowers (Handique, 1986). Anthesis takes place between 8.00 to 16.30h. Anther dehiscence commenced an hour after anthesis. Pollen production is very abundant to the extent of 170, 240 to 232, 320 pollen grains per anther. Stigma receptivity is from 4 to 64h after anthesis. Under natural conditions, pollination could be continual process and lasts upto 72h.

Fruit: Berry, yellow or green in colour, globose, 2.5cm in diameter. Each fruit contains 200-400 small brown seeds (Fig. 3).

Seed: Smooth, brown, 3mm in diameter and compressed (Fig. 4).

Seed dispersal: Although seed is the primary mechanism for spread, it also sprouts from the roots and possibly from the root crown. Cattle eat fruit and disseminate the seed *via* foeces. The seed is spread long distances as a contaminant in hay, grass seed, and sod and by moving water. The seed may be dispersed by equipment or human and other animal movement. The seeds contain a sticky substance that adheres to the undersides of mowers, tires, front grills, blades and shoes. Contaminated mowers, vehicles, and other equipment can transfer seed to uninfected pastures. The seed may also adhere to animal fur, providing another means for tropical soda apple seed dispersal.

Propagation: It is propagated through seeds.

Seed storage: Seeds germinated well upto eight months of storage and then, decreased with an increase in storage period (Mitra and Kushari, 1987a). The seed placed in dry, dark storage in the laboratory for upto 12 months remained viable (Mullahey *et al.*, 1996; Patterson *et al.*, 1997).

Seed germination: In the laboratory, seed viability and germination rates were high for seed sown under variable conditions. Germination is epigeal (Mullahey and Cornell,1994). Germination rates of 69 (Akanda *et al.*, 1996) to 82.5 per cent (Brown *et al.*, 1996; Mullahey *et al.*, 1993) have been reported for tropical soda apple in the laboratory and scarification of seed coat may improve germination (Akanda *et al.*, 1996; Brown *et al.*, 1996). In the laboratory, germination of seed removed from fruit was 82.5 per cent. Germination increased to 90 per cent when seed passed through the digestive tract of cattle within 24h, indicating that scarification or partial digestion of the seed coat may increase germination rates. Seed taking longer than six days to pass through the digestive tract may no longer germinate (Brown *et al.*, 1996). Although scarified tropical soda apple seed germinated more quickly than unscarfied seed, per cent germination was reduced, presumably from injury to the embryo (Akanda *et al.*, 1996). Germination rates increased upto 82 per cent when seed was subjected to dormancy breaking acid treatments and exposed to the same alternating temperatures (Suryawanshi *et al.*, 2001).

Fig. 1. Plant

Fig. 2. Flower

Fig. 3. Fruit

Fig. 4. Seeds

Economic part used: Fresh berries

Alkaloids: Maiti *et al.* (1964) were the first to report the presence of solasodine (5.40 per cent) in berries. Chandra *et al.* (1970) reported the variation in alkaloid content from 0.90 to 7.40 per cent. Khanna and Murthy (1974) screened the population and found that the solasodine content ranged from 0 to 4 per cent. The range of 0.90 to 2.20 per cent of solasodine was recorded by Singh *et al.* (1978). Nandhakumar (1983) recorded the highest content of 2.70 per cent. Paturde *et al.* (2002) reported that the variety Arka Mahima contained high amount of solasodine (2.78 per cent).

Medicinal uses: Steroids produced by this plant have been used to treat cancer, Addison's disease, rheumatic arthritis, chronic asthma, leukemia, obesity, palsy and skin diseases. It is also used in manufacturing of oral contraceptive tablets, nematicidal and bactericidal properties.

Agricultural uses: Seenivasan *et al.* (2000) reported that cabbage dimond back moth was effectively controlled by spray of three per cent.

Past work done on Seed Science and Technology

Pre-sowing seed treatment

According to Suryawanshi *et al.* (1998), top of paper media was found to be the best for conducting germination test. Freshly harvested seeds show little germination which gradually increases with storage and reaches its optimum between seven and nine months storage. A storage period of atleast one month is necessary to ensure better and quicker germination. The seeds washed with water give a higher germination than unwashed seeds. By manipulating the conditions of seed treatment and regulating temperature, very high rate of germination can be achieved and the total duration of germination can also be reduced. Scarification of seeds with five per cent nitric acid improves the germination of the seeds. The seeds scarified with H_2SO_4 (20%) at the rate of 20ml/2g of seed and mechanically scarified seeds increased the rate of germination, but not the over all percentage (Akanda *et al.*, 1996) (Table 1).

Table 1. Effect of mechanical and acid scarification on rate and total germination in tropical soda apple

Treatment	Time to 50 per cent germination	Total germination[a]
	D	%
Un-scarified (control)	13.9	70.8
Mechanical	9.6	64.1
Sulphuric acid (SA) (20%)	8.5	55.4
LSD (0.05)	2.1	3.7

[a] initial TZ viability = 89%

Seeds soaked in tap water (or) hot water increased the rate of germination to 26 per cent and seeds soaked in KNO_3, GA_3 and ethephon increased it to 53 per cent (Akanda *et al.*, 1996) (Table 2).

Table 2. Effect of mechanical and acid scarification on rate and total germination in tropical soda apple

Treatment	Time to 50 per cent germination	Total germination[a]
	D	%
Un scarified (control)	14.9	70.4
Tap water	10.8	75.4
Hot water	11.3	73.3
H_2SO_4 (20%)	8.5	55.4
KNO_3 (0.2%)	7.8	81.5
GA_3 (100μM)	8.0	77.8
Ethephon (0.69μM)	7.8	75.8
LSD (0.05)	1.3	7.2

[a] initial TZ viability = 89%

Seed treatment with X rays and EMS (Ethyl Methane Sulphonate) reduced the germination percentage and survival of seedlings with increasing radiation dose. Whereas effects with SA treatments were comparatively less pronounced than X rays and EMS (Ram and Kaul,1991) (Table 3).

Table 3. Effect of mutagenic treatments on seed germination and seedling survival (%) of *Solanum khasianum*

Treatment	Germination % of control	RRL-20-2 Lethality (%)	Survival (%) of control	Germination % of control	RRL-SL-2 Lethality (%)	Survival % of control
Control	100	0	100	100	0	100
X-rays (KR)						
5	77.5	22.5	72.0	92.0	8.0	90.0
10	68.7	31.3	65.6	75.5	24.5	75.0
20	52.5	47.5	50.6	62.6	37.6	61.5
30	47.7	52.3	42.7	45.4	54.6	43.7
EMS (%)						
0.1	75.0	25.0	66.7	88.5	11.5	81.5
0.2	66.5	35.5	59.0	75.5	24.5	72.0
0.5	60.2	39.8	55.0	64.0	36.0	58.0
1.0	52.0	48.0	49.0	58.9	41.1	55.5
SA (%)						
0.1	100.2	0	82.6	102.1	0	98.7
0.2	102.5	0	90.0	102.7	0	87.7
0.5	105.1	0	100.0	100.2	0	88.4
1.0	98.7	0	90.0	98.5	0	92.5

Seed treatment with X rays and EMS significantly reduced the pollen fertility depending on the dose. SA affects though less pronounced followed a similar pattern (Ram and Kaul,1991) (Table 4).

Table 4. Percentage of pollen fertility following various mutagenic treatments in the M_1 generation

Sl. No.	Treatments	RRL-20-2	RRL-SL-6
1.	Control	96.5	95.0
2.	**X rays (KR)**		
	5	95.0	92.00
	10	85.0	86.00
	20	82.0	71.50

	30	68.0	63.00
3.	**EMS (%)**		
	0.1	85.0	92.00
	0.2	95.0	91.00
	0.5	85.0	82.00
	1.0	81.0	79.00
4.	**SA (%)**		
	0.1	95.0	94.00
	0.2	91.0	89.00
	0.5	92.0	86.00
	1.0	86.0	84.00

Seeds were soaked in ethrel (100, 250 and 500ppm) for 4h and seed treatment followed by foliar spray of ethrel solution at two and four leaf stages stimulate leaf development, branching and increased the alkaloid content of the berries. Repeated treatments are more efficacious than seed treatment alone. The flowering was not affected by seed treatment, but the repeated treatments proved inhibitory. Fruit development was delayed in the treated plants but yield remained unaltered (Chauhan and Kharbteng,1992).

Seeds treated with nitric acid at low concentrations (10%) stimulated the seed germination and seedling growth (Mitra and Kushari, 1987).

The seeds irradiated with gamma rays at the rates upto 5.0KR increased the percentage of long style flowers and enhanced berry numbers / plant (Chauhan and Ravindran, 1979). Soaking of seeds in ethrel at 500ppm and 700ppm concentrations for 4h, stimulated the germination whereas the higher concentration of 1000 (or) 1250ppm inhibited the seed germination (Chauhan, 1978).

Seed germination : One of the serious problem in the commercial cultivation is the low and prolonged germination of seeds in pans and nursery beds. The reason for low germination is that the mucilage covering, containing mostly glycoalkaloid, in the form of dry membrane on the seeds might be preventing rapid and uniform seed germination.

Seed storage: Seeds maintained the germination capacity upto eight months of storage and then decreased with an increase in storage period upto 12 months (Mitra and Kushari, 1987). Germination percentage of seed is 75 per cent.

REFERENCES

Akanda, R. U., J. J. Mullahey and D. G. Shilling 1996. Environmental factors affecting germination of tropical soda apple (*Solanum viarum*). *Weed* Sci., 44(3):570-574.

Babu, C. R. and F. N. Hepper 1979. Taxonomy and nomenclature of *Solanum khasianum* and some of its relatives. *Kew. Bull.,* 34: 407-411.

Brown, W. F., J. J. Mullahey and R. U. Akanda 1996. Survivability of tropical soda apple seed in the gastro-intestinal tract of cattle. **In:** Proc. of the tropical soda apple symposium; January 9-10, 1996 (Eds.) F. L. Bartow and F. L. Gainesville, University of Florida, Institute of Food and Agriculture Science: p. 35-39.

Chandra, V., B. Singh, A. Singh and L. D. Kapoor. 1970. Variation in the solasodine content of fruits of *Solanum khasianum* at different stages of development in Lucknow. *Indian For.,* 96: 352-360.

Chauhan, Y. S. 1978. Effect of ethrel on the germination of *S. khasianum* Cl. seeds. *Indian J. Pharmaceutical Sci.,* 40(2):61-63.

Chauhan, Y. S. and J. S. Kharbteng. 1992. Ethrel effect on growth, flowering, fruiting and alkaloid content in *S. khasianum* Cl. *Recent Advances in Medicinal, Aromatic and Spice Crops, 2* : 473-480.

Chauhan, Y. S. and S. Ravindran 1979. Gamma ray effects on heterostyly in *Solanum khasianum* Clarke. *Curr. Sci., 48(15)* : 698.

Gadwal, V. R. 1977. Commercial cultivation of *Solanum* crop (Glaxo strain). **In:** Cultivation and utilization of medicinal and aromatic plants (Eds.) Atal, C. K. and B. M. Kapur, Regional Research Laboratory. Jammu – Tawi. pp:31-33.

Handique, A. K. 1986. Breeding behaviour of *Solanum khasianum* Clarke. *Indian Drugs,* 20: 292-293.

Khanna, K. R. and A. S. Murthy. 1974. Inheritance of solasodine in *Solanum khasianum* Clarke. *Indian J. Genet. Plant Breed.,* 34: 1200-1203.

Maiti, P. C., S. Mookerjee, R. Mathew and A. N. Henry. 1964. A rich source of solasodine from *Solanum khasianum* Clarke var. Chatterjeanum Sen Gupta. *Curr. Sci.,* 33: 730.

Mitra, S. and D. P. Kushari 1987. Effect of pretreatment with nitric acid on the seed germination and seedling success of Mogor variety of *Solanum khasianum. Environmental Ecology, 5 (4)* : 628-632.

Mitra, S. and D. P. Kushari 1987a. Seasonal effect on seed germination of *Solanum viarum. Proc. of National Seminar on Medicinal and Aromatic Plants,* Tamil Nadu Agricultural University, Coimbatore.

Mullahey, J. J. and J. Cornell 1994. Biology of tropical soda apple (*Solanum viarum*) an introduced weed in Florida. *Weed Technology.* 8(3): 465-469.

Mullahey, J. J., M. Nee, R. P. Wunderlin and K. R. Delaney 1993. Weed alert! Tropical soda apple (*Solanum viarum*): a new weed threat in subtropical regions. *Weed Technology.* 7(3): 783-786.

Mullahey, J. J., R. U. Akanda and D. G. Shilling 1996. Integrated management strategies to control *Solanum viarum.* **In**: Proceedings of second international weed control congress; June 25-28, 1996, Copenhagen, Denmark. (Ed.) O. R. Corvallis: International Weed Science Society, p. 1009-1014.

Nandhakumar. D. 1983. Intervarietal hybridization in steroid bearing *Solanum viarum* and *S. incanum* and induction of autotetraploidy in *S. viarum,* Ph. D. thesis. University of Agricultural Sciences, Bangalore.

Patterson, D. T., M. McGowan, J. J. Mullahey and R. G. Westbrooks 1997. Effects of temperature and photoperiod on tropical soda apple (*Solanum viarum* Dunal) and its potential range in the U.S. *Weed Sci.,*. 45(3): 404-408.

Paturde, J. T., S. T. Wankhade, P. P. Khode and A. H. Khan 2002. Performance of Arka Mahima (Tetraploid) against Arka Sanjeevini (Diploid) varieties of wild brinjal (*Solanum viarum* D.) under different plant spacings and fertility levels. *Agric. Sci. Digest*, 22(1): 67-68.

Ram, G. and B. L. Kaul 1991. Biological effect of X rays, ethyl methane sulphonate and sodium azide in *Solanum khasianum* Clarke. *Indian J. Forestry,* 14(4): 275-281.

Saini, A. D. and R. C. Biswas. 1967. Studies on the physiology of *Solanum khasianum* Clarke. II. Effect of different sowing dates on growth and development. *Indian J. Plant Physiol.,* 10: 36-43.

Seenivasan, N. R., P. Sridhar and Gnanamurthy 2000. Efficacy of some plant extracts against (*Plutella xylostella)* on cabbage. *South Indian Hort.*, 51 (1-6):183-185.

Singh, B., K. R. Khanna and A. S. Murty 1978. A note on genetic variability for glyco alkaloids in *Solanum khasianum* Clarke. *Indian J. Hort.,* 35: 145.

Suryawanshi, Y. B., R. B. Patil and N. D. Moholkar 1998. Studies on germination procedure in some medicinal species. *Seed Tech. News.,* 28: 57-58.

Suryawanshi, Y. B., R. B. Patil and N. D. Moholkar 2001. Study on seed germination procedures in some medicinal plant species. *Seed Res.,* 29(2): 141-144

3.38. *Solanum nigrum*

Common name: Black night shade

Botanical name: *Solanum nigrum* L.

Family : Solanaceae

Vernacular names: Makoi, *Makoy*, Ghamai (Hindi); Kakamachigida (Kannada); Tudavalam (Malayalam); Kangni, Ringni (Marathi); Manathakkali, Milaku thakkali (Tamil); Kakamanchi (Telugu).

Origin and distribution: It is native to Europe. It is distributed in Asia and northern Africa, the Caucasus, Central Asia, Western Siberia, the Far East. Distributed also in Scandinavia, Middle and Atlantic Europe, Mediterranean, Iran, India, China, Japan, North Africa; an adventive plant in North America.

Habitat: It is mainly a weed of moist environment and thrives in areas of low rainfall only where the land is under irrigation. It is best adapted to soils of high fertility, especially those high in nitrogen and phosphorus. It has been naturalized in both temperate and tropical regions. This species colonize moist environments, only occurring in areas of low rainfall when the land is subjected to irrigation. However, it is susceptible to frost. This plant prefers rich, cultivated, well moistened soils, but can even grow on saline soils and in stony places. It is also found near roads, houses, waste dumps and bush thickets along rivers.

Habit: It is an erect to spreading, 15-100cm tall, branching, annual to short lived (2 - 5 years) perennial herb to sub-shrub (Fig. 1).

Root: Roots are mainly fibrous with a shallow taproot.

Stem: Stems are round to angular and covered with curved multicellular hairs sometimes tipped with a glandular head, upto 70cm high and branched.

Leaf: Simple leaves are 2-8cm long by 1-5.5cm wide, green to purple in colour, alternate, ovate or ovate-lanceolate to lanceolate in shape with wavy toothed margins, decurrent on the petiole, glabrous or sparsely pubescent.

Flower: Inflorescences are simple, with pedicels closely clustered and forming an extended cymes or raceme like cluster of two to ten flowers. Flowers are 4-18mm in diameter and bell shaped, with a star shaped, white corolla with a translucent to yellowish basal star. The calyx consists of ovate sepal lobes that are 1.2-2.5mm long and a star shaped corolla of five white lobes that are 1.8-7.5mm long. Stamens five, anthers are in a conical group. Gynoecium composed of two fused carpels (Fig. 2).

Flowering, pollination and fruit set: This plant flowers in June - September, bears fruit in July - November. Flowering occurs 5-9 weeks after germination and continues until the plant dies. Early germinating plants take several weeks more to flower compared to late germinating plants. It is an entomophilous.

Fruit: Fruits are broadly ovoid and have dull purple to black or yellowish green berries that are 6-10mm broad. Berries may remain on the plant or fall to the ground when ripe. Fruits are without sclerotic granules on the surface (Fig. 3). Each berry contains 15 to 96 bone coloured seeds (Ogg and Rogers, 1989; Defelice, 2003).

Seed: Seeds are tan, obovate, flat in cross section, 1.7-2.4mm long (Fig. 4). The potential of seed production vary in a range of 600 to 680 seeds per plant (Keeley and Thullen, 1983; Holm *et al.* 1991; Defelice, 2003).

Seed dispersal: The ripe berries are extremely succulent, often brightly colour and/or aromatic, and are attractive to a variety of birds and animals. They are therefore, effectively dispersed by birds all over the world, as well as by various animals such as rodents, lizards, rabbits, cattle and man. There is evidence that the seeds pass through the digestive tracts of both birds and animals undamaged (Burgert *et al.,* 1973; Weller and Phipps, 1978/79). The seeds can also be dispersed by water.

Propagation : By seeds and shoot cuttings.

Seed collection: Seeds may be collected from ripe fruits dried under shade and stored in room condition.

Fig. 1. Plant Fig. 2. Flower

Fig. 3. Fruit Fig. 4. Seeds

Seed dormancy and germination: Roberts *et al.* (1978) opined that the seeds germinated rapidly at an alternate temperature of 25/30°C and treating the seeds with 0.2 per cent KNO_3 or 500ppm GA_3. Wagenvoort and Vanopstal (1979) recommended alternate temperature for improving the germination of seeds. According to Tyagi and Sharma (1977), seeds washed with two per cent acetic acid resulted in good germination. According to Horowitz and Givelberg (1982), seeds did not germinate in dark, but in light 98 per cent germination was recorded on fifth day between 20 and 35°C temperature with optimum of 25°C.

Long survival of dormant seeds enables a plant population to build up a bank of seeds of different ages in the soil. Seeds can remain viable in the soil for a long time. Viability of seeds in the soil has been reported from 18 to 39 years depending on soil disturbance. The mechanism of seed dormancy is associated with impermeable seed coat (Suthar *et al.,* 2009). Minimum temperature for seed germination is 10-12°C, maximum is 34-36°C and optimum is 24-26°C. Seeds germinate from depths of 0.5-1cm. Alternating temperatures are required to break dormancy and lower light intensities or darkness are generally more favourable for germination. There is vigorous recruitment from soil stored seed in burnt and unburnt areas after rain. Levels of primary seed dormancy among populations, enabling it to maximise its chance of survival by regulating germination timing to favourable conditions.

Economic part used : Dry leaves, fruits and seeds

Alkaloids: Solanigrin, solamargine and solasodine.

Medicinal Uses *:* Infusion of herb applied to anthrax pustules. The herb also used as a diuretic and laxative. Freshly prepared extract of herb is effective in cirrohosis of liver. Juice of fresh leaves produces dilation of the pupils. Berries used in anasarca and heart diseases. They are employed as a domestic remedy in fevers, diarrhoea and eye troubles.

Past work done on Seed Science and Technology

Seed dormancy and germination: Suthar *et al.* (2009) observed 72 per cent germination when the seeds were scarified with needle pin while with sand paper, 78 per cent seed germination was observed followed by 33 per cent germination with mortar pestle scarification (Table 1). It was observed that sand paper scarification method was the best for breaking dormancy. The macroscleried layer of mature seed coat is mainly responsible for imparting impermeability to the seeds.

Table 1. Effects of mechanical scarification on seed germination of *Solanum nigrum*

Method	Germination (%)
Unscarified seeds	52 ± 1.5
Sacrification with needle pin	72 ± 1.3
Sand paper	78 ± 0.9
Mortar pestle	33 ± 1.0

Scarification with diluted sulphuric acid gave 48 per cent of germination at 1.0min. and then, it decreased with increasing in time. Similarly, germination was the highest, 38 per cent at 30sec. with concentrated sulphuric acid and decreased later on (Table 2). It was observed that chemical scarification with both dilute and concentrated sulphuric acid showed adverse effect as compared to that of control (unscarified seeds), which could be due to adverse effect caused by the acid to the embryo.

Table 2. Effect of chemical scarification with dilute and concentrated H_2SO_4 on seed germination of *Solanum nigrum*

	Method	Germination (%)
Unscarified seeds		53 ± 1.8
Scarified with	0.15 min.	23 ± 1.2
a. Dilute H_2SO_4	0.30 min.	34 ± 1.8
	1.0 min.	48 ± 2.8
	2.0 min.	32 ± 1.6
b. Conc. H_2SO_4	0.15 min.	34 ± 1.2
	0.30 min.	38 ± 0.8
	1.0 min.	20 ± 2.6
	2.0 min.	10 ± 2.6

Low imbibition was observed in unscarified seeds as compared to sand paper scarified seeds at all pH treatments. But, in both seeds, there was no significant variation in imbibition percentage in acidic pH as well as alkaline pH range. The rate of germination of scarified and unscarified seeds also showed variation. The maximum percentage germination occurred in petriplate irrigated with only distilled water. The percentage rate germination was low at acidic as well as alkaline pH in both the sets of scarified and unscarified seeds (Table 3).

Table 3. Effect of pH on imbibition and seed germination of *Solanum nigrum*

		pH values							
Parameter	**Seeds**	**4.1**	**5.1**	**6.1**	**6.5**	**D/W**	**8.1**	**9.2**	**10.3**
Percentage of imbibition (6h)	Scarified	8.39	8.31	8.59	8.43	8.51	8.13	8.21	8.17
	Unscarified	6.83	6.31	6.91	6.83	6.91	6.78	6.78	6.61
Percentage of germination	Scarified	71 ± 1.2	73 ± 0.8	75 ± 1.4	78 ± 1.2	82 ± 1.6	76 ± 3.2	73 ± 0.6	63 ± 1.2
	Unscarified	31 ± 0.6	38 ± 0.8	40 ± 1.2	38 ± 1.2	48 ± 1.6	36 ± 0.8	28 ± 1.8	22 ± 2.0

Both scarified and unscarified seeds showed an increase in rate of imbibition with increase in time interval. After soaking, the scarified seeds for 5.5h, the maximum imbibition was observed and it remained constant afterwards. Similarly, the rate of imbibition of unscarified seeds was low and maximum imbibition was observed at 6.0h and then remained constant (Table 4).

Table 4. Rate of imbibition on scarified and unscarified seeds of *S. nigrum*

	Percentage imbibition in gram	
Time (h)	**Unscarified seeds**	**Scarified seeds**
0.5	5.31	5.81
1.0	5.38	5.93
1.5	5.48	6.46
2.0	5.50	6.49
2.5	5.53	6.54
3.0	5.58	6.58
3.5	5.60	6.63
4.0	5.81	6.71
4.5	5.93	7.13
5.0	6.31	7.41
5.5	6.68	7.42
6.0	6.73	7.42
6.5	6.74	7.42
7.0	6.74	7.42

The maximum percentage of seed germination was observed in all the three sets in sand paper scarified seeds as compared to unscarified seeds. In scarified seeds, the maximum percentage of seed germination was observed in the seeds, sown 1.0cm deep in the soil followed by seeds, sown superficially and 2.0cm deep in the soil. Similar trends were observed in case of unscarified seeds with low values. Thus, a suitable depth of sowing of the seeds is 1.0cm below the soil surface (Table 5).

Table 5. Effect of sowing depth on seed germination of *Solanum nigrum*

Sowing depth	Germination percentage	
	Scarified	**Unscarified**
Superficial	68 ± 7	38 ± 3
1cm	78 ± 8	52 ± 4
2cm	54 ± 4	44 ± 2

Table 6. Effect of temperature on seed germination of *Solanum nigrum*

Temperature (°C)	Seeds	Germination (%) / day				
		1	2	3	4	5
5	Scarified	-	-	7 ± 0.8	9 ± 0.8	3 ± 0.3
	Unscarified	-	-	1 ± 0.9	3 ± 0.2	2 ± 0.2
10	Scarified	-	-	8 ± 0.6	7 ± 0.2	4 ± 0.4
	Unscarified	-	-	2 ± 0.4	2 ± 0.6	1 ± 0.6
27	Scarified	-	4 ± 0.8	28 ± 0.6	38 ± 1.3	14 ± 1.8
	Unscarified	-	1 ± 0.3	8 ± 1.2	6 ± 0.3	4 ± 0.8
32	Scarified	-	-	14 ± 1.2	22 ± 0.8	18 ± 0.2
	Unscarified	-	-	8 ± 0.3	4 ± 0.3	3 ± 0.4
40	Scarified	-	-	6 ± 0.6	18 ± 0.7	3 ± 0.3
	Unscarified	-	-	5 ± 0.2	3 ± 0.3	1 ± 0.7
50	Scarified	-	-	8 ± 0.2	3 ± 0.4	7 ± 0.3
	Unscarified	-	-	3 ± 0.1	4 ± 0.8	2± 0.2
60	Scarified	-	-	3 ± 0.2	8 ± 0.7	2 ± 0.2
	Unscarified	-	-	1 ± 0.3	3 ± 0.3	2 ± 0.4

Unscarified seeds showed lesser percentage of germination as compared to sand paper scarified seeds at all temperature treatments. The maximum percentage of germination was observed at 27°C followed by 32°C in scarified seeds in which the lowest percentage seed germination was observed at low temperature (3 per cent after five days of incubation at 5°C) as well as at high temperature (2 per cent after five days of incubation at 60°C). Variation was observed in the germination rate of scarified and unscarified seeds and the maximum percentage of germination was noted on fourth day. The rate of percentage germination was low for first two days and on last day (Table 6). Therefore, the temperature (27°C) increased the germination of scarified seeds.

Seed extraction methods

Traditional: Seeds and pulp were squeezed from the berries and thoroughly washed with tap water. The water cleaned seeds were sun dried for one day, then stored in a chamber environmentally controlled at 25°C and 30 per cent relative humidity. Similar drying and storage conditions were used in all other extraction methods.

Concentrated hydrochloric acid: The volume of seed pulp obtained by the traditional method above was measured and subjected to acid treatment using a ratio of 158ml pulp: 8.9ml concentrated HCl. The digestion mixture was stirred every 10min. for 30min. at room temperature. The seeds were thoroughly washed with tap water, dried then stored.

Fermentation: The seed and pulp obtained by the traditional method was left to ferment for three days at 25°C in an incubator. The mixture was stirred three times a day. The seeds were then, dried and stored.

Sodium carbonate: The volume of seed and pulp obtained as previously described was measured and an equal volume of ten per cent sodium carbonate added. The mixture was incubated for two days at 25°C and stirred every eight hours. The seeds were subsequently dried and stored.

Seed extraction using concentrated HCl gave the highest germination rates (92%) followed by a fermentation method (88%) using ten per cent Na_2CO_3 or traditional method gave germination rates of 85 and 83 per cent respectively. The traditional method proved superior for long term storage needs with minimum seed deterioration. Concentrated HCl method is used where seeds are extracted for immediate use. The traditional method had the least deleterious effect on seed viability after storage for eight months in a controlled environment at 25°C and 30 per cent relative humidity (Agong, 1993).

The average 1000 seed weight differed significantly over time between and within extraction methods (Table 7 and 8).

Table 7. Effect of seed extraction procedures on the viability (%) categorized as bright red (BR), red (R) and without colour change (W) following tetrazolium solution test

Method	January			April			August		
	BR	R	W	BR	R	W	BR	R	W
Conc. HCl.	95a	2a	3a	64a	9a	27a	52a	15a	33a
10% Na_2CO_3	91b	4b	5b	61a	7a	28a	35b	20b	45b
Fermentation	93b	5b	2c	63a	11b	26a	61c	8c	31c
Traditional	94a	4b	2c	64b	4b	20b	64d	7d	29d

Values followed by a common letter in each column do not differ significantly at $p = 0.05$.

(Agong, 1993)

Seed storage and burial: Ozer (1972) reported that seeds buried in the soil can remain viable for 30 years or more. Uremis and Uygur (2005) recorded that seeds recovered seven years after burial and seed longevity was determined using germination tests. Burial depth, 15 and 30cm had no significant effect on seed viability. Seed longevity was 5.38 per cent.

Table 8: Effect of seed extraction procedures on average 1000 seed weight

Method	January	April	August
Conc. HCl.	0.682a	0.678a	0.677a
10% Na_2CO_3	0.680b	0.672b	0.665b
Fermentation	0.680b	0.665c	0.663b
Traditional	0.683a	0.682d	0.680c

(Agong, 1993)

Suthar *et al.* (2009) observed the highest percentage of seed germination in the month of June followed by July, August, October and December in case of sand paper scarified seeds and June in case of unscarified seeds, while there was not much variation in other months. The scarified seeds showed higher germination as compared to that of unscarified seeds at every month (Table 9).

Table 9. Effect of months and scarification on seed germination (per cent) of *S. nigrum*

Months	Scarified seeds (%)	Unscarified seeds (%)
October	68 ± 1.2	43 ± 0.8
November	63 ± 1.3	42 ± 0.3
December	68 ± 1.2	46 ± 0.4
January	62 ± 0.8	43 ± 0.8
February	58 ± 0.7	38 ± 0.7
March	53 ± 0.3	37 ± 1.3
April	68 ± 0.8	43 ± 0.8
May	63 ± 1.1	46 ± 0.9
June	73 ± 0.8	54 ± 1.5
July	71 ± 1.2	52 ± 1.1
August	69 ± 0.3	50 ± 0.7
September	49 ± 0.2	38 ± 0.9

Gunasekaran, (2003) undertaken the experiments at the Department of Seed Science and Technology, Tamil Nadu Agricultural University, Coimbatore in *Solanum nigrum* and his findings are given below.

The seeds attained physiological maturity at 40DAA when seed dry weight, germination and vigour index were highest. The seeds extracted by HCI and Na_2CO_3 methods gave maximum seed recovery.

Regarding the medium for germination test, top of paper was found to be the best medium. The first and final count of germination test could be taken at seventh and fourteenth day, respectively. Seed treatment with halogen mixture @ $3g.kg^{-1}$ of seed and packing in 700 gauge polythene bag proved to be the best for seed storage. Seed extracted by manual and fermentation maintained higher germination and seedling vigour over storage when compared to HCI and Na_2CO_3 methods.

From the accelerated ageing test, it was confirmed that seed treatment with halogen mixture and packing in 700 gauge polythene bag and seed extraction by fermentation and manual methods were found to maintain higher vigour and viability in storage.

Seed production plasticity: Plants tolerate wide range of habitats, able to start flowering in early stage of growth and potentially capable of producing huge number of seeds under favourable growing conditions and few seeds under unfavourable conditions.

Seed dormancy: The seeds go through seasonal dormancy cycle during period of a year. They germinate only when the environmental condition is favourable for germination and its subsequent growth. Therefore, it can build up a seed bank for a long period of time.

Growth habit: Although it is a late germinating weed, it is able to grow fast and occupy opportunity space between crop stands. It thrives very well when the crop is less competitive or crop canopy is sparse.

REFERENCES

Agong, S. G. 1993. Seed extraction procedures for long term conservation of *Solanum nigrum*. *Seed Sci. & Technol., 21(2)* : 447-451.

Burgert, K. L., O. C. Burnside and C. R. Fenster. 1973. Black nightshade leaves its mark. *Nebr. Univ. Coll. Agric. Q. Serving Farm*, 20: 8-10.

Defelice, M. S. 2003. The black nightshade, *Solanum nigrum* L. *Weed Tech.*, 17:421-427.

Gunasekaran, S. 2003. Seed development and maturation, extraction methods, germination test and storage in *Solanum nigrum* Linn. M.Sc (Ag.) Thesis, TNAU, Coimbatore

Holm, L G., D. L. Pluckneet, J. V. Pancho, and J. P. Herberger. 1991. The World's Worst Weeds: Distribution and Biology. Malabar, FL: Krienger Publishing (The University Press of Hawaii, Honolulu). 609 p.

Horowitz, M. and A. Givelberg. 1982. Effect of high temperature on germination and dormancy of *Solanum nigrum* seeds. *Phytoparasitica,* 10(4): 279-280.

Keeley P. E. and R. J. Thullen 1983. Influence of planting date on the growth of black nightshade *(Solanum nigrum). Weed Sci.,* 31:180–184.

Ogg, A. G., Jr. and B. S. Rogers 1989. Taxonomy, distribution, biology and control of black nightshade (*Solanum nigrum*) and related species in the United States and Canada. *Rev. Weed Sci.*, 4:25-58.

Ozer, Z. 1972. Yabanci ot tohumlarinin yasam muddetleri. *Ataturk Universitesi Ziraat Fakultesi Dergisi,* 3: 233-239.

Roberts, H. A., Lockett and M. Patricia. 1978. Seed dormancy and field emergence in *Solanum nigrum*. *Weed. Res.*, 18: 231-241.

Suthar, A. C., V. R. Naik and R. M. Mulani. 2009. Seed and seed germination in *Solanum nigrum* Linn. *American-Eurasian J. Agric. Environ. Sci.*, 5 (2): 179-183.

Tyagi, M. C. and B. Sharma. 1977. Glycoalkaloid inhibits seed germination in *Solanum viarum*. *Seed Res.*, 5(2): 182-183.

Uremis, F. I. and N. Uygur. 2005. Seed viability of some weed species after seven years of burial in the Cukurova region of Turkey. *Asian J. Plant Sci.*, 4: 1-5.

Wagenvoort, W. A. and N. A. Vanopstal. 1979. The effect of constant and alternating temperature, rinsing, stratification and fertilizer on germination of some weed species. *Scientia Hort.*, 10:15-20.

Weller, R. F. and R. H. Phipps 1978/1979. A review of black nightshade (*Solanum nigrum* L.). *Prot. Ecol.*, 1: 121-139

3.39. *Solanum surattense*

Common names: Febrifuge plant, Prickly brinjal, Jacquin's nightshade, Dwarf wild brinjal, Spiny nightshade

Botanical name: *Solanum surattense* Burm. F Syn. *S. armatum* Forssk., S. *virginianum* Jacq., *S. xanthocarpum* Schrad. & Wendl. *S. Jacquini* Will., *S. diffusum* Roxb.

Family : Solanaceae

Vernacular names: Katvaedana, Kantakar (Assamese); Kantakari (Bengali); Bhoyaringani (Gujarati); Bhatakataiya, Chhotikateri, Berkateli, Kateli, Katai, Kateri, Ringani (Hindi); Nelagulla, Kiragulla (Kannada); Kath (Kashmiri); Kantakari chunda, Kandan kathri (Malayalam); Bhauringani, Kataringani (Marathi); Chakada Bhoji, Bhejibegun, Ankranti (Oriya); Kandyali, Mahori, Warumba (Punjabi); Kantakari, Nidigadhika (Sanskrit); Kandangatri, Kandankatri, Kandanghathiri, Mullu kaththiri (Tamil); Mulaka, Chinnamulaka, Vakudu, Pinnamulaka, Nelamulaka, Vankuda (Telugu).

Origin and distribution: It occurs throughout India and also distributed in Sri Lanka, Malaysia and tropical Australia.

Habitat: It is commonly found in dry situations as a weed along the roadsides and wastelands.

Habit: It is a prostrate herb, 19-40cm long with scattered stellate hairs and prickly stem with woody base (Fig. 1).

Stem: Stem branched much and younger ones clothed with dense, cylindrical, 0.7-1.4cm in width, greenish brown, stellate and tomentose hairs. Prickles are compressed straight, glabrous and shining, often 1-3cm long.

Leaf: Leaves ovate or elliptic, 1.1-2.5cm long, lamina 1.4-7cm long, 0.4-4.4cm width, sinuate or sub-pinnatifid, obtuse or sub-acute, stellately hairy on both sides, armed on the midrib and often on the nerves with long yellow sharp prickles. Petiole is long, stellately hairy and prickly.

Flower: Inflorescence axillary cymose, peduncle 1-2cm long. Flowers purple, 0.8-2cm long, 0.3-0.7cm width, villous. Sepals five, 0.3-0.5cm long, 0.2-0.3cm width, green, margin hairy, apex acute, pubescent. Petals five, 0.4-0.7cm long, 0.2-0.4cm wide, margin hairy undulate, apex obtuse, pubescent. Stamens five, filament upto 0.1cm long, smaller than anthers, Anthers 0.4-0.6 (0.7)cm long. Style 0.7-1.1cm long, partially hairy, stigma capitate, ovary blackish, rounded (Fig. 2).

Flowering, pollination and fruiting*:* Flowers and fruits appear during summer. It is an entomophilous (Sharma, 1993).

Fruit: Fruit is typically berry, 0.7-1cm long, globose, yellow with white or green streaks (Fig. 3).

Seed : Seeds discoid, minutely reticulate, whitish yellow in colour, numerous, shiny, 2.5mm in diameter and glabrous (Fig. 4).

Fig. 1. Plant

Fig. 2. Flower

Fig. 3. Fruit

Fig. 4. Seeds

Propagation : By seeds and cuttings.

Seed dispersal: This is easily dispersed through animals as the fruits have hooked bristles and two strong hooked beaks. Seed dispersal takes place through birds (77%), crab eating fox (19%) and lizards (4%).

Seed dormancy and germination : Germination rate of seeds collected directly from fruit (control), bird droppings, the foeces of crab eating foxes and lizards were 64, 64, 53, and 80 per cent, respectively.

Economic part used : Dried roots, fruits, flower, seed

Alkaloids: Steroidal alkaloid solasodine is the principal alkaloid. Solasonine is present in fruits. The presence of diosgenin in the plant has been reported.

Medicinal uses : It cures cough, asthma, chest pain and fever, diuretic, tooth and gum disorders and bronchitis.

REFERENCE

Sharma, O. P. 1993. Plant taxonomy. Tata McGraw-Hill Publishing Company Limited, New Delhi. pp.483.

3.40. *Solanum torvum*

Common name: Devil's fig, Turkey berry, Susumber, Gully bean, Thai eggplant

Botanical name: *Solanum torvum* Sw. Syn. *Solanum hermanii*, *Solanum ferrugineum* Jacq., *S. ficifolium* Ort.

Family: Solanaceae

Vernacular names: Bhi-tita (Assamese); Tita bagoon (Bengali); Bhurat, Bhankatiya, Katai (Hindi); Sundekkayi (Kannada); Anachunda (Malayalam); Marang (Marathi); Brihati (Sanskrit); Sundaikkai (Tamil); Kottuvastu (Telugu).

Origin and distribution: It is native to West Indies and distributed throughout tropical Africa, Asia, Australia and the Pacific Islands including Hawaii, Guam and American Samoa.

Habitat: It grows best in full sunlight and does well in light shade or shade for part of the day, but cannot survive under a closed forest canopy. Turkey berry single plants, groups and thickets are most frequently seen on roadsides, vacant lands, brushy pastures, recently abandoned farmland, landslides and river banks.

Habit: It is evergreen, usually 2 or 3m in height and 2cm in basal diameter, pubescent with stellate hairs, but may reach 5m in height and 8cm in basal diameter (Fig. 1).

Stem: The shrub usually has a single stem at ground level, but it may branch on the lower stem. The stem bark is gray and nearly smooth with raised lenticels. The inner bark has a green layer over an ivory color. The twigs are gray green and covered with star shaped hairs. The spines are short and slightly curved and vary from thick throughout the plant, including the leaf midrib to entirely absent.

Leaf: Leaf is opposite or one per node, broadly ovate with the broader entire or deeply lobed. The petioles are 1 to 6cm long and the blades are 7 to 23 by 5 to 18cm and covered with hairs dense on lower leaf surface, sparser on upper surface; glandular hairs on pedicels; prickles 3-7mm long, scattered on stems and leaf surfaces, sparse to absent on mature growth.

Flower: Lower flowers perfect, upper ones staminate, actinomorphic, numerous (50-100) in congested, branched corymbs, at first terminal, soon lateral by sympodial growth, peduncle to first fork 1-2cm long, pedicels 5-10mm long; calyx 3-4mm long, the lobes apiculate, 2-3mm long; corolla white, stellate, 2.5cm in diameter, the lobes lanceolate, 10mm long, apex acute; stamens five, inserted on corolla tube; filaments 1mm long; anthers yellow, attenuate, 6-7mm long, opening by small apical pores; ovary globose, pubescent; style one, erect, 8-10mm long in perfect flowers in staminate flowers (Wagner *et al.* 1999) (Fig. 2).

Flowering, pollination and fruiting: Flowering in year round and fruiting in May. Pollination is an entomophilous (Sharma, 1993).

Fig. 1. Plant

Fig. 2. Flower

Fig. 3. Fruit

Fig. 4. Seeds

Fruit: Berry globular, 10-15mm diameter, drab yellow, drying black, thin fleshed with 210-220 seeds (Fig. 3).

Seed: Seeds are ovate to broadly ovate, flat, broadly elliptic or nearly circular, infrequently C shaped in outline, 1.5-2mm long, brown seeds (Fig. 4).

Seed dispersal: The seeds are spread by fruit eating birds (PIER, 2003). They may also be spread on equipment or in contaminated soil.

Propagation: By seeds, cuttings

Economic part used: Roots, fruits

Alkaloids: Solasodine.

Medicinal uses: Stomach ache, antihelmenthetic.

REFERENCES

PIER (Pacific Islands Ecosystems at Risk). 2002. Invasive plant species: *Solanum torvum*. Available: http://www.hear.org/pier (Accessed: April 22, 2003).

Sharma, O. P. 1993. Plant taxonomy. Tata McGraw-Hill Publishing Company Limited, New Delhi. pp.483.

Wagner, W. L., D. R. Herbst, and S. H. Sohmer. 1999. Manual of the flowering plants of Hawaii. 2 vols. Bishop Museum Special Publication 83, University of Hawaii and Bishop Museum Press, Honolulu, HI

3.41. *Sphaeranthus indicus*

Common name: East Indian Globe Thistle

Botanical name: *Sphaeranthus indicus* L. Syn. *S. hisutus* Willd., *S. mollis* Roxb.

Family: Asteraceae

Vernacular names: Kamadarus (Assamese); Surmuriya, Chhagal Nadi, Mudmudiya, Murmuriya, Mundi, Gorakh mundi (Bengali); Mundi, Gorakhmundi (Gujarati, Hindi and Kannada); Mirnagnee, Atookamanni, Mirangnee, Atakkamaniyan (Malayalam); Mundi, Baras Bondi, Gorakh mundi (Marathi); Buikadamba, Murisa, Bokashungi (Oriya); Gorakhmundi, Mundi, Ghundi, Khamadrus (Punjabi); Mahamundi, Mundi, Hapus (Sanskrit); Kotook, Karandai, Kottakarantai, Visnukkarantai (Tamil); Boddatarupa, Bodasaramu, Bodataramu, Bodatarapu (Telugu); Mundi (Urdu).

Origin and distribution: It is native to India and distributed throughout India, Sri Lanka, Africa and Australia.

Habitat: It is found abundantly in damp situations in the plains all over India, ascending to an altitude of 1500m in the hills, especially as a *rabi* season weed in the rice fields.

Habit: A spreading aromatic annual herb, about 30-60cm tall with glandular, hairy, winged stems, wings toothed (Fig. 1).

Root: It is much branched, strongly scented, and annual erect with branched tappering tap roots (Vikani *et al.*, 2008). The root shows on its outer side metaderm, a typical brown colored tissue. It consists of suberized cells, arranged irregularly and forms a protective layer. Radial groups of pericyclic fibers and few stone cells are seen alternating with radially arranged secretory canals in the secondary cortex. Phloem is parenchymatous and radially arranged. Medullary rays are pitted, lignified and about two to five seriate.

Stem : It is a much branched, strongly scented with winged cylindrical stem and the wings toothed. The stem shows cork with two to three layers of parenchymatous cells covered with papillose cuticle having trichomes and can be distinguished by the presence of a discontinuous ring of lignified pericyclic fibers and a well developed ring of bicollateral vascular bundle surrounding the pith. Medullary rays are pitted, lignified and about unitetraseriate.

Leaf: Leaves are sessile, decurrent, 2-7cm long, 1-1.5cm wide, obovate-oblong, rounded or sub-acute, glandular hairy, spinous serrate or dentate, narrowed at the base and greenish brown in colour. The leaf is dorsiventral and shows abundant trichomes of varying types on both the epidermis. Simple trichomes are three to four celled, thick walled and measure 130.8–145.2μm in length and 29.0–43.5μm in width. Trichomes are straight knee shaped, with a swollen base and with collapsed cell at the middle or at the apex. Midrib shows three to four collateral vascular bundles associated with a group of sclerenchymatous cells on either side.

Flower: Flowers are borne in terminal, solitary, globose, clusters of heads. Heads of flowers are purple, bracts are short slender and acuminate. In each head, the outer flowers are females, few or many, fertile, the central flowers bisexual, fertile or sterile, involucre narrow, bracts paleaceous, spathulate, acute, ciliate; receptacle small, naked. Corolla of female flowers are purple, slender, tubular, minutely two to three toothed; corolla of hermaphrodite flowers are purplish white, tubular or funnel-shaped, four to five toothed, anther base sagittate, auricles acute or tailed, style armed, filiform, sometimes connate (Fig. 2).

Flowering: Flowering and fruiting occurs during November to January in Indian conditions.

Fruit : Fruits are oblong and have compressed achene in which pappus is absent. Odor of herb is slightly aromatic, but disappears on long storage (Fig. 3 and 4).

Propagation: By seeds

Economic part used: Root, bark, leaves, flowers, and seeds.

Fig. 1. Plant

Fig. 2. Flower

Fig. 3. Seeds

Fig. 4. Fruit

Alkaloids: Sphaeronthanolide.

Medicinal uses: The plant is bitter, stomachic, stimulant, alterative, pectoral, demulcent and externally emollient. It is used in epilepsy, hemicrania, jaundice, liver disorders and gastric disorders. A paste of the herb mixed with oil is good for pruritus and painful swellings. The mots are diuretic, expectorant, febrifuge, anthelmintic and stomachic. They are useful in strangury, diabetes, leprosy, fever, pectoralgia, cough, gastric complaints, hernia, haemorrhoids, helminthiasis and dyspepsia. Oil prepared by using the root is useful in scrofula. The leaf is good for skin diseases and is considered as a nervine tonic. The flowers are highly esteemed as alterative, depurative, refrigerant and tonic, and used as blood purifier in skin diseases.

REFERENCE

Vikani, K. V., R. D. Dangar, N. S. Kapadia and M. B. Shah 2008. A pharmacognostic study on *Sphaeranthus indicus*. *J. Nat. Remed.*, 8:61–67.

3.42. *Stevia rebaudiana*

Common name: Sweet honey leaf, Sweet leaf, Honey leaf, Sweet herb of Paraguay, Candy leaf

Botanical name: *Stevia rebaudiana* Bertoni

Family. Asteraceae

Vernacular names: Madhu Patra (Hindi); Sarkarai Thulasi, Seeni Thulasi, Inippu Thulasi (Tamil)

Origin and distribution: It is native to Paraguay, Bolivia, Argentina and Brazil, where it has been cultivated for centuries by the Native American Guarani. It is grown commercially in many parts of the world and especially in Japan, China, Korea, Malaysia, Thailand, Israel and much of South America. It has been successfully cultivated in recent years in many areas of Indian states: Rajasthan, Maharashtra, Kerala and Odisha.

Habitat: The natural climate is semi-humid and sub-tropical with a temperature extremes from 21 to 42°C, averaging 24°C with annual rainfall of 1375mm (Brandle *et al.*, 1998). The soil should be in the pH range 6.5-7.5; well drained red soil and sandy loam soil. Saline soils should be avoided to cultivate this plant.

Habit: It is an herbaceous short lived perennial that reaches about 50-75cm in height with a spread of around 45-50cm (Fig. 1).

Stem: The plant has an extensive tap root system with brittle stems.

Leaf: Sessile, oppositely arranged lanceolate to oblanceolate leaves, 3-4cm long, serrated above the middle, trichome structures on the leaf surface are two distinct sizes, one large and one small.

Flower: The tiny white florets are perfect, borne in small corymbs of 2-6 florets. Corymbs are arranged in loose panicles (Fig. 2).

Flowering and pollination: The tiny white flowers appear late in the summer. Oddone (1997) considers it to be self-incompatible and insect pollinated.

Fruit: It is five ribbed spindle shaped achene with a feathery pappus.

Seed: Contained in slender achenes about 3mm length, each achene has about 20 persistent pappus bristles (Fig. 3).

Propagation: The seeds germinate poorly and propagation is usually done from cuttings.

Seed propagation: The seeds are collected during late summer. The seeds are sown indoors in early spring or frost free climate. At about 8-10 weeks from sowing, the plants may be transplanted. Excessive rainfall during pollination can affect both seed yield and germination (Carneiro, 1990). Seed is best stored at 0°C, but even under low temperature conditions germination will still decline 50 per cent over three years (Shuping and Shizhen, 1995). Sealing of storage containers or using lower temperatures did not prevent the decrease in germination over time.

Fig. 1. Plant Fig. 2. Flower

Fig. 3. Seeds

Propagation through cuttings: Propagation by stem cuttings, which roots easily, but it requires high labour inputs.

Economic part used: Leaves

Alkaloids: The dry leaves contain 5-10 per cent steviosides and 2-4 per cent rebaudiosides (Dzyuba, 1998).

Medicinal uses: It is an ideal, natural, herbal and calorie free bio-sweetener with no side effect. Its white crystalline compound is the natural herbal sweetener with no calories and is over 100-300 times sweeter than table sugar. This plant extract is also used to treat high blood pressure and there is hypertensive effect on both systolic and diastolic blood pressure, which is dose dependent and it also have antiviral, antifungal and antibacterial property (Chan *et al.*, 1998). It is always a slight lowering in arterial blood pressure at low or normal doses, changing to slight rise in arterial pressure at very high doses (Machado *et al.*, 1986).

Past work done on Seed Science and Technology

A study was undertaken to investigate the low seed germination of seeds. Influence of pollination and effect of light and darkness during germination were evaluated. The results revealed that there were two types of seed; black and tan. Black seed weighed more than tan seed, 0.300 vs. 0.178mg, and viability of black seed based on tetrazolium chloride was much higher than tan seed, 76.7 vs. 8.3 per cent (Table 1). Germination in the dark was higher for black as compared to tan seed (83.7% vs. 16.0%) while light increases the germination of black seed, but not tan seed (Fig. 1). This suggests that tan seeds represent non-viable seed that are produced without fertilization. There was no significant difference in black seed germination among the four pollination treatments suggesting that incompatibility is not a factor in these clones (Table 2). However, all pollination treatments increase seed germination of black seed over the control suggesting that some active manipulation of the blossoms is necessary to achieve pollination. It would appear that many of the black seeds in the control were misclassified tan seed (Goettemoeller and Ching, 1999).

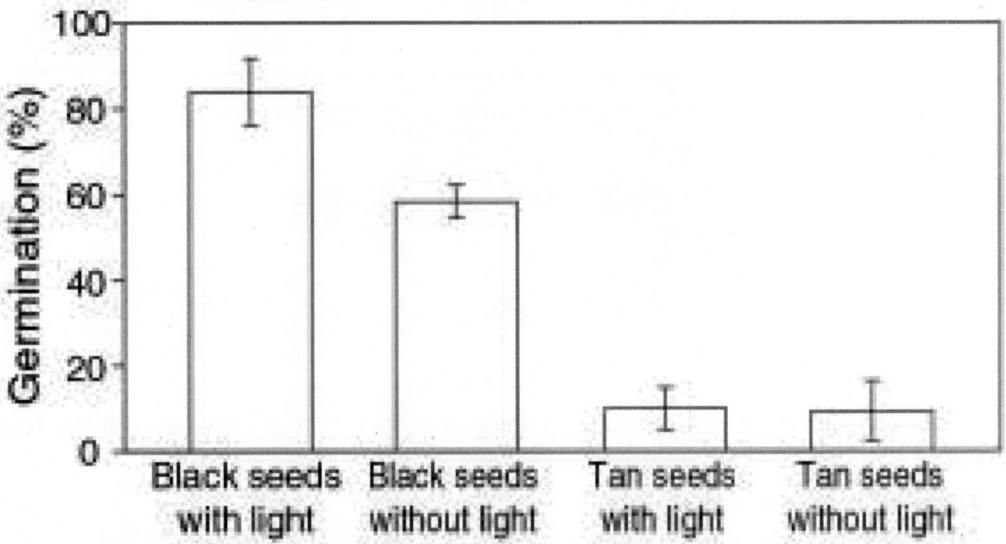

Fig. 1. Average germination rate for cross- pollinated "Chinese" stevia (± SD).

Table 1. Seed weight and viability of stevia

Seed color	Seed weight (mg/1000 seeds ± SD)	Seed viability (% ± SD)
Black	300 ± 5	76.7 ± 23.1
Tan	178 ± 8	8.3 ± 10.4

Table 2. Germination of black stevia seeds under light derived from various pollination treatments.

Plant pollination treatment	Germination (% ± SD)
Selfing (hand)	93.3 ± 0.6
Crossing (hand)	92.0 ± 2.0
Crossing (bumble bees)	78.3 ± 14.5
Crossing (wind)	68.3 ± 22.5
Control[z]	36.3 ± 25.5[y]

[z] Control germination = 62, 36, 11 [y] Plants in isolation

REFERENCES

Brandle, J. E., A. N. Starratt and M. Gijzen 1998 . *Stevia rebaudiana*: its agricultural, biological and chemical properties. *Canadian J. Plant Sci.,* 78: 527-536.

Carneiro, J. W. P. 1990. *Stevia rebaudiana* (Bert.) production of seed. M.Sc. Thesis, State University of Maringa, Brazil (English abstract).

Chan, P. X., D. Y. Liu, J. C. Jen Y. J. Tomlinson, B. W. P. Huang and J. T. Cheng 1998. The effect of stevioside on blood pressure and plasma catecholamines in spontaneously hypertensive rats. *Life Sci.,* 63 (19): 1679-1684.

Dzyuba, O. O. 1998. *Stevia rebaudiana* (Bertoni) Hemsely - a new source of natural sugar substitute for Russia (Russian). *Rastitel Nye Resursy,* 34 (2): 86-95.

Goettemoeller, J. and A. Ching 1999. Seed germination in *Stevia rebaudiana.* **In:** Perspectives on new crops and new uses (ed.) J. Janick, ASHS Press, Alexandria, VA. p. 510–511.

Machado, E., A. M. Chagas and D. S. Reis 1986. *Stevia rebaudiana* (Bert.) Bertoni in the arterial pressure of the dog. Third Brazilan Seminar on *Stevia rebaudiana* (Bert.) (Summaries), Angelucci, E. (Coordinator). p.11.

Oddone, B. 1997. How to grow stevia. Technical manual. Guarani Botanicals, Pawtucket, CT

Shuping, C. and S. Shizhen 1995. Study on storage technique of *Stevia rebaudiana* seed (English abstr.). *Acta Agronomica Sinica* 21 : 102-105

3.43. *Trichopus zeylanicus*

Common name: Aarokya pachai

Botanical name: *Trichopus zeylanicus* subsp. *travancoricus* (Bedd.) Burkill

Family : Trichopadaceae

Vernacular names: Jeevani (Hindi); Aarokya pacha, Saasthan kizhangu (Malayalam); Aarokya pachai (Tamil).

Origin and distribution: Endemic to Western Ghats. Located in the hills of Kerala and Tamil Nadu. In Kerala, it is locally abundant in dense forests of Bonacaud, Kallar in Thiruvananthapuram district and Kulathupuzha and Achankoil in Kollam district. In Tamil Nadu, it is locally common in Keeriparai, Pechiparai and Mahendragiri hills in Kaniyakumari district and Kalakad tiger reserve forests and Agasthiarmalai hills in Tirunelveli district.

Habitat: In dense, shady moist places and slopes of semi-evergreen to evergreen forests.

Habit: A perennial erect herb. This plant can at once be recognized in the forest by the bunch of leaves projecting above the ground. Since this is a perennial rhizomatous herb that produces a rosette of 10-15 (weighing 100-200g) evergreen leaves (with two or three flush in every year) (Fig. 1).

Root: Grey, numerous, fibrous, arising at the base of the rhizome.

Rhizome: Rhizome is short, stout, about 2cm long, brown.

Stem: Several, slender, 4-12cm long, cylindrical, glabrous, rigid, purplish brown arising from underground rhizome.

Fig. 1. Plant Fig. 2. Flower

Fig. 3. Fruit

Leaf: Single in each stem, triangular ovate to ovate - lanceolate, base truncate to sub - cordate or rounded, apex acute and minutely apiculate, margin entire, coriaceous, glabrous, dark green occasionally with purplish shade, with five nerves emerging from the base of the leaves. Leaf stalks 3-7cm long, rigid, purplish brown, glabrous with a central groove.

Flower: Bisexual, 3-5, clustered at the base of the leaves on about 3cm long slender stalks, 1cm across, dark brown (Fig. 2).

Flowering, pollination and fruiting: Flowering and fruiting occurs during November to March. It is autogamous (Kubitzki, 1998).

Fruit: Ellipsoid, distinctly three angled, indehiscent, 1-2.5cm long, glabrous, dark brown (Fig. 3).

Seed : Six, oblong, with grooves on the upper side.

Propagation : Rhizomes and seeds

Economic part used : Fruits, leaves

Medicinal uses : Fruits are used by the local tribal to increase strength and energy of the body. This plant reveals strong anti-fatigue properties. It shows potent anti-stress, energy boosting, anti-hepatotoxic and immune modulating property. It is used for gaining evergreen health and energy and preventing the ageing related disorders. Paste made from equal quantity of leaves and tender fruits, is consumed once a day for body pain, fever and whooping cough.

REFERENCE

Kubitzki, K. 1998. The families and genera of vascular plants. Vol. III: Flowering plants monocotyledons Liliance (Except Orchidaceae). Springer verlog berlin Heidelberg, New York.

3.44. *Vitex negundo*

Common name : Chast tree

Botanical name: *Vitex negundo* L.

Family : Verbenaceae

Vernacular names: Nirgundi, Nishinda, Samalu (Bengali); Nagoda, Shamalic (Gujarati); Mewri, Nirgundi, Nisinda, Sambhalu, Sawbhalu (Hindi); Bilenekki (Kannada); Indrani (Malayalam); Nirgunda (Marathi); Banna, Marwan, Maura, Mawa, Swanjan Torbanna (Punjabi); Nirgundi, Sephalika, Sindhuvara, Svetasu-rasa, Vrikshaha (Sanskrit); Chinduvaram, Nirnochchi, Nochchi, Notchi, Vellai nochchi (Tamil); Sindhuvara, Vavili, Nalla vavili, Tellavavili (Telugu).

Origin and distribution: This plant is indigenous to the Mediterranean countries and Central Asia. It can be grown in all over the country.

Habitat: It is found in wastelands upto 2000m in the Himalayas.

Habit: It is an aromatic large shrub. Sometimes grow like a small tree too (Fig.1).

Stem: Stem colour is bluish green or purple. Stem shape is square.

Leaf: Each leaflets looks like an eye. Underside of the leaf has white powdery hair. Leaves are bitter to taste.

Flower: Flowers are in bunches and are light lavender in colour (Fig. 2).

Flowering, pollination and fruiting: Flowering and fruiting occurs during August to March. It is pollinated by wasps.

Fruit: The fruits are ovoid or obovoid, 4mm in diameter, four seeded succulent drupes, black or purple when ripe (Fig. 3).

Seed : Seeds are black in colour when ripe and globose in shape.

Propagation : Seeds and micropropagation

Fig. 1. Plant

Fig. 2. Flower

Fig. 3a. Mature fruit

Fig. 3b. Dry fruit

Seed germination: Seeds of this plant germinated better when they were previously soaked in water for 15 to 24h. Soaking for more than 24h was detrimental to the germination (Ma and Liu, 1986).

Economic part used: Whole plant.

Medicinal uses: The leaves are used to treat cough and asthma, guinea worms, epilepsy, foetid ears, joint pains, aches etc. Leaves are used as digestive agent and are good for eyes. Roots are used for joint ache, inflammations, flatulence, wounds and ulcers, breathing problems, malaria and leprosy. Flowers are used for diarrhoea, cholera, fever, hemorrhages and cardiac disorders. Seeds are boiled and eaten to prevent the spread of toxins from poisonous bites of animals.

Agricultural uses

The plant products notchi (*Vitex negundo* L.) leaf powder @ 1:100 protected the pea seeds against storage environment and biotic factors significantly better than captan and activated clay by keeping the viability at about 70 per cent after 8 months of storage (Paramasivam, 1990).

Thambidurai and Jayaraj (2000) observed the plant extract of notchi @ 5 per cent for the effective control of chilli thrips (50%) compared to application of dimethoate (29.58%).

Prakasam *et al.* (2001) suggested that combination of different extracts of *Azadirachta indica, Ocimum sanctum* and *Vitex negundo* var. *purpurescens* could be suitably applied to carrot for effective and safe management of post harvest rot disease of carrot.

Antonysamy (2004) reported that aggergatum onion treated with three per cent of panchagavya along with Notchi, *Calotropis, Nerium, Aloe* and pungam in cow's urine for seven days increased the yield of bulbs by 17.5t.ha^{-1}.

Nandhakumar (2010) observed higher germination in maize at five days after sowing with *Albizia lebbeck* as green leaf manure and seed soaking with *Vitex negundo* leaf extracts.

Other uses: Leaves are kept along with the cereal and millet seeds during storage, to prevent the occurrence of pests. The pungent smell of leaves resists the pest attack. Mostly planted as hedges in gardens and in home gardens.

REFERENCES

Antonysamy, 2004. Organic aggregatum onion better packaging and quality. The Hindu News paper. May 20. pp.10.

Ma, C. G. and D. Y. Liu 1986. Effect of experimental soaking of the seed of the tree species. *Forest Sci. Technology*, 12: 10-13.

Nandhakumar, M. R. 2010. Utilisation of leafs and leaf extracts of tropical trees for eco friendly vrkshayurvedic farming in maize (*Zea mays* L.) M.Sc (Ag.) Thesis, AC &RI, TNAU, Madurai.

Paramasivam, V. 1990. Studies on development, maturation, quality and storability of pea (*Pisum sativum*) seeds M.Sc. Thesis, Department of Seed technology, TNAU, Coimbatore.

Prakasam, V., S. Abraham and C. Kannan 2001. Management of post harvest fungal diseases of carrot using botanicals. *South Indian Hort.*, 49 (Special): 271-274.

Thambidurai, G. and S. Jayaraj 2000. Efficacy of certain botanical pesticides in the management of major pests of brinjal, bhendi and chillies. MSc Thesis, TNAU, Madurai.

3.45. *Withania somnifera*

Common name: Winter cherry, Indian ginseng

Botanical name: *Withania somnifera* (L.) Dunal

Family: Solanaceae

Vernacular names: Ashvagandha (Assamese, Bengali, Oriya); Asgandha, Ghodaasun (Gujarati); Asgandh, Aswagandha, Punir, Rasbhari (Hindi); Angarberu, Hiremaddina-gida, Kanchuki (Kannada); Asagandh (Kashmiri); Amukkuram (Malayalam); Asagandha, Askagandha, Askand, Ghoda, Tilli (Marathi); Asgandh (Punjabi); Hayagandhj, Vjjigandhj (Sanskrit); Amukkira, Amukkirakilangu (Tamil); Pennerugadda, Vajigandha, Asvagandhi (Telugu); Asgand (Urdu).

Origin and distribution: It is native to Australia, East Asia and Africa. It is distributed in south and east Africa, Congo, Madagascar, from Palestine upto north India covering Israel, Jordan, Egypt, Sudan, Iran and Pakistan. In India, the plant grows well throughout the drier parts and in subtropical and semi temperate regions, including the states of Maharashtra, Madhya Pradesh, Gujarat, Rajasthan, Uttar Pradesh, Haryana and Punjab extending to Himachal Pradesh and Jammu and Kashmir from plains to the height of 1700m.

Habitat: As it is a late rainy season crop, it requires dry weather conditions. Rainfall of 650-750mm is best suited. One or two late winter rains are enough for its roots to develop fully. Well drained sandy loam soils or light red soil with a pH of 7.5-8.0 with good drainage is best for its cultivation. This also helps to uproot the plant without causing any damage to them.

Habit: Small or middle sized erect evergreen shrub, slightly hairy plant, upto 1.5m height, stem and branches covered with star shaped hairs. It grows as a stout shrub that reaches a height of 170cm with long tuberous roots (Fig. 1).

Leaf: Leaves are simple, petiolate, elliptic ovate to broadly ovate, entire, exstipulate, acute, cuneate or oblique, glabrous and upto 10cm long. Leaves on vegetative shoots alternate and large, those on floral branches opposite, arranged somewhat laterally in pairs of one large and one small leaf, having on their axil a cymose cluster of 5-25 inconspicuous pale green flowers.

Flower: The flowers are bisexual, greenish or lucid yellow, shortly pedicellate, 4-6mm in diameter. Pale green to yellow flowers, small 1cm long, and few flowers borne together in short axillary clusters. Calyx gamosepalous, enlarged with inflated fruit; sepals five partite, persistent, lobes linear, acute. Corolla is five lobed gamopetalous; lobes spreading or recurved, acute, pubescent and greenish yellow. Stamens are five, attached near the base of the corolla, epipetalous. Anthers are oblong, dehiscing longitudinally. Filaments are linear, slender. Gynoecium is syncarpous, composed a minute swollen ovary subtended by a long slender style. Stigmas are shortly bifid (Fig. 2).

Flowering, pollination and fruiting: It flowers throughout the year. Optimum flowering and fruiting is observed during August – November and February - April. The flowers open in the morning during summer and in the afternoon during winter. The flowers are highly protogynous as stigma becomes receptive much before anther dehiscence. In summer (May and June), 65-75 per cent pollen grains are fertile, while in winter (October, November, February and March), the fertility increases (85-96%). An average of 3600 pollen grains is present for each ovule. Flexible floral morphology, particularly heterostyly, first prompts the flowers for cross pollination, but if failed the length of stamens increases to make them suitable for self pollination. It has highly flexible mating systems and is facultative xenogamous. Out crossing is brought about by insects particularly large and small ants and bees. Autogamy and geitonogamy bring about high fruit and seedset percentage. However, seed germination percentage is very low (Singh, 2009).

Fruit : The fruit is a berry, 7mm across, red, globose, smooth, enclosed in an inflated, membraneous, somewhat five angled, pubescent, persistent calyx. The fruits turn orange red in colour when they mature, containing numerous small, smooth, discoid seeds (Fig. 3).

Seed : Seed is 2.5mm diameter, yellow, small, smooth, reniform in shape somewhat scurfy (Fig. 4).

Propagation: It is propagated by division, cuttings or seed. Seed is the best way to propagate them.

Seed propagation: The seeds are directly sown in the main field by broadcasting and covering with a thin layer of soil. Alternately, they may be sown in sunken beds at a distance of 5cm in between. The seeds are sown during October-January or from July to September. In six to seven days, 80 per cent of the seeds germinate. The seedlings are transplanted when they are six weeks old.

Seed development and maturation: The seed attains physiological maturity at 35 days after anthesis when the colour of seeds changes from orange to deep red. Harvestable maturity of seed is reached at 42 days after anthesis when germination and vigour are at peak.

Seed collection: The fruits are collected during November - December and the fruit is crushed to obtain the seeds.

Fig. 1. Plant

Fig. 2. Flower

Fig. 3. Fruit

Fig. 4. Seeds

Germination improvement: The seeds are washed to remove mucilage and dried under shade. The seeds are soaked for 24h before sowing. Seed sown on moist sand will germinate in 14-21 days at 20°C. Mechanical scarification of seeds with sand for six minutes followed by soaking in GA_3 500ppm solution for five hours significantly improved the germination of seeds. Prior to sowing, the seeds are treated with dithane M-45 @ 3g/kg of seeds.

Seed testing: Bharath (2008) carried out the studies on standardization of seed testing procedures and storage studies and his findings are given below.

- The better performance was observed in 25°C in between paper method in the presence of light for 24h. The germination percentage (84.90) and the speed of germination (0.320), root length (6.30cm), shoot length (12.10cm), seedling vigour index (1563) and seedling dry weight (17.80mg) were higher.
- The number of days fixed for first count and second count (Fifth and Nineth day respectively). But, lower values of these were found in sand medium at 15°C in dark without KNO_3.

- The germination and other seed quality parameters were decreased in the ten months of storage. The germination decreased to the extent of 22.00 per cent and KNO_3 treated seeds maintained best throughout the period of storage.
- The GA_3 and cow's urine treated seeds recorded reduction in their seed quality from fourth month onwards. The control seeds showed better results than cow's urine treated seeds from seventh month onwards for all the seed quality parameters.

Micro-propagation: Addition of BAP (0.3mg/l) and K (0.2mg/l) has been found to give a good response of shoot proliferation with a regeneration of 85 per cent (Kulkarni *et al.*, 2000).

Economic part used : Roots, barks, leaves, fruits and seeds

Alkaloids: Among the various alkaloids, withanine is the main constituent. The other alkaloids are somniferine, somnine, somniferinine, withananine, pseudo-withanine, tropine, pseudo-tropine, choline, cuscohygrine, isop-elletierine, anaferine and anahydrine. Two acyl steryl glucoside *viz.*, sitoindoside VII and sitoindoside VIII have been isolated from root. The leaves contain steroidal lactones, which are commonly called withanolides.

Medicinal uses: It increases the health and longevity. It is used as a sedative, anthelmintic and sexual vitality. Fruits, leaves and seeds have been traditionally used for the Ayurvedic system as aphrodisiacs, diuretics and for treating memory loss.

Past work done on Seed Science and Technology

Mehta and Sen (1990) reported that in Ashwagandha exposing the seed to eight hours light and 16h darkness recorded 63.3 per cent germination, whereas 10h light and 14h dark resulted in minimum germination of 15 per cent. They also added that germination was maximum in blue light and minimum in red light while there was no germination in complete darkness.

The seeds recorded highest germination after soaking for 24h in 400ppm GA_3 (Limbale *et al.,* 2001). Soaking of seeds in one per cent sodium nitrate solution for 24h significantly reduced the number of days taken for germination (9 days), enhanced the germination percentage (92.13), vigour index (670.73), shoot length (30.73cm), root length (15.99cm) and dry root yield (131.03kg.ha^{-1}) compared to other pre-sowing seed treatments (Kattimani *et al.*, 2001) (Table 1).

Verma *et al.* (2001) found that the seeds treated with 100ppm GA_3 resulted in improved percentage of germination, speed of germination, emergence and coefficient of velocity of germination, and a reduced mean germination time, compared with the control seeds.

Table 1. Influence of seed vigour tests with different pre-sowing seed treatments in ashwagandha

Treatments	Days taken for germination (No)	Germination (%)	Vigour index	At harvest		Dry root yield (kg ha^{-1})
				Shoot length (cm)	Root length (cm)	
Water	12.0	27.93	104.98	20.63	9.10	72.01
Potassium nitrate 0.5%	10.00	57.20	229.56	23.66	11.45	107.69
Potassium nitrate 1.0%	10.00	67.87	397.24	24.86	13.73	114.35
Ammonium nitrate 0.5%	11.00	50.37	265.15	25.63	13.06	72.01
Ammonium nitrate 1.0%	11.00	78.77	400.12	29.10	11.89	71.04
Cobalt nitrate 0.5%	11.00	61.40	314.16	25.87	12.66	90.02
Cobalt nitrate 1.0%	11.00	56.07	301.12	23.86	12.38	92.02
Sodium nitrate 0.5%	9.33	67.97	366.57	24.38	11.32	103.35
Sodium nitrate 1.0%	9.00	92.13	670.73	30.0	15.99	131.03
Calcium nitrate 0.5%	10.67	56.40	380.80	28.71	13.02	110.02
Calcium nitrate 1.0%	11.00	67.70	369.64	24.62	11.96	97.69
Zinc nitrate 0.5%	11.00	83.13	509.06	24.42	13.95	93.35
Zinc nitrate 1.0%	11.00	55.47	342.04	24.82	11.93	90.02
Seed sown as such (Control)	13.67	25.73	102.24	18.90	12.14	71.01
S.Em±	0.146	3.264	5.105	0.139	0.351	1.87
CD at 5%	0.425	9.491	14.844	0.405	0.102	5.20

The combination of sand scarification of seed for 6min. and soaking in GA_3 500ppm for 5h each recorded the highest germination per cent (83.5) than the other treatments and control (Table 2) (Vakeswaran, 2001).

Table 2. Effect of seed treatments on germination and vigour parameters on ashwagandha seeds

Treatment		Treatment duration	Germination (%)	Root length (cm)	Shoot length (cm)	Dry matter production (mg 10 $seedlings^{-1}$)	Vigour index
Sand scarification		2min.	70 (57.1)	2.87	2.91	19.5	408
		4min.	72 (58.0)	2.88	2.91	20.3	418
		6min.	77 (61.3)	2.88	2.93	21.0	445
H_2SO_4	1%	2min.	68 (56.0)	2.86	2.90	19.7	396
	3%	2min.	73 (58.6)	2.86	2.93	19.7	423

	5%	2min.	73 (58.6)	2.85	2.91	19.5	401
H_2SO_4	1%	4min.	69 (56.5)	2.87	2.92	19.7	400
	3%	4min.	69 (56.5)	2.88	2.94	20.5	443
	5%	4min.	76 (60.6)	2.87	2.91	19.7	382
H_2SO_4	1%	6min.	66 (54.3)	2.85	2.90	19.3	383
	3%	6min.	66 (54.3)	2.86	2.92	19.3	399
	5%	6min.	69 (56.5)	2.81	2.87	18.5	354
KNO_3	5%	5h	62 (52.1)	2.85	2.92	19.5	390
	10%	5h	67 (55.2)	2.86	2.93	19.5	392
GA_3	100 ppm	5h	68 (55.5)	2.87	2.93	19.7	406
	200 ppm	5h	70 (56.7)	2.89	2.93	20.8	435
	300 ppm	5h	75 (59.7)	2.89	2.95	20.8	440
	500 ppm	5h	75 (59.7)	2.90	2.96	21.5	466
Thiourea	5%	5h	64 (53.1)	2.86	2.90	19.5	370
	10%	5h	66 (54.0)	2.86	2.91	19.7	377
Sand scarification + soaking in GA_3	500 ppm	6min. + 5h	84 (66.0)	2.91	2.98	20.8	490
Water		12h	72 (58.0)	2.87	2.93	20.8	417
		24h	76 (60.6)	2.88	2.93	21.0	441
Control			62 (51.9)	2.84	2.89	19.0	355
SEd			0.074	NS	NS	0.48	3.17
CD (P=0.05)			1.488			0.84	6.32

(Figures in parenthesis indicate arcsine values)

REFERENCES

Bharath, V. B. L. K. 2008. Standardization of seed testing procedures and storage studies in selected medicinal crops. M.Sc. (Agri.) Thesis University of Agricultural Sciences, Dharwad.

Kattimani, K. N., Y. N. Reddy and B. R. Rajeswar Rao. 2001. Influence of pre-sowing seed treatments on seedling vigour, root length and dry root yield of ashwagandha *Withania somnifera* under semi-arid tropical climate of Hyderabad. CIMAP, Field station, Boduppal, Uppal, Hyderabad.

Kulkarni, A. A., S. R. Thengane and K. V. Krishnamurthy. 2000. Direct shoot regeneration from node, internode, hypocotyls and embryo explants of *Withania somnifera. Plant cell, tissue organ culture* 62(3): 203-209.

Limbale, S. D., S. T. Ghule, V. H. Bhuse and R. D. Wash. 2001. Stimulation of germination and dormancy studies in some important medicinal plants. National Seminar on Production, Processing and Marketing of Medicinal, Aromatic and Dye yielding Crops, p.188.

Mehta, M. and D. M. Sen 1990. Effect of different photoperiod and light of different wave length on seed germination of *Withania somnifera* (Linn.) Dunal. *Sci. Cul.*, 56 (7): 300-301.

Verma, S., R. K. Sharma and D. K. Shrivastava. 2001. Seed germination, viability and invigoration studies in medicinal plants of commercial value. **In:** Proc. National Seminar on the Frontiers of Research and Development in Medicinal Plants, Lucknow, India, 16-18, September, 2001.

Singh, V. 2009. Phenology and reproductive biology of *Withania somnifera* (L.) Dunal (Solanaceae) *J. Plant Reproductive Biology* 1(1):81-86.

Vakeswaran, V. 2001. Investigation on certain aspects of true seeds and synthetic seed development in Ashwagandha (*Withania somnifera* Dunal). Ph.D. Thesis, Tamil Nadu Agricultural University, Coimbatore.

PART - IV

MEDICINAL CLIMBERS

4.1. *Abrus precatorius*

Common names: Indian liquorice, Rosary pea, Precatory bean, Jequerity bean, Crab's eye, Prayer beans, Buddhist's rosary bead,

Botanical name: *Abrus precatorius* L. Syn. *Glycine abrus* L., *Abrus abrus* (L.) W. Wight

Family: Fabaceae

Vernacular names: Liluwani, Raturmani (Assamese); Chunhali, Kunch (Bengali); Chanothi, Gunja (Gujarati); Ganchi, Gunchi, Rati (Hindi); Ganji, Gul-Ganju, Guluganji, Madhuka (Kannada); Kath (Kashmiri); Kunni, Kunnikuru (Malayalam); Chanoti, Gunchi, Gunja (Marathi); Gunja, Runji (Oriya); Labrigunchi, Ratak (Punjabi); Gunja (Sanskrit); Gundumani, Kuntumani (Tamil); Guruginia, Guruvenda (Telugu); Ghongcha, Ratti (Urdu).

Origin and distribution: Indigenous to Indonesia, but also growing in Surinam. Common in forest tracts of India and Myanmar, ascending in the outer Himalayas. A seashore species in the Indian archipelago, it is cosmopolitan in the tropics and is found throughout India on hedges and bushes in exposed areas.

Habitat: Full sun or partial shade, well drained moist soil and frost free spots.

Habit: A beautiful, much branched, slender, perennial, deciduous, woody, prickly twining or climbing herb (Fig. 1).

Stem: Stem cylindrical, wrinkled; bark smooth textured brown.

Leaf: Leaves stipulate, pinnately compound; leaflets 7-24 pairs, 0.6-2.5 x 0.4-1.2cm, turgid, oblong, obtuse, truncate at both ends, appressed hairy; corolla pink; stamens monodelphous; filaments sub-equal.

Flower: Flowers in axillary racemes, shorter than leaves, fascicled on the swollen nodes, pink or pinkish white; calyx lobes short, appressed hairy (Fig. 2).

Flowering, pollination and fruiting: Flowering and fruiting occurs during August-January. The essential organs cannot regain their original position, because the petals fall off as soon as the flower has received a single visit. So in this case one single visit of the insect is enough for cross pollination. Instances of the same kind of explosive arrangement are afforded by the flowers (Mishra, 2009).

Fruit: Pods 1.5-5.0 x 0.8-1.5cm, turgid, oblong, appressed hairy, with a sharp deflexed beak, silky textured, three to five seeded (Fig. 3a and 3b).

Seed: Seeds are elliptic to sub-globose, 0.5cm in diameter, smooth, glossy, shining red with black blotch around the hilum (Fig. 4).

Fig. 1. Plant

Fig. 2. Flower

Fig. 3a. Immature pod

Fig. 3b. Pod with seed

Fig. 4. Seeds

Propagation: The species is easily propagated through seeds.

Seed propagation: Soak the seeds in hot water for 5min. and then in cold water for 24h are found to germinate better. Dibble the seeds in polybags. In 7- 10 days, the seeds begin sprouting, the germination per cent being 70.

Economic part used: Leaves, roots and seeds are used.

Alkaloids: The seeds are very toxic due to the phyto-chemical abrin (consisting of 5 glyco-proteins); a single seed, if broken can cause blindness or even death, if ingested. Abrin is a ribosome inactivating protein (it blocks protein synthesis) and is one of the most deadly plant toxins known.

Medicinal uses

The seeds are poisonous

Dried leaf: Anti-spasmodic.

Dried seed: Anthelmintic, anti-diarrhoeal, inhibits intestinal motility.

Fresh root: Anti-convulsant.

Seed: Anti-bacterial, anti-spasmodic, insecticide.

REFERENCE

Mishra, S. R. 2009. Understanding plant reproduction. Discovery Publishing House Private Limited, New Delhi. p. 355.

4.2. *Argyreia nervosa*

Common names: Baby wood rose, Elephant creeper, Elephant climber, Elephant vine, Woolly morning glory, Silver morning glory, Silky elephant glory.

Botanical name: *Argyreia nervosa* (Burm. F.) Bojer., Syn. *A. speciosa* (L. F.) Sweet, *Convolvulus nervosus* Burm. F., *C. speciosus* L. F., *Lettsomia nervosa* (Burm. F.) Roxb., *Rivea nervosa* (Burm. F.) Hallier F.

Family: Convolvulaceae

Vernacular names: Bijarka (Bengali); Samudarsoka (Gujarati); Samudra Shokh (Hindi, Marathi); Samudra paala, Chandrapada (Kannada, Telugu); Samudra somk (Konkani); Samudra paccha (Malayalam); Bryddhotareko (Oriya); Vridha darukha, Murva (Sanskrit); Kadarpaalai (Tamil).

Origin and distribution: It is native to India and introduced to numerous areas World wide, including Hawaii, Africa, Deccan and the Caribbean. It is found to grow throughout India especially along coastal areas of Western Ghats, habitations near moist forests and among the way side thickets in coastal regions.

Habitat: This species enjoys warmer temperature and is tolerant of arid conditions, usually planted in open sun or partial shade. It is a vigorous, fast growing heavy climber that needs strong support and should not be planted mixing with other slender vines. This plant thrives in well drained soil and also grows well in sandy and gravelly red soil. It requires less water hence, watering can be done once in two or three days.

Habit: It is a huge climbing vine around 10m height. In fact, it is a lovely climber with shiny stems and crowded leaves and large, pretty pinkish purple flowers. It is soft to touch because of the presence of velvetty hair. These vines have large, heart shaped leaves and clusters of 5-7.5cm flowers, which turn into attractive pods that really have the appearance of a wood rose and thus, is also known as 'baby wood rose'. This is woody, lactiferous climber that can reach upto 15m in height (Fig. 1).

Stem: Stems are softly hairy.

Root: Root brown, cylindrical, 1-1.5cm thick; smooth, round wood is scant, flexible and smooth, latex oozes at cuts.

Leaf: Leaves are green, simple, arranged in alternate manner, broadly ovate or heart shaped, upto 25cm long and 11cm broad with prominent venation, smooth from above, densely white hairy beneath with long stalks.

Flower: Inflorescences are axillary cymes. Flowers pink, 1.2-2cm long. Peduncle is 0.5-10cm, tomentose. Bracts are large and foliaceous, hairy outside, smooth inside, resembling outer sepals. Two outer sepals concealing the rest, white silky or in fruit hirsute, 8.7-10mm inner oblong, 7.5mm white silky. Corolla is pinkish purple in colour, funnel shaped, silky pubescent on the outside; corolla tube slightly inflated, urn shaped, upto 5cm long. Stamens far exerted. Ovary two celled (Fig. 2).

Flowering and pollination: The plant can start growing flowers as early as 18 months from seed. For this to occur, there must be sufficient watering and adequate room for roots to grow; it can take upto five years for the first signs of flowering to become visible (Ashutosh *et al.,* 2011). Usually, this plant flowers during July-December and March-April. The Hawk moth pollinates flowers at night and usually looks for pale colored flowers, and requires a lot of energy because of the speeds they are capable of flying at, some close $30mh^{-1}$. Having a pale pinkish color and sufficient sugar content in its nectar makes it a perfect candidate for pollination by this species of moth. The bee is drawn to the plant by the scent of nectar. In contrast to the hawk moth, the bee is an active pollinator during the daylight hours. During daylight hours, wasps may move from one flowering plant to another as the bright colors and olfactory cues attract them. It does not contribute to pollination in this plant as often as bees, but it still may unwittingly participate from time to time.

Fruit: Berry, red, fleshy, globose, surrounded by the enlarged sepals which are red within. The fruits are imitating woody capsules, in fact these are indehiscent berries; size of a single fruit measures upto 2.5cm in diameter (Fig. 3).

Seed: Seeds are black, sub-globose, 4-4.5mm diameter, grey, minutely tomoentose round the large areole, hairy, embedded in burly pulp. There are 3-5 seeds, commonly four per flower (Fig. 4).

Seed dispersal: The berries are dispersed by frugiverous birds.

Propagation: It can be propagated by scarified seeds and cuttings.

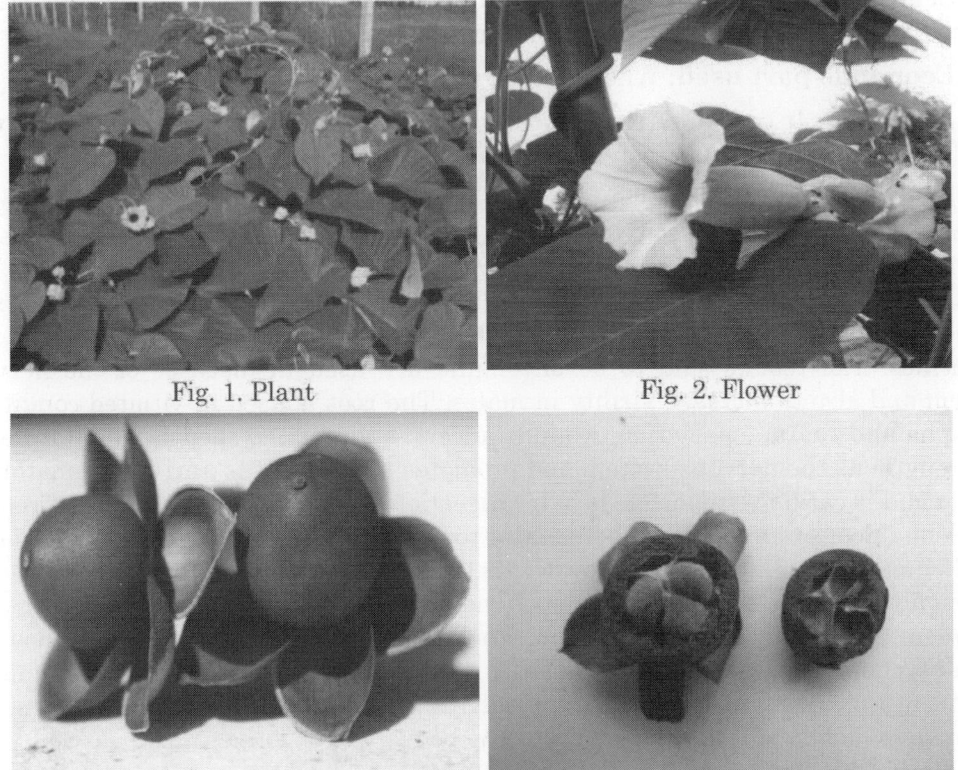

Fig. 1. Plant

Fig. 2. Flower

Fig. 3. Fruit

Fig. 4. Seeds

Seed propagation: Two fresh seeds are sown directly (mother beds are not required in this case as it has recorded over 80 per cent of germination) in each polybag or container. Sowing should not be more than 2.5cm deep. Germination starts within twenty days. Soaking seeds in cold water for 24h can hasten germination.

Seed collection: The seeds are found in the pods of dried flowers. These cannot be harvested until the pods are completely dried.

Economic part used: Whole plants, leaves, fruits, seeds and roots.

Alkaloids: It contains triterpenoids, flavonoids, steroids, argyreioside, 1- triacontanol, epifriedelinol, ß-sitosterol and lipids.

Medicinal uses: The plant is widely used for psychotropic and ornamental purposes. The roots and seeds have been used as a support of the nervous system and mild aphrodisiac. The whole plant is reported to have purifying properties. It is also taken to maintain healthy joints. All parts are reported to possess antiseptic properties. Its roots and leaves are extensively used in the preparation of herbal medicines. The root is said to be bitter and are used in diseases of the nervous system and also to increase virility in males. The root is used in vitiated conditions of kapha and vaata, emaciation, wounds, ulcers, anorexia, dyspepsia and flatulence. It strengthens the nervous system and promotes memory. It is anti-inflammatory in action and is cardiac supporter. It helps digestion. It is also helpful in throat related problems. Root extraction is effective as a tonic and the root paste is applied in case of rheumatic swellings. Root powder is used in elephantiasis. The leaves are commonly used in skin diseases such as boils, tumours, sores and other skin disorders. The leaves are also used to expel guinea worms. The seeds are known to have psychoactive properties. It increases sperm count and decreases inflammation in the uterus. It is also helpful in resolving urine related problems and is also effective in diabetes. It also strengthens the body.

Other uses: This ornamental creeper can bring height, dimension, colour and shade to the gardens. It is often planted for its foliage and bright flowers. Sprawling on compound walls brings natural look to the house.

REFERENCE

Ashutosh, M., A. A. Kumar and P. A. Ranjan 2011. A literature review on *Argyreia nervosa* (Burm. F.) Bojer. *International J. Res. Ayurveda Pharmacy*, 2(5):1501-1504.

4.3. *Aristolochia indica*

Common names: Serpent root plant, Indian birthwort

Botanical name: *Aristolochia indica* L.

Family: Aristolochiaceae

Vernacular names: Isharmul (Bengali and Hindi); Arkmula (Gujarati); Isvaberusa (Kannada); Garudakkoti (Malayalam); Sapsund (Marathi); Ishvari (Sanskrit); Utamanielaye, Isvaramuli, Adagam, Iswara kodi, Peru marunthu, Perunkizhangu, Thalaichuruli (Tamil); Esvaraveru (Telugu).

Origin and distribution: It is native to India and it is more common in southern and eastern India.

Habitat: It is commonly found in lower hilly regions and in plains.

Habit: It is a twining shrub, semiwoody, having more or less swollen nodes (Fig. 1).

Root: Roots are stout, brown outside, creamy white inside, strongly aromatic. Branches are stout, vertically grooved, hairless.

Stem: Stems are woody below, slender and soft above. Main rootstock is woody.

Leaf: The leaves are about 9cm x 5cm in size, ovate-elliptic or panduriform, glabrous. It is stout and glabrous.

Flower: Small, irregular, often offensively smelling, protogyny in nature, greenish-white, in small bunches arising from the axils of the leaves. Perianth is globose with a purple dilated and trumpet shaped mouth with a strap shaped brown purple appendage or lip behind. Stamens six, fused to form a structure about 4-5mm diameter. Anthers sessile. Stigma six. Ovary 6-locular. Ovules numerous in each locule (Fig. 2).

Flowering, pollination and fruiting: Flowering and fruiting occurs during December- February (Neelima *et al.*, 2011). It is pollinated by insects (Gonzalez and Stevenson, 2000).

Fruit: Fruit is a sub-globose capsule, about 5cm long, longer than broad, opening from below upwards into six valves. Capsules consist of a thickened seed bearing section about 3cm long and a much thinner section also about 3cm long which splits into a number of thread like structures attached to the valves of the capsule (Fig. 3a and 3b).

Seed : Numerous, heart shaped, flat, about 8mm across, surrounded by distinct white papery wing, brown, hairless (Fig. 4).

Propagation : By seeds and cuttings.

Economic part used : Dried leaves, stems and roots.

Alkaloids: Aristolochic acid and aristolochine (root).

Medicinal uses : In small doses, it promotes digestion and regulates menstruation; it is a tonic, stimulant and curing fevers.

Fig. 1. Plant

Fig. 2. Flower

Fig. 3a. Mature fruit

Fig. 3b. Dry fruit

Fig. 4. Seeds

REFERENCES

Gonzalez, F. and D. W. Stevenson 2000. Perianth development and systematics of *Aristolochia*. *Flora*, 195: 370–391.

Neelima, M., G. P. Prasad, G. Sudarsanam, G. P. Pratap and B. Jyothi 2011. Ethnobotanical studies in Rapur forest division of Nellore district in Andhra Pradesh. *Life Sci. Leaflets*, 11: 333-345.

4.4. *Cephaelis ipecacuanha*

Common name: Ipecac, American Ipecac, Ipecac Spurge, White Ipecacuanha

Botanical name: *Cephaelis ipecacuanha* (Brot.) A. Rich.

Family : Rubiaceae

Vernacular names: Ipecacuanha (Hindi).

Origin and distribution: It is native to South America, mainly Brazil. In India, it is cultivated in Assam and Bengal.

Habitat: It prefers moist, shady woods. It prefers well drained, humus rich soil in shade with ample moisture and humidity.

Habit: It is a small trailing evergreen shrub with a slender stem growing to 30cm which grows partly underground and is often procumbent at the base, the lower portion being knotted (Fig. 1).

Root: Roots are slender, spreading horizontally and descending obliquely into the ground, from 10-15cm long, simple, or divided into a few diverging branches, about as thick as a goose-quill, ringed, when fresh pale-brown, when dry umber colored, blackish umber colored, or grayish brown; the cortical integument with a reddish, resinous, glittering fracture, and readily separating from a central woody axis. Mature roots moniliform, in bunches, fibrous rootlets are given off from the knots, and some of them develop an abnormally thick bark, in which much starch is deposited.

Stem: The stem is simple, suffruticose, from 60-90cm long, ascending, often rooting near the ground, smooth and cinereous at the base, and downy and green near the apex.

Leaves: The leaves are rarely more than four or six on a stem, oblong-ovate, acute, roughish with hairs, from 7.5-10cm, 2.5-5.0cm broad; those at the top of the stem are opposite, those toward the base alternate. Petioles short and downy. The leaves are opposite, petioled, entire, smooth, dark green, and usually crowded near the top of the plant. At the base of each pair of leaf stalks there is a pair of whitish, laciniate, cut stipules. Stipules erect, appressed, membraneous, deciduous, and four to six cleft.

Flower: The flowers are small, white, in semiglobose heads, of 8, 12, or more; the involucre is one leaved, spreading, deeply four to six- parted, with obovate, acuminate, ciliated segments. Bracts to each flower one, obovate - oblong, acute, and downy. Calyx is minute, obovate, whitish, adhering to the ovary, with five bluntish, short teeth. The corolla is white, funnel shaped, the tube cylindrical, downy on the outside and at the orifice, the limb shorter than the tube, with five ovate, reflexed segments. Stamens six; filaments filiform, white, and smooth; anthers linear, longer than the filaments, projecting a little beyond the corolla. The stamens and pistils are dimorphous, *i.e.,* some flowers bear long stamens and short pistils, and conversely, other flowers short stamens and long pistils with heterostyly. Ovary with a fleshy disc at the apex; style filiform; stigma two, linear (Fig. 2).

Flowering, pollination and fruiting: Flowering occurs during January and February. The berries ripen during September - October. The peculiar structure of the flower requires the intervention of insects for the purpose of fertilization; when cultivated in hot houses it is therefore necessary to transfer mechanically some pollen to the stigma, if the plant is expected to bear fruit (Meyer, 1883).

Fruit : The berry is ovate, obtuse, about the size of a kidney bean, at first purple, afterward violet black, two celled, two seeded, with a longitudinal, fleshy partition, nucules plano-convex and furrowed on the flat side (Fig. 3).

Fig. 1. Plant

Fig. 2. Flower

Fig. 3. Fruit

Seed : Seed is white, solitary in each pyrene and similar to it in form, testa membranous, embryo small, straight, at the base of the abundant horny endosperm. About 1/3 of the seeds are sterile.

Seed dispersal: Transportation of seeds by birds (Sick, 1993)

Propagation : Propagated by greenwood cuttings in late spring in sandy soil or by root cuttings during harvesting.

Seed treatment: Veeraragavathatham *et al.* (1988) found the seeds treated with 200ppm of KH_2PO_4 solution for 24h exhibited maximum germination when compared to control.

Economic part used : Root bark, Rhizome

Alkaloids: Emetine, cephaelin and psychotrine.

Medicinal uses : It cures amoebic dysentery and loss of appetite, diaphoretic, emetic.

REFERENCES

Meyer, A. 1883. *Psychotria Ipecacuanha. Archiv. d. Pharm.,* 221: 721-745.

Sick, H. 1993. Birds in Brazil. Princeton University Press, New Jersey.

Veeraragavathatham, D., R. Balakrishnan and M. D. Abdul Khader 1988. A note on seed germination in Ipecac (*Cephalis ipeceucanha*). *South Indian Hort.,* 36 (4) : 212- 213.

4.5. *Chonemorpha fragrans*

Common name: Frangipani vine

Botanical name: *Chonemorpha fragrans* Moon Alston. Syn. *Chonemorpha grandiflora,* (Roth) M.R. and S.M.

Family: Apocynaceae

Vernacular names: Garbhedaro (Hindi); Chandra hoovina balli (Kannada); Paal valli (Malayalam); Murva, Morala (Sanskrit); Velutha kaaka kodi (Tamil); Chaga (Telugu).

Origin and distribution: In India, this has been reported from the Western Ghats, North East India, Uttar Pradesh and Andaman and Nicobar Islands.

Habitat: Prefers to grow in moist deciduous to semi-evergreen forests.

Habit: Perennial, very heavy climber, hence can only be used in stronger supports such as larger trellises, columns or can be allowed to ramble over tall, strong trees on the high compound walls. It is a disease free, quick growing, deciduous climber, prefers partial shade to full sun. Loamy soil is the best soil and does not thrive in poor soils. All parts of the plant exude milk when injured (Fig. 1).

Stem: Bark is thin, grey, numerously lenticelled.

Leaf: Simple, arranged in opposite manner, upto 20cm across, broadly ovate to orbicular, base cordate or rounded and the lamina densely brownish hairy and velvetty with prominent veins.

Flower: Large cymose clusters of exuberant bisexual blooms with delicious, large upto 10cm across. Large, pure white, twisted back corolla with yellow throat, very fragrant (Fig. 2).

Flowering and fruiting: Flowers almost throughout the year, but profusely during May – July. Fruiting period is July-September.

Fruit: Pired follicles, 20-40cm long, linear, woody, with smooth surface (Fig. 3).

Seed: Seeds many, flat, shortly beaked with long white silky coma.

Propagation: By seeds and stem cuttings.

Fig. 1. Plant

Fig. 2. Flower

Fig. 3. Fruit

Seed propagation: Seeds are collected directly from the fruits when the fruits develop dehiscing spilt (without waiting for the complete dispersal of the seeds). Collection of fresh seeds is carried out during July-September. These seeds are required to be sown within two months from the collected date for the best result of germination. Sowing is done in sandy, raised mother beds. It takes 15-20 days for germination. The obtained per cent of germination is 70. Seedlings of 5cm height can be transplanted to the containers or polythene bags.

Vegetative propagation: Soft wood cuttings of pencil thickness measuring 10-15cm with minimum two nodes are selected. Stem cuttings are planted in raised, pure sand beds in mist chambers during January - March. Sprouting starts within 20 days but rooting is initiated only after 30 days. Three month old cuttings can be transplanted in the containers or directly to the main field.

Economic part used: Roots

Medicinal uses: Roots are reportedly used as a laxative, astringent, carminative and expectorant. Roots are primarily used to treat intermittent fevers, skin diseases, constipation, cardiac debility, bronchitis, jaundice, cough and diabetes. They are also useful in vitiated conditions of vata and kapha, syphilis, stomach ache, gas in stomach and bowels, colic, helminthiasis, hyperdipsia, urethrorrhoea and hyperacidity. It is also used in the treatment of diarrhoea, polyuria, boils, leprosy, eye diseases, vomiting and poisoning (Yoganarasimhan, 2000).

Other uses: Eye catching and decorates the garden fences and side walls with large shiny, velvet hairy leaves with prominent raised veins.

REFERENCE

Yoganarasimhan, S. N. 2000. Medicinal Plants of India, Volume 2, Tamil Nadu. Bangalore: Cybermedia, p. 346.

4.6. *Cissus quadrangularis*

Common name: Veldt Grape, Devil's Backbone, Edible stemmed wine, Bonesetter, Adamant creeper

Botanical name: *Cissus quadrangularis* L., Syn. *Vitis quadrangularis*

Family: Vitaceae

Vernacular names: Harjora (Assamese); Har, Harbhanga, Hasjora, Horjora (Bengali); Hadsankal, Hadsand, Chodhari (Gujarati); Hadjod, Hadjora, Hadsankari, Hadsarihari, Kandvel, Hadjor (Hindi); Mangaravalli, Vajravalli (Kannada); Peranta, Changala peranda (Malayalam); Kanvel (Marathi); Hadabhanga (Oriya); Asthisamhara, Asthisirikala (Sanskrit); Pirandai (Tamil); Naleru, Nulleratiga, Vajravalli (Telugu); Sandubooru (Tulu); Hrajora, Hadsankal (Urdu).

Origin and distribution: It is probably native to India or Sri Lanka, but is also found in Africa, Arabia and South East Asia. It is found throughout the hotter parts of India alongside hedges, neighboring countries like Pakistan, Bangladesh, Sri Lanka and Malaysia.

Habitat: It is commonly found in tropical and subtropical xeric wood. It is widely adaptable and can be grown throughout the year. It is usually grown on fences and in between tree plantations. It can be cultivated in plains coastal areas, jungles and wastelands upto 500m elevation. When grown in fields for cultivation, it is supported with bamboo sticks. It is also grown in hanging baskets and pots. It grows profusely, it may take seven months to one year, once it attains good growth it starts flowering.

Habit: It is a fleshy succulent shrubby climber (Fig. 1).

Stem: Reaches a height of 1.5m and has quadrangular sectioned branches with internodes 8 to 10cm long and 1.2 to 1.5cm wide. Along each angle is a leathery edge. Each has a tendril emerging from the opposite side of the node. The surface is smooth, glabrous, buff colored with greenish tinge, angular portion reddish-brown; no taste and odour.

Fig. 1. Plant

Fig. 2. Flower

Fig. 3a. Mature fruit

Fig. 3a. Ripe fruit

Leaf: Leaves are simple 2.5-5cm long, broadly ovate or reniform, sometimes 3-7 lobed, denticulate, glabrous, cordate, rounded, truncate or cuneate at the base; petioles 6-12mm long; stipules small broadly ovate, obtuse.

Flower: Flowers are in umbellate cymes, opposite to the leaves, bisexual, tetramerous, sepal like a cup and slightly lobed, petals four, white, with four stamens (Fig. 2).

Flowering and pollination: Plant flowers in the month of June - December (Guhabaksh *et al.*, 2001). It is pollinated by beetle.

Fruit: Globular berries are red when ripe with a single seed, dark purple to black, scarcely 6mm long and apiculate (Fig. 3a and 3b).

Seed: Seeds four with prominent cord like raphe extending to a chalazas knot.

Propagation: It is mainly propagated by stem cuttings. They are usually cut between the two nodes and then planted. The pot can contain soil mixed with little cow dung for planting. It may take one month also to get new leaves, as it grows it may just spread across, at that time it is essential to be tied to a rope.

Economic part used: Stem, root

Alkaloids: Phytochemical studies on methanol extract revealed the presence of triterpenes including α- and β-amyrins, ß-sitosterol, ketosteroids, phenols, tannins, carotene and vitamin C. Several other constituents such as flavonoids quercetin and kaempferol and stilbene derivatives, quadrangularins A, B, C and many others e.g. resveratrol, piceatanon, pallidol, perthenocissi and phyto-sterols have been isolated from plant.

Medicinal uses: The plant is used both, internally as well as externally. As it is a healing herb and styptic one, it is used to promote the healing of bone fractures. For this purpose, externally, the crushed stems are used as a poultice over bone fractures, along with the juice of its roasted stems, orally is given with ghee. The powdered stem is mixed with pulses and fried in sesame oil as a remedy for several vata diseases.

REFERENCE

Guhabaksh, D. N., D. C. Pal and P. Sersuma 2001. A Lexicon of Medicinal Plants in India. Vol. 1, Naya Prokash, Kolkatta. p. 443-445.

4.7. *Embelia ribes*

Common name: False Pepper, Bidang

Botanical name: *Embelia ribes* Burm. f.

Family: Myrsinaceae

Vernacular names: Vidang, Biakol-lata (Assamese); Biranga, Vidang (Bengali); Bavding (Gujarati); Bhabhiranga, Baberang, Vayavidang (Hindi); Vayu-vilanga, Vayuvidanga, Vaayu vilana, Hulimeese (Kannada); Babading (Kashmiri); Vizhalari, Vizalari, Vizhal (Malayalam); Vavading, Vavding (Marathi); Bidanga, Vidanga (Oriya); Babrung, Vavaring (Punjabi); Jantughna, Vidang (Sanskrit); Vivilangam, Vaivelangam (Tamil); Vaividungalu, Vidangamu (Telugu); Baobarang, Babrang (Urdu).

Origin and distribution: It is widely distributed throughout Pakistan and India. In India, it is commonly found in Meghalaya and Assam. It is recorded from Itanagar, Kimin, Hunli, Dibang Valley, Nechephu and Dirang areas (West Kameng district), Joram and Hija areas near Hapoli in Lower Subansiri district of Arunachal Pradesh. The species was also reported from upper Assam and from Cherapunji and

Jowai areas of Khasi/Jaintia hills and Mao areas of Manipur, Kawnpui - Aizawl, Serchhip areas of Mizoram in both high and low altitudes of north east India. This species is reported to be vulnerable in the Western Ghats of Tamil Nadu and Karnataka states of India and at a lower risk in Kerala state of peninsular India. It is found throughout India upto the altitude of 1500m. It grows under diffused sunlight conditions. In North East region, it is commonly distributed in Arunachal Pradesh, Meghalaya and Mizoram in the altitudinal zone of 500 to 2500m. It is a red listed climbing shrub found in the semi-evergreen to evergreen forests of India, Sri Lanka, Malaysia and China (Guhabakshi *et al.*, 2001).

Habitat: It can be cultivated in all types of soils. However, the well drained light cotton soils or sandy loam soils are best suited for its cultivation.

Habit: A large semi climber shrub with long drooping branches (Fig. 1).

Root: Root is deeply penetrating in the soil and takes help of trees for their climbing habit.

Stem: Stem is long slender brittle, bark studded with lenticels.

Leaf: Leaves are simple, coriaceous, asymmetric alternate, elliptic ovate lanceolate, smooth leaves gland dotted, short and obtusely acuminate, broad, entire upper surface is perfectly glabrous and shiny, lower surface is buff with nodulated glands. It is about 7.5cm long and 4cm broad (Asadulla *et al.*, 2012).

Flower: Flowers are greenish white seen in terminal or axillary lax panicled racemes and petiole length is 7.5-10cm (Fig. 2). It is polygamo dioecious with low male female ratio (1:5). Male flowers were short lived compared to female flowers. Though the female flowers have anthers, they do not produce pollen.

Flowering, pollination and fruiting: Flowering during March - April and fruiting during June - October. It is pollinated by bees and wasps.

Fruit: It is found in bunches. The fruit is globular and wrinkled, berries, 3-4mm in diameter, varying colour from dull red to nearly black, a short pedicel is often present, the pericarp is brittle enclosing a single seed covered with a membrane. Mature fruits turn brown or black (Fig. 3).

Seed: Globose and seeds remain dormant for a long period (Fig. 4).

Seed dispersal: Fruits are dispersed by birds.

Propagation: The plant is propagated by seeds and cuttings.

Seed maturity: Generally, all seeds do not mature at the same time and maturation starts from the apical towards bottom region. In case of seed germination trials, only fully mature seeds having brown black color initiate germination. This plant can give 80 to 90 per cent viability only with matured seeds (Shankar *et al.*, 2012).

Seed collection: The fresh seeds are collected during March - April.

Fig. 1. Plant

Fig. 2. Flower

Fig. 3. Fruit

Fig. 4. Seed

Seed germination: The seeds are sown directly in raised beds. Within 20-25days, about 50 per cent of the seeds show germination. Then, the nursery is maintained for 60 days. Then, the seedlings attain tall as a thumb and transplanted in to the field. The seedlings are irrigated thrice a day.

Vegetative propagation: The seeds often fail to germinate therefore, vegetative propagation through stem cuttings is advisable. Pencil sized stem cuttings bearing three nodes and the cut ends treated with IBA are planted 3cm deep in polybags filled with pure sand and kept in shade. It is watered regularly. The rooting commences only after 35-40 days and with in three weeks one or two axillary buds above the leaf scars develop and grow into branches. After the formation of healthy root system, it is ready for transplanting.

Economic part used: Fruits, leaves and root bark.

Alkaloids: The plant contains embelin, quercitol, and fatty ingredients; an alkaloid, christembine, a resinoid, tannins and minute quantities of volatile oil.

Medicinal uses: *Embelia ribes* traditionally used as appetizer, carminative, laxative, anthelmintic and liver diseases. It posses many pharmacological actions like analgesic, anti-fertility, antioxidant, antiandrogenic and many more.

REFERENCES

Asadulla, S., Ramandang and Rajasekharan 2012. Taxonomy and flower stages in *Embelia ribes* Burm F. *International J. Res. Pharmacy Chemistry*, 2(2):488-494.

Guhabakshi, D. N., P. Sensarma and D. C. Pal 2001. A lexicon Medicinal Plants of India. Calcutta, India, Naya Prokash, Calcutta pp.135-136.

Shankar, R., G. S. Lavekar, S. Deb, B. K. Sharma and M. S. Rawat 2012. Distribution, conservation and folk uses of Vaibidang (*Embelia ribes* Burm. f.) *International J. Biodiversity Conservation,* 4 (13):525-529.

4.8. *Gloriosa superba*

Common name: Glory lily, Superb lily, Tiger lily. Tiger claws, Meadow saffron

Botanical name: *Gloriosa superba* L.

Family: Liliaceae

Vernacular names: Karihari Languli (Hindi); Agnishike, Akkathangiballi, Karadikannina godde, Nangulika (Kannada); Dudhio Vaccbonag (Gujarati); Indai, Karianag, Khadyanag (Marathi); Kanvali kizhangu, Kalappai kizhangu, Thanthondri kilangu, Venthati kilangu, Karthigai kizhangu (Tamil); Adavinabhi, Kalappa gadda (Telugu).

Origin and distribution: It is native to tropical Asia and Africa. It is found growing throughout tropical India, from north west Himalayas to Assam and the Deccan Peninsula, extending upto an elevation of 2120m. In Karnataka, it is generally found growing along the western ghats. It is also found growing in Madagascar, Sri Lanka, Indochina and on the adjacent islands. It is state flower of Tamil Nadu and national flower of Zimbabwe. In Tamil Nadu, it is grown in 5000ha in Erode, Dindigul, Perambalur, Ariyalur and Salem districts.

Habitat: Hot moist climate is best suited for its cultivation. It can be grown in any type of soil including rocky gravelly soil having good drainage. Sandy loam soil having a pH of 6-7 is considered to be the most suitable.

Habit: It is an herbaceous, climbing perennial, growing between 3.5 to 6m in length, but usually trained at 1.2m above ground level (Fig. 1).

Root: Tubers are cylindrical, large, "V" shaped with the two arms equal or unequal in length pointed towards end; brownish externally with a yellowish internal surface.

Stem: The vines are tall, weak stemmed with tuberous roots that support-themselves by means of cirrhosed tips. One to four stems arise from a single V shaped fleshy cylindrical tuber.

Leaf: The leaves are stalk less, lance shaped, alternate or opposite or in whorls of upto three. Leaves are 5 to 15cm long by 4 to 5cm wide with parallel veins and tips ending in spiral tendrils which are used for climbing.

Flower: Flowers solitary, large, borne on long, spreading pedicels, actinomorphic, hermaphrodite, perianth segments six, free, lanceolate, keeled within at base, long persistent; hypogynous, the anthers extorse, medifixed and versatile, opening by longitudinal slits; ovary superior, the carpels cohering only by their inner margins, the ovules numerous, the style deflected at base and projecting from the flower more or less horizontally. Large, showy, long stalked flowers are made up of six long refluxed petals usually with wavy margins. Flower size: 6 to 10cm long by 1 to 2.5cm wide. The flower colour, usually very bright ranging from red with yellow margins to very pale yellow forms with a mauve or purple stripe; pale white forms also occur. In the bud stage, the petals hang down over the ovary and on maturity; they assume an erect position, leaving the ovary with its stigma exposed at right angles. There are five long stamens, each with a long anther that displays profuse orange - yellow pollen. The ovary is three celled ellipsoidal capsule (Fig. 2).

Flowering, pollination and fruiting

Flowering and fruiting occurs during July to October in Indian conditions. It exhibits herkogamy. The flowers have deflexed stigma which requires assisted pollination. Hand pollination is done between 7-11am every day. Pollens are collected using brush and dusted on the just opened flowers to ensure maximum seed set.

Studies conducted by Narain (1976) in glory lily showed that stigma became receptive after anthesis with stigmatic hairs becoming shiny at the time of receptivity. The receptivity was maximum between 9.00 to 10.30am and continued for the next 27 to 32h. However, the receptivity of the stigma even before anthesis and continuance upto three days had been recorded by the Mamtha (1989) in the same crop. She also observed good pod set with artificial cross pollination. The reason for low fruit set in self pollination was indicated due to self incompati-bility mechanisms or deflexed nature of anther and stigma thus preventing self pollination as suggested by Le Roux and Robertse (1997).

Fruit: The fruit is oblong, 6 to 12cm by 2 to 2.5cm and contains about 20 globose red seeds in each valve. The fruit is loculidal capsule. The fruits had three equal lobes; fruits are green, dried to pale and then, black colour dehisced into three sections (Fig. 3).

Seed: There are numerous seeds in a capsule and the seeds are warty and dorsally compressed. Seeds are oval in shape, testa spongy, 30-150 seeds per capsule, pale orange, attached to the sutures (Fig. 4).

Fig. 1. Plant

Fig. 2. Flower

Fig. 3a. Immature fruit

Fig. 3b. Dry fruit

Fig. 4. Seeds

Propagation: The climber is propagated through seeds and 'V' shape tubers.

Seed propagation: The fresh seeds are collected during October. Then, the seeds are soaked in water for 24h and sow in beds. In 20 to 30 days, 50 per cent of the seeds germinate. Seedlings raised from the seeds are able to produce only tubers and fail to set flowers and seeds in the first year. In case of plants raised from tubers, fruits and seeds can be obtained in the same year of planting.

Seed treatment: The seed germination can be improved by soaking in 4000ppm thiourea and the early germination could be achieved by soaking in GA_3 (Supari *et al.*, 1993).

Vegetative propagation: The tubers are cut into divisions with at least two eye buds and weighing not less than 50 to 60g. Plant the tubers horizontally finger deep in furrows opened at 60cm apart. The ideal planting season is during March-April. The dormant tubers start sprouting from the month of May till August.

Economic part used: Seed

Alkaloids: The seeds contain alkaloid called Colchicine and thiocolchicocide (Chaudhuri and Thakur, 1998).

Medicinal uses: It was widely used in the treatment of gout, ulcers, leprosy, piles, inflammation, abdominal pains, intestinal worms, thirst, bruises, infertility and skin problem.

Other uses: Inducing polyploidy in plants.

Past work done on Seed Science and Technology

Investigations to study the causes of low seed set in this species have revealed that the species is (i) both self and cross pollinated and (ii) seed set is dependent upon both pollinator activity and the time of pollination (Raina and Gupta, 1999; Narain, 1976). Although there are no self or cross incompatibility barriers, the hercogamous and attractively coloured flowers favour cross pollination. The peculiar structure of the large flowers with six perianth lobes bent backwards, six radiating anthers and the style bent almost 90° at the point of attachment to the ovary, does not make them suitable for pollination by small insects. Only large insects like bumble bees and birds like *Nectarinia zeylonica* and *Nectarinia asiatica* with long beaks have been reported to be visiting these flowers. This limits the possibility of good cross pollination, although wind is another factor which would be aiding its pollination.

Table 1. Chronology of flower opening on a branch in *G. superba* L.

No. of flower starting from base on a branch	Stage of flower development (days after opening of first flower)			
	I	II	III	IV
1	0	3.20	5.80	9.00
2	2.20	5.20	7.40	10.80
3	4.20	7.55	9.05	12.50
4	6.00	9.25	11.25	14.75
5	7.80	10.96	13.30	16.63
6	8.90	12.20	14.70	17.50

Wherer, I – Bud-opening stage;

II – Stigma receptive stage (Perianth colour is crimson at the tip, yellow in the middle and greenish at the base);

III – Perianth colour is crimson upto the middle portion and yellow towards base;

IV – Perianth colour is entirely crimson with lobes starting to dry (Gupta and Raina, 2001).

To overcome this problem, the species has developed the mechanism of sequential opening of its flowers. An average of six flowers develops fully on a branch and they open in a sequential manner. The first flower opens towards the base of the branch with the subsequent flowers opening away from the first flower. No two flowers on a branch were observed to be at the same stage of development at any given time.

The pollination studies revealed that, dusting of foreign pollen and stigma one day after anthesis resulted in higher pod set of 88 per cent, while pollination of the day of anthesis resulted in 80 per cent and bud pollination one day before anthesis resulted in 68 per cent pod set (Table 2). These results clearly revealed that the stigma was receptive even before anthesis and continued to attain the maximum receptivity one day post anthesis. Among the different pollination methods, hand pollination was the best as revealed through the superiority in terms of pod and seed yield parameters accompanied with better seed quality parameters (Table 3 and 4) (Venudevan *et al.*, 2011).

Table 2. Studies of stigma receptivity and method of pollination to improve pod and seed set and seed quality in glory lily

Treatment	Stigma receptivity (pod set %)
Pollination one day before anthesis	68 (55.55)
Pollination on the day of anthesis	80 (63.43)
Pollination one day after anthesis	88 (69.79)
Pollination two days after anthesis	76 (60.66)
Pollination three days after anthesis	60 (50.76)
Mean	74.4 (59.60)
SEd	1.151
CD at 5 %	2.441

(Figures in parentheses indicate transformed values)

Table 3. Studies of stigma receptivity and method of pollination to improve pod and seed set and seed quality in glory lily

Treatment	Pod set (%)	Days to pod maturity	Length of pod (cm)	Girth of pod (cm)	Weight of pod (g)
Open pollination	64 (55.13)	63.7	7.7	4.9	4.49
Self pollination	24 (29.33)	64.4	7.5	4.6	3.74
Hand cross pollination	96 (78.46)	62.2	8.4	6.8	4.92
Hand self pollination	44 (41.55)	66.3	7.4	4.0	4.37
Natural cross pollination	80 (46.14)	63.1	7.9	6.0	3.93

Mean	61.64 (51.70)	63.94	7.78	3.88	4.29
SEd	1.262	1.369	0.146	0.079	0.088
CD at 5 %	2.675	2.903	0.310	0.169	0.187

(Figures in parentheses indicate transformed values)

Table 4. Studies of stigma receptivity and method of pollination to improve pod and seed set and seed quality in glory lily.

Treatment	Weight of seed pod (%)	100 seed weight (g)	No. of seeds pod^{-1}	Shelling percentage	Germi-nation (%)	Vigour index
Open pollination	2.01	3.39	65.0	44.76 (41.99)	52 (46.14)	1170
Self pollination	1.60	3.41	62.0	48.40 (44.08)	51 (45.57)	1132
Hand cross pollination	3.04	3.35	74.0	61.78 (51.81)	54 (47.29)	1252
Hand self pollination	1.92	3.43	59.0	43.57(41.30)	50 (45.00)	1100
Natural cross pollination	2.24	3.37	67.0	56.99 (49.01)	53 (46.72)	1213
Mean	2.30	3.39	65.4	51.10 (45.63)	52.00(46.14)	1173
SEd	0.044	0.071	1.234	0.606	0.670	26.309
CD at 5 %	0.094	NS	2.616	1.285	1.398	54.880

(Figures in parentheses indicate transformed values)

REFERENCES

Chaudhuri, P. K. and R. S. Thakur 1998. A review article on *Gloriosa superba*. *Curr. Res. medicinal and aromatic plants*, 16: 51-64.

Gupta, L. M. and R. Raina 2001. Significance of sequential opening of flowers in *Gloriosa superba* L. *Curr. Sci.*, 80(10): 1266-1267.

Le Roux, L. G. and P. J. Robertse 1997. Aspects relating to seed production in *Gloriosa superba* L. *South African J. Bot.*, 63: 191-197.

Mamtha, H. 1989. Studies on the problem of fruit set and it's important in *Gloriosa superba* L. M.Sc (Hort.) Thesis, University of Agricultural Sciences, Bangalore.

Narain, P. 1976. Studies on pollination mechanism and breeding system in Glory lily. *J. Hort.*, 33: 194-199.

Raina, R. and L. M. Gupta 1999. Increasing seed yield in glory lily (*Gloriosa superba*) experimental approaches. *Acta Horti.*, 502: 175-179.

Supari, M. R., A. A. Farooqui, T. G. Prasad 1993. Influence of various pre-sowing treatments and growth regulators on seed germination in *Gloriosa superba*. *Indian J. Forestry*, 16(2) : 123-126.

Venudevan, B., S. Sundareswaran, A. Vijayakumar and K. Rajamani 2011. Studies on improving seed set and quality in Glory Lily through pollination methods. *Madras Agric. J.*, 98 (1- 3): 33-35.

4.9. *Gymnema sylvestre*

Common name: Small Indian Ipecacuanha, Ram's horn, Miracle fruit

Botanical name: *Gymnema sylvestre* R.Br. ex Schult., Syn. *Periploca sylvestris* Willd., *Gymnema melicida* Edgew.

Family: Apocynaceae

Vernacular names: Gadalshingi (Bengali); Dhuleti, Kaavalee (Gujarati); Chhota dudhilata, Gudmar, Gurmar, Medhashingi (Hindi); Kadhasige, Sannagera, Sannagerasehambu (Kannada); Chakkarakkolli, Madhunasini (Malayalam); Kavali, Bedaki, Bedakuli, Kalikardori, Kaoli (Marathi); Meshasringi (Oriya); Ajaballi, Ajaghandini, Karnika, Kshinavartta, Madhunasini (Sanskrit); Sirukurinchaan, Tadigam, Amudupushpam, Ayagam, Kogilam (Tamil); Bodaparta, Podapatra (Telugu); Gurmar, Gurmar booti, Gurmar patta (Urdu).

Origin and distribution: It is native to India and commonly found in the Deccan Peninsular extending to the parts of northern and western India.

Habitat: It is a tropical and sub tropical climber and is known to perform in areas with well distributed high or average rainfall. It is common in hills of evergreen forests. It can grow in variety of soils, but grows in well drained red loamy soils or in medium black soils. The crop can also be grown in dry areas and it does not tolerate water logging. It thrives better on humus rich loamy soils with good shade.

Habit: It is a large, more pubescent lactiferous woody climber, much branched, running over the tops of tall trees (Fig. 1).

Leaf: Leaves are 3-5cm long and upto 3cm broad, ovate-elliptic, acute or shortly acuminate, pubescent on both sides; base rounded or heart shaped with 6-13mm long pubescent petioles.

Flower: The flowers are pedunculate or nearly sessile cymes, peduncles densely pubescent, shorter than petioles (6-13mm long, pubescent) arising from between them, sometimes producing successive umbels or whorls of flowers. The pedicles are 3-12mm long pubescent; bracts minute, ovate, oblong, hairy and ciliate. The sepals are pubescent, divided at base or nearly so, 2mm long, oblong and obtuse, the petals are yellow, 4-5mm across, tubes campanulate 1.5mm long about equal to lobes; lobes thick; ovate deltoid spreading recurved, glabrous. The corona of five processes inserted on the corolla tube, alternate with its lobes, free at the short deltoids sub-acute tips above the sinus, the adenate portion decurrent, channeled with strongly ciliated margins. The apex of the style is thick, sub-hemispherical, much exerted beyond the anthers, pearly white (Fig. 2).

Flowering, pollination and fruiting: Flowering occurs in October–January, while fruits mature from March to May (Anon., 2008). The pollination behaviour is entomophilous.

Fruit: Follicles are upto 7.5cm long and 1cm broad, rigid, lanceolate, and attenuated into a beak, usually single (Fig. 3).

Fig. 1. Plant

Fig. 2. Flower

Fig. 3. Fruit

Seed: Seeds are about 1.3cm long, narrowly ovoid - oblong, flat, with a thin, broad, brown and glabrous marginal wing.

Seed dispersal: The tufts of hairs attached to the micropylar end of the seeds effectively disperse the seeds at larger distances.

Propagation: This plant can be propagated by seeds, cuttings, and apical shoots.

Seed propagation : Seeds are collected from freshly harvested fruits and soaked overnight in water. Then, the seeds are sown in the seed pans containing soil mixed with sand. The seed pans are watered daily and the seeds germinate in about 15 day's time. Within 7-10 days, about 40 per cent seed germination can be observed.

Seed development and maturation: Seeds attain physiological maturity 100 days after anthesis with colour change of pods from green to dark brown. Collect the seeds during February-April.

Seed germination : Seed setting is poor in this species and the seeds show a maximum germination of 50 - 55 per cent when sown in soil mixed with vermicompost. Leaching the seeds in tap water for 12h followed by soaking in KNO_3 2g/l for 6h recorded higher germination.

Seed grading: Density grading using South Decota air blower at 0.5" water pressure improves seed germination.

Seed storage: Seed treatment with carbendazim 50% WP @ 2g/kg of seeds and halogen formulation (Bleaching powder + $CaCO_3$ + Arappu leaf powder @ 5:4:1) @ 3g/kg seed and packed in moisture vapour proof container prolongs viability.

Vegetative propagation

Seed germination is poor; hence, plants are preferably raised vegetatively through cuttings. Cuttings can be obtained throughout the year in moist humid conditions of South India. In North and Central India, cuttings are preferably planted in February-March. Terminal and axillary cuttings with three to four nodes from one year old plants are the best planting material. The semi-hardwood cuttings are prepared to the length of 15-20cm from the apical shoots.

These cuttings are planted in polybags and transplant on sufficient rooting. Alternately these cuttings may also be planted directly near a tree. The rooting would initiate in about six days and the success per cent is 60. The stem cuttings are dipped in IBA @100ppm for six minutes before planting in the nursery to promote rooting. Before planting in polythene bags the cuttings are also treated with one per cent Bavistin. In the mist propagation technique, the apical shoots are buried and covered with soil leaving a light weight on it. It takes about 30-40 days to produce roots. The pots/polybags/beds are not over watered at any time, as water logging is very harmful for this plant. Once it is about 10-15cm, it can be trans-planted near a large tree that will support this climber. The cuttings are planted at the onset of rains with a spacing of 2m within and between the rows in less than 60cm deep cubic pits.

Economic part used: Dry leaves

Alkaloids: The produce contains 7-9.6 per cent of gymnemic acid as active principle. Besides, alanine, aminobutyric acid, isoleucine, valine, adenine, choline, gymnamine (alkaloid) and many other ingredients are isolated from leaves (Anon., 2008).

Medicinal uses: It is hypoglycaemic, astringent, stomachic, antiperiodic, diuretic, tonic and refrigerant. It is used as a destroyer of glycosuria, and is believed to neutralize the excess sugar present in the body. It is also used to treat other urinary disorders.

Agricultural uses: Seenivasan *et al.* (2000) reported that cabbage dimond back moth was effectively controlled by spray of three per cent leaf extract of *Gymnema sylvestris.*

Past work done in seed science and technology

Hara Kumar (1997) conducted a field and laboratory studies in *Gymnema sylvestre* to study the seed development and maturation, germination potential and storage of seeds under ambient conditions. The findings of this experiment are follows.

(a) The seeds attained physiological maturity at 100 days after anthesis associated with the change of colour of capsule from green to dark brown and with a thin film of waxy layer formed on upper portion of capsule before it split longitudinally and starts to dehisce.

(b) The seeds collected from first umbel exhibited better quality in terms of vigour and viability than that of other umbels exhibiting polymorphism.

(c) For processing the bulk seeds of *Gymnema,* an air pressure of 1.25cm of water pressure under vertical column through blower can be adopted.

(d) Regarding the media for germination, sand was adjudged as the best medium for testing the germination of *Gymnema.* The germination of the seeds starts at fifth day of sowing and attains maximum at 12th day and thereafter, the seedling growth slows down. Hence, 12th day has been fixed as the final count day.

(e) The freshly harvested seeds recorded about 38 per cent germination. To improve the germination potential, seeds can be subjected to leaching with water for 12h followed by soaking in 0.2 per cent KNO_3 for 6h. This treatment improves the germination to the extent of 75 per cent.

(f) The seeds can be safely stored with maximum vigour and viability under ambient conditions in polyethylene bag with or without thiram treatment for a period of three months.

REFERENCES

Anonymous 2008. Agro-techniques of selected medicinal plants, Vol. 1. National Medicinal Plants Board, New Delhi.

Hara Kumar, C. 1997. Seed technological studies in *Gymnema sylvestre* R.Br. M.Sc. (Ag.) Thesis, TNAU, Coimbatore.

Seenivasan, N. Philip Sridhar, R. and Gnanamurthy 2000. Efficacy of some plant extracts against *(Plutella xylostella)* on cabbage. *South Indian Hort.*, 51 (1-6): 183-185.

4.10. *Hemidesmus indicus*

Common name: Indian Sarsaparilla

Botanical name: *Hemidesmus indicus* (L.) R. Br. Syn. *Periploca indica* L.

Family : Asclepiadaceae

Vernacular names: Vaga Sariva (Assamese); Anantamul, Shvetashariva (Bengali); Sariva, Upalasari (Gujarati); Anantamul, Dudhli, Kapuri, Pitmari (Hindi); Namada veru, Bili Namadaberu, Anantamool, Sugankha palada gidda, Sogade (Kannada); Anant mool (Kashmiri); Nannari, Nannar, Narunenti (Malayalam); Anantamul (Manipuri); Khadki Rasna, Upalsari, Anantmul (Marathi); Onotomulo (Oriya); Anantmool, Ushbah (Punjabi); Sariba (Sanskrit); Nannari, Narukku moolam, Naru neendi (Tamil); Tella Sugandhi, Sugandhiphala(Telugu); Ushba (Urdu).

Origin and distribution: It occurs over the greater part of India, from the upper Gangetic plain Eastwards to Assam and in some places in Central, Western and South India.

Habitat: Garden loam soil rich in organic matter are suited for this plant. It is a dry land weed, common on roadside and waste places.

Habit: It is a vine, which trails on the ground and climbs by means of tendrils growing in pairs from the petioles of the evergreen leaves. The vine emerges from a long, tuberous rootstock, and can reach upto 1-3m (Fig. 1).

Stem: Stem wiry

Root: The tuberous root is dark brown, coma silvery white, tortuous with transversely cracked and longitudinally fissured bark. It has a strong central vasculature and a pleasant smell and taste.

Leaf: Leaves are simple, highly variable, linear, lanceolate to orbicular, opposite.

Flower: Flowers are 5-15, in axillary fasciles. Calyx lobes imbricate, herbaceous to thick, glandular within. Corolla yellow/purple, (sub) rotate; lobes valvate, thick, rugose within. Stamens connive around styles; filaments incurved, distinct; pollinia sub-erect; pollinial bags spherical, closely appressed to the brownish, sheathy caudicle. Corona single, coralline (Fig. 2).

Flowering: Flowering and fruiting occurs during October - December. It is having distinct floral morphology especially, the presence of gynostegium, pollen either in granular tetrads or waxy pollinia and highly evolved system of insect pollination.

Fruit: Ovaries are conic. Follicles are terete (Fig. 3).

Seed : The seeds are coma brownish white, oblong and covered in tiny silvery white hairs.

Propagation : It can be propagated through seeds, stem cuttings and tubers. However, seed propagation gives better results.

Seed propagation: Fresh seeds are collected throughout the year. The seeds are soaked in water for 24h and sow in raised beds at the onset of monsoon for better growth. Sixty per cent of the seeds germinate within a fortnight. The seedlings are transplanted to polythene bags when they are four leaved.

Fig. 1. Plant Fig. 2. Flower

Fig. 3. Fruit

Vegetative propagation: Root cuttings with a length of 10cm are planted in polybags or raised beds. They sprout in 15-20 days with 40 per cent success. Sprouting percentage of stem cuttings treated with boric acid (250 and 100ppm) and IBA 100ppm for 12h was 20 per cent.

Micro-propagation: Patnaik and Debata (1996) developed a protocol for *in vitro* propagation from nodal segments on MS supplemented with NAA (0.054ppm) + K (1.5ppm).

Seed dormancy and germination : Collect fresh seeds any time of the year (since fruiting is observed throughout the year). Soak them in water for 24h and sowing in raised beds at the onset of monsoon for better growth. Sixty per cent of the seeds germinate within fortnight.

Economic part used : Root

Alkaloids: β-sitosterol, 2-hydroxy-4-methoxy benzaldehyde, α-amyrin, β-amyrin and its acetate, hexatriacontane, lupeol octacosonoate, lupeol and its acetate.

Medicinal uses : It is used for venereal diseases, herpes, skin diseases, arthritis, rheumatism, gout, epilepsy, insanity, chronic nervous diseases, abdominal distention, intestinal gas, debility, impotence and turbid urine.

REFERENCE

Patnaik, J. and B. K. Debata 1996. Micropropagation *of Hemidesmus indicus* through axillary bud culture. *Plant Cell Reports* 15: 427-430.

4.11. *Ipomoea mauritiana*

Common name: Giant Potato, Large Forest Ipomoea

Botanical name: *Ipomoea mauritiana* Jacq. Syn. *Ipomoea digitata, Ipomoea insignis*

Family : Convolvulaceae

Vernacular names: Vidarikand, Bhuyikohada (Hindi and Marathi); Bhumi chakkara gedde (Kannada); Mutalakkilannu, Paalmuthukku (Malayalam); Palmudamgi (Tamil); Bhuchakra gedda, Nelagummudu (Telugu).

Origin and distribution: This species is native to Tropical America, now globally distributed in the Pan tropics. Within India, it has been recorded in the moist tropical region. In Kerala, Kasaragod, Cannanore, Wayanad, Calicut, Malappuram, Thrissur, Palghat, and Trivandrum districts; almost throughout the state.

Habitat: This is usually found to grow in open areas of moist deciduous forests. It has been recorded from riparian vegetation, low land forests and in grasslands throughout the warmer parts of India.

Habit: This is a tuber bearing vine with beautiful bell shaped, pinkish purple flowers (Fig.1).

Root: Tuberous roots

Stem: This vine has stems that can grow to 10m.

Leaf: Leaves are orbicular (round) in outline but palmate, 5-9 lobed, upto 20cm long and 18cm wide and the base of leaves is usually cordate. Leaf stalk upto 10cm in length, individual lobes are ovate-lanceolate with sharp apex and entire margins.

Flower: Inflorescence about 15-18cm long. Sepals about 8-10mm long. Flowers are pink or reddish purple, with a darker center, 5-6cm across. Corolla trumpet shaped, about 35-50 x 70-80mm. Sepals about 10 x 5-7mm. Staminal filaments attached to the corolla tube about 5mm from the base, filaments about 12mm long clothed in hairs on the basal section. Anthers about 4mm long. Ovary about 1-3mm long, seated on a fleshy disk about 0.5-0.75mm high. Style about 25-26mm long. Stigma globular, about 2mm diameter, Ovules two per locule (Fig. 2).

Fig. 1. Plant

Fig. 2. Flower

Flowering, pollination and fruiting: It blooms intermittently throughout the year but will be peak during May to July. Fruiting occurs during July - October. Pollination is mostly through the agency of insects (Pooja, 2004).

Fruit: Fruits are dehiscing capsules, upto 1.5cm long.

Seed : There are four seeds per capsule, black in colour, hard seed coat and are covered with cottony hair.

Propagation : This is propagated through seeds and cuttings of tubers.

Seed propagation: Collection of fresh seeds is carried out in the months of June and July. These seeds should be sown within two months from the collected date for better result of germination. The seeds soaked in water for 18h and planted in coir dust media induced rapid and highest germination per cent of seeds (Priyadarshani, 2012). Usually the seeds are sown directly in containers or can be raised mother beds. It takes 15-20 days for germination. The obtained per cent of germination is 80. In case of raised mother beds, seedlings of 5-10cm height can be transplanted to containers. Six month old seedlings can be planted in the field.

Vegetative propagation: Tubers of three to five year old mother plants are cut into two to three long pieces and planted in containers. These cut tubers start sprouting within 20 days.

Economic part used : Tubers

Medicinal uses : This is an indispensable tuber for women health. Its tubers are used to moderate menstrual discharge and often prescribed to increase secretion of mother's milk. It is advised in case of debility after delivery and to increase weight. This can be used against diseases of spleen and liver, complaints related to digestion and general debility of men. It is reported to be used in oligospermia (low sperm count). It can also reduce accumulation of fat.

Other uses: This is ideal and most rewarding climber to decorate wire fences, beautify trellises and to dress up walls and pillars. It can be used as colourful living screens. It adds splendid vertical dimension to the garden that offers privacy and provide breezing shade. Spill of pink flowers with shower of green foliage assign the hanging pots a very delightful look.

REFERENCES

Pooja, 2004. Angiosperms. Discovery Publishing House, New Delhi. pp.298.

Priyadarshani, N. D. N. 2012. Seed germination of kiribadu (*Ipomoea mauritiana*) as influenced by soaking time and sowing media. In: Proc. Abstracts of Jaffna University International Research Conference (JUICE-2012). p 12.

4.12. *Operculina turpethum*

Common name: Trivit

Botanical name: *Operculina turpethum* (L.) J. Silva Manso

Family : Convolvulaceae

Vernacular names: Nisoth, Nisotar and Pitohari (Hindi); Tigade balli (Kannada); Trikolpa konna (Malayalam); Nishottara (Marathi); Trivrit, Nishotra and Triputa (Sanskrit); Sivadhai (Tamil); Tigada (Telugu).

Origin and distribution: The plant is found throughout tropical dry and moist deciduous regions in central and peninsular India.

Habitat : The natural habitats are hedges along coastal areas and wayside thickets. In India, it is found wild throughout upto an altitude of 900m above sea level. Occasionally planted in gardens and rarely cultivated.

Habit: It is a large climbing shrub that exudes milky juice when injured. It is the fastest growing, hardy climber and can grow upto a height of 15m (Fig. 1). It has no tendrils, support should be given for climbing.

Root: Roots are long, slender, fleshy and extensively branched.

Stem: Stems are very long, fast growing, purplish green, twining with three twisted, sharp wings.

Leaf: Leaves measure 5-10cm long and 3-7cm broad, ovate or oblong, more or less pubescent on both sides, especially when young, minutely reticulate veined.

Flower: Inflorescence is a cyme with few flowers. Peduncles are stout, 2.5-5cm long. The two bracts are large, caducous and often pinkish. In flower, the outer sepals are upto 2.2cm long, accrescent, much angled in fruit, broadly ovate, mucronate, concave, and pubescent; the three inner sepals are smaller, barely 2cm long, very thinly membraneous, and glabrous. Corolla is white, 3.8–5cm long and campanulate (Fig. 2).

Flowering, pollination and fruiting: Flowering and fruiting occur twice a year from September to November and then, from March to May. Pollination is mostly through the insects, on account of the flowers being mostly of large size and bright red, violet, blue or white coloured (Pooja, 2004).

Fruit: Fruits appear as cones (due to the twisted sepals) latter unfolds as lotus flowers (Fig. 3).

Seed : Seeds are four in number, pure black in colour, enclosed within the bloated, transparent, floral parts (Fig. 4).

Fig. 1. Plant

Fig. 2. Flower

Fig. 3. Fruit

Fig. 4. Seed

Propagation : The plants can be raised by vegetative propagation as well as from seeds.

Seed propagation: Seeds are sown in sunken mother beds, irrespective of any seasons. Sowing should not be more than 2.5cm deep. It starts germination within 15 days. Two month old seedlings are transplanted to the containers or field. Ninety per cent germination can be observed.

Seed collection: April-May is the best season for seed collection when fruits mature and the seeds are sown in nursery beds immediately after collection.

Seed treatment: Soaking of seeds in water for 24h followed by mechanical scarification by rubbing seed coat with sand paper gives 95 per cent germination.

Vegetative propagation

In case of drooping down, new plantlets arise from the nodes when the climbers touch the ground. New plantlets can be jiggled out and detached for propagation.

In case of cuttings, matured stem cuttings (measuring 8mm broad including wings) bearing four nodes are planted (although the internodes are long) in sandy mother beds in such a way that underground portion as well the aerial portion should have at least two nodes each. Soil medium should have more proportion of coarse sand. After planting, the bed is covered with polythene cover to enhance the root initiation by creating more humidity. Sprouting starts within ten days, root initiation starts after 20 days. Two to three month old sprouting are transplanted in containers or into the field.

Economic part used : Roots.

Alkaloids: The active principle of the leaves is a cardiotonic substance named oleandrin. It has anti-inflammatory and stimulant properties. The roots, bark and seeds contain cardio-active glycosides, formerly designated as neriodorin, neriodorein and karabin, which are anti-inflammatory, stimulant and good pain relievers. Plant contains β-sitosterol, α and β-turpethin, coumarin, scopoletin, lupeol, betulin.

Medicinal uses: Roots are used in chronic constipation as an effective laxative. Drug of this climber is also used to remove excess mucous from lungs and to expel worms from the digestive system. As having hot potency, it is good as anti-inflammatory and to reduce fever caused by inflammations. It is used to relieve flatulence and colic and in case of obesity, it is used to decrease fat. Curative property against abdominal disorders and rat bite poisoning has also been reported.

Other uses: It can be used as a focal point in garden due to its dense cool foliage and bright white flowers along with quick growing habit without any major pest problem.

REFERENCE

Pooja, 2004. Angiosperms. Discovery Publishing House, New Delhi. Pp. 298.

4.13. *Piper longum*

Common name: Long pepper, Indian long pepper

Botanical name: *Piper longum* L. Syn. *Piper sarmentosum, Piper latifolium, Chavica roxburghii, Chavica sarmentosa*

Family: Piperaceae

Vernacular names: Piplamor (Bengali); Pipari (Gujarati); Pipal, Pipali (Hindi); Kandan Lippili (Kannada); Pipli (Konkani); Pippali (Marathi); Tippali (Malayalam); Pippali, Magadhi (Sanskrit); Tippili, Pippili, Sirumulam, Kandan tippili (Tamil); Pippallu (Telugu); Pipul (Urdu).

Origin and distribution: It is native to India and widely distributed in India, Nepal, Indonesia, Malaysia, Sri Lanka, Timor and Philippines. In India, it is cultivated in West Bengal, Lower elevations of Assam, Amaravathi region of Maharashtra, Anamalais of Tamil Nadu and Kerala and parts of Odisha.

Habitat: The plant grows well under hot humid climate at an elevation of 100-1000m. It is commonly found in tropical rain forests. It can be successfully grown even in areas receiving heavy rainfall with high relative humidity. It can be grown as an under crop in coconut and arecanut gardens with 20-25 per cent shade intensity. It flourishes well in rich, well drained loamy soil. Laterite soils rich in organic matter content with good moisture holding capacity are also suitable.

Habit: A small aromatic plant, slender, perennial, trailing on ground, also climbing with woody roots growing to height of 90 - 120cm and the young shoots are downy (Fig. 1).

Leaf: Leaves of sterile shoots long petioled, ovate, deeply cordate. Fertile branches erect, 30-60cm tall, minutely puberulous with short petioled, narrowly ovate, elliptic or oblong leaves, acute at apex or gradually tapering into a short acumen; thinly membraneous. Lower leaves are 6-10cm long, 5cm wide, broadly ovate, deeply cordate with big lobes at base. Upper leaves oblong- ovate, cordate, dark green, shining above, pale on lower surface, stipules conspicuous, 1.3cm long, but soon falling.

Flower: Spikes of flowers are solitary, unisexual, bracts of male spikes narrow, bracts of female circular, densely packed flowers. The male spikes are larger, slender and are 2.5-7.5cm long, stamens two, while the female spikes are 1.25-2.0cm long when in flowering stage, growing to about 3cm wherein fruiting stage (Fig. 2).

Flowering, pollination and fruiting: The plant flowers in rains and fruits in early winters. It is wind and water pollinated crop.

Fruit: The fruits are berry, small, ovoid, sunk in fleshy spike which is 2.4-4.0cm, ovoid, oblong, blackish green, shining (Fig. 3).

Propagation: The plant is propagated by suckers or rooted vine cuttings.

Vegetative propagation: Best time for preparing the stem cuttings is March - April. 15-20cm long 2-3 noded vines are kept in poly bags or the cuttings are planted in shaded nursery beds with the upper most bud exposed. The rooted seedlings are planted during May last.

Suckers arising from the base of a mature plant may also be used for propagation. The suckers are planted individually. These plants are prone to many diseases during the early stages of development and hence, care should be taken to prevent their occurrence.

Fig. 1. Plant

Fig. 2. Flower spike

Fig. 3. Dried spike

Economic part used: Leaves, dry spikes and roots

Alkaloids: Longumine, piperine, piplatin.

Medicinal uses: The drugs are used for diseases of respiratory tract, cough, bronchitis, asthma, as a counter irritant and analgesic when applied for muscular pains and inflammation. It is used as a snuff in conia and drowsiness and internally as carminative and as a sedative in insomnia and epilepsy. It is also used as a general tonic and haematinic, as cholagogue in obstruction of bile duct and gall bladder, as an anthelmintic and in dysentery and leprosy. Long pepper is used as an alternative tonic. It is also used in sciatica and paralysis. The dried whole plant is used for chronic fever, abdominal diseases, worms from both the mouth and anus, for tumours, anaemia and to prevent conception. The fruit is used as carminative to treat heart pain and to promote blood circulation. It is given to mothers after childbirth. It is used for chronic coughs, bronchitis and to prevent conception. The fruit is given to women after parturition to check haemorrhage and to ward off fever. Dried root is used for tetanus, body pain, malaria, rheumatism and also as a carminative. Dried seeds are used as a stimulant, carminative, diuretic and vermifuge. They are also used to stimulate menstruation and to check nausea. The seed is used against alcoholism and for bronchitis. Unripe dried fruit is used for stomach abdominal pain, diarrhoea, cold dysentery, headache and toothache. It is also used to treat vomiting and acid regulation. Leaves are used to treat bronchial asthma. Root and fruit are antidotes to snake bite and scorpion stings.

4.14. *Pueraria tuberosa*

Common names: Indian kudzu, Nepalese kudzu

Botanical name: *Pueraria tuberosa* (Roxb. Ex Willd.) DC. Syn. *Hedysarum tuberosum*

Family: Fabaceae

Vernacular names: Shimia batraji (Bengali); Vidarikand (Gujarati); Sural, Bilaikand, Bharda, Tirra, Bankumra Vidarikand (Hindi); Gumadigida (Kannada); Palmutukku (Malayalam); Ghorbel, Badra (Marathi); Bhukushmandi, Vidari, Vidarika (Sanskrit); Darigummadi, Niala-gumodi (Tamil); Darigummadi (Telugu).

Origin and distribution: It is native to India, Pakistan and Nepal. It is distributed widely in India and Pakistan. In India, it is widely distributed in central, southern and eastern regions including tropical Himalayas. It is fairly common in the lower Ghats of Kerala and Tamil Nadu.

Habitat: Dry deciduous to moist deciduous forests. It prefers well drained moist soils.

Habit: It is a large woody deciduous climbing shrub, one metre height with huge tuberous roots (Fig. 1).

Root: Tubers are globose or pot-like, about 25cm across, white inside, starchy and mildly sweet (Fig. 2).

Stem: The woody stems grow upto 12cm in diameter.

Leaf: Leaves are divided into three leaflets. The leaflets are egg shaped, with round base and unequal sides. They are 18cm long and 16cm wide. Leaves alternate, three foliate, base obtuse and unequal sided, apex acuminate, margin ciliate, glabrous above, densely pubescent below. Leaf stalks upto 18cm long, lateral nerves 7-8 pairs.

Flower: Flowers bisexual, about 1.5cm across, bluish purple to pale violet, shortly stalked, densely arranged in 15-25cm long axillary panicles (Fig. 3).

Flowering and fruiting: Flowering occurs between December and February and fruiting occurs during March to June.

Fig. 1. Plant

Fig. 3. Flower

Fig. 2. Tuber

Fruit: Pods are flat and 5-7cm long, densely clothed with long, silky, bristly brown hairs.

Seed: Seeds are 3-6 in number.

Propagation: Tuber, seeds, divisions and layers.

Seed propagation: Pre-soak the seeds for 12h in warm water and sow in a warm greenhouse in early spring. Germination takes place within two weeks.

Economic part used: Tuber, root

Medicinal uses: The root is used as a demulcent and refrigerant in the treatment of fevers. It is peeled and bruised into a cataplasm and used to reduce swellings. It is crushed and rubbed on the body in the treatment of fevers and rheumatism. It is also emetic, galactogogue and tonic.

4.15. *Tinospora cordifolia*

Common name: Heart leaved moonseed plant

Botanical name: *Tinospora cordifolia* (Willd.) Hook. F. & Thomson

Family: Menispermaceae

Vernacular names: Siddhilata, Amarlata (Assamese); Gulancha (Bengali); Galo (Gujarati); Gilo, Giloya, Gurcha (Hindi); Amrita balli (Kannada); Amrita, Gilo (Kashmiri); Chittamrutu (Malayalam); Gulvel, Shindilakodi (Marathi); Guluchi (Oriya); Guduchi, Madhuparni, Amrita, Chinnaruha, Vatsadaani, Tantrika Kundalini, Chakralakshanika (Sanskrit); Amirthavalli, Seenthil, Amirthai, Somavalli, Gundali, Amirthakodi (Tamil); Manapala, Tippatiga (Telugu); Gilo (Punjabi and Urdu).

Origin and distribution: This is native to India and found throughout tropical India ascending to an altitude of 300m and in south Asia, Indonesia, Philippines, Thailand, Myanmar, China and in Sri Lanka.

Habitat: It prefers wide range of soil, acid to alkaline and it needs moderate levels of soil moisture.

Habit: A large climber with large succulent, glabrous, deciduous stems, throwing aerial roots like a banyan tree (Fig. 1).

Root: Long thread like, aerial, arise from branches.

Stem: The stem structure is fibrous and the transverse section exhibits a yellowish wood with radially arranged wedge shaped wood bundles, containing large vessels, separated by narrow medullary rays. The bark is creamy white to grey, deeply left spirally and stem contains rosette like lenticels.

Leaf: The leaves are membraneous and cordate in shape. 5-10cm long, 7-9 nerved, petiole slightly shorter than leaves.

Flower: Flowers are in axillary position, 2-9cm long raceme on leaflet branches, unisexual, small, less than 2mm in size and yellow in colour. Minute, male and female separate, male flowers grouped in axils of bract, female solitary (Fig. 2).

Flowering, pollination and fruiting: Flowering occurs in May-June, while fruiting is witnessed in September–October. It is cross pollinated by wind.

Fruit: Fruits are fleshy and single seeded. Drupe, size of pea, red in colour, ovoid, glossy, succulent (Fig. 3).

Seed : The seeds are fleshy, curved and pea sized.

Propagation : The plant is propagated either through seeds or through cuttings.

Seed propagation: The seeds are soaked in cold water for 24h for better germination. These seeds are sown in polybags during May-July. It takes 10 to 12

days for the seeds to germinate with the germination success being 80 to 90 per cent. Seeds sown without pre-soaking show only 30 to 35 per cent germination successes. The seedlings are maintained in the nursery for about 45 days.

Fig. 1. Plant

Fig. 2. Flower

Fig. 3. Fruit

Vegetative propagation: During June-July, 10-15cm long, pencil thickness semi-hard wood cuttings having four to eight nodes are obtained from older stems and planted in polybags in a slanting manner with two nodes buried in the rooting medium. Cuttings should be sown within 24h of their removal from the mother plant. Nearly all the cuttings root in about 20 to 30 days. About 90 per cent of the cuttings sprout in 45 days.

Economic part used : Dried stem part and roots

Alkaloids: The plant contains various alkaloids *viz.*, diterpenoid lactones, steroids, glycosides, aliphatic compounds, polysaccharides. Some constituents have been isolated from plant mainly they are tinosporone, tinosporic acid, cordifolisides A to E, syringen, berberine, giloin, gilenin, crude giloinin, arabinogalactan polysaccharide, picrotene, bergenin, gilosterol, tinosporol, tinosporidine, sitosterol, cordifol, heptacosanol, octacosonal, tinosporide, columbin, chasmanthin, palmarin, palmatosides C and F, amritosides, cordioside, tinosponone, ecdysterone, makisterone A, hydroxyecdysone, magnoflorine, tembetarine, syringine, glucan polysaccharide, syringine apiosylglycoside, isocolumbin, palmatine, tetrahydropalmaitine, jatrorrhizine respectively (Singh *et al.*, 2003).

Medicinal uses: It is a tonic and has alterative, diuretic and aphrodisiac properties. It is a febrifuge used in malarial and chronic fever. It is also a liver tonic. The plant is used in general debility, loss of appetite, fevers, urinary disorders, diabetes, rheumatism and dyspepsia.

REFERENCE

Singh, S. S., S. C. Pandey, S. Srivastava, V. S. Gupta, B. Patro and A. C. Ghosh 2003. Chemistry and medicinal properties of *Tinospora cordifolia* (Guduchi) *Indian J. Pharmacology*, 35; 83-91.

4.16. *Tylophora indica*

Common names: Indian ipecac, Indian ipecacuahna

Botanical name: *Tylophora indica* Merr. Syn. *Asclepias asthmatica, Cynanchum indicum, T.asthmatica*

Family : Apocyanaceace

Vernacular names: Antamul (Assamese); Antamul, Anantamul (Bengali): Damnivel (Gujarati); Janglipikvam, Antamul (Hindi); Antamula, Adumutadhagida, Nipaladaberu (Kannada); Nansjera-patsja, Vallippala (Malayalam); Atkari, Khadari, Khodiki, Pitthakaadi, Pitthamaari, Pitvel, Raasna, (Marathi); Mendi, Mulini (Oriya); Antamul, Arkaparni, Lataksiri, Shwasaghni (Sanskrit); Kagitham, Asthma kodi, Nanjaruppan, Nanju murithaan, Kodipaalai, Kondaisani, Mirkkurinja, Nayppala, Tellavidavela (Tamil); Kukkapala, Tellayadala,Verripala (Telugu).

Origin and distribution: It is native to India, found in plains, forests, hills of southern and eastern India. The species is common in peninsular India, extending to Bihar, Odisha, West Bengal, and north-eastern states.

Habitat: It is found growing wild mostly in forests in plains and hills upto a height of 1000m. It forms dense patches in the forests in moist, humid conditions in open hill slopes and narrow valley. The plant grows in the area with lesser rainfall. It grows in wide range of well drained soil and prefers scanty localities.

Habit: It is a profusely branching perennial climber, which attains a length of 15m or more. It has short, knotty, 3-4cm thick rootstock (Fig. 1).

Root: The roots are numerous, fine, upto 15cm long, very brittle and consist of fibres or fibrils.

Leaf: Leaves are opposite, ovate to orbicular, elliptic–lanceolate, shinning green, leathery cordate at base, 3-10cm long and petioles are upto 12mm long.

Flower: Flowers are small (1-1.5cm across), greenish yellow outside and purplish within. They are arranged in umbellate racemes. Calyx is coarsely hairy outside. Corolla is greenish yellow or greenish purple, 5-6mm long (Fig. 2).

Flowering, pollination and fruiting: The flowering appears in the second year of growth. Flowering and fruiting occur from October to December. It is an example of entomophilous pollination.

Fruit: A follicle, 5-10cm long, 1-2cm broad, in pairs, pointed at tip with many fine ridges (Fig. 3).

Seed : Seeds are ovate and elongate into 2-2.5cm long coma.

Propagation : Plants can also be raised through seeds or using stem cuttings, ground layering and root coppices from the mother plants in March.

Fig. 1. Plant Fig. 2. Flower

Fig. 3. Fruit

Seed propagation: The seeds have a high germination percentage (more than 90 per cent), but fruit set is rare. The mature follicles containing seeds can be collected at the end of the cold season. Seeds start germinating in ten days and germination is completed within three weeks.

Vegetative propagation: Vegetative propagules as cuttings from stem coppices or ground layering can be prepared in spring season. Success rate of these propagules is high. Vegetative propagules sprout in three to four weeks.

Tissue culture: Thomas and Philip (2005) have reported high frequency shoot organogenesis from leaf derived callus.

Economic part used : Dried roots, Leaf.

Alkaloids: The roots and leaves contain pharmacologically active alkaloids like tylophorine, tylophorinine and anti cancerous tylophorinidine (Mulchandani *et al.*, 1971; Gopalakrishnan *et al.*, 1980).

Medicinal uses : Root or leaf powder is used in diarrohea, dysentery and intermittent fever. It is an expectorant and administered in respiratory infections, bronchitis and whooping cough. Dried leaves are emetic, diaphoretic and expectorant. It is believed to have laxative, expectorant, diaphoretic and purgative properties. It has been used for the treatment of various respiratory problems and inflammatory conditions like asthma including allergies, bronchitis and colds, as well as dysentery and osteoarthritis pain. It is becoming increasingly popular for the treatment of asthma (Sunil and Priya, 2012).

REFERENCES

Gopalakrishnan, C., D. Shankaranarayan, S. K. Nazimudeen and L. Kameshwaran 1980. Studies of pharmacological effects of extracts and total alkaloids of *Tylophora indica*. *Indian J. Med. Res.* 71: 940-948.

Mulchandani, S. B., S. S. Iyer and L. P. Badheka 1975. Structure of tylophorinindine. A new potential antitumor alkaloid from *Tylophora indica*. *Chem. India*. 19: 505-506.

Sunil, K. and S. Priya 2012. *Tylophora indica* - an Indian ipecacuahna: A review, *International J. Phytotherapy Res.,* 2(2): 1-14.

Thomas, D. and B. Philip 2005. Thidiazuron induced high frequency shoot organogenesis from leaf derived callus of a medicinal climber, *Tylophora indica*. *In vitro cellular and Development Biology- Plant*, 41: 124-128.

PART - V

PERENNIAL MEDICINAL PLANTS

5.1. *Aegle marmelos*

Common names: Bengal quince, Indian quince, Golden apple, Holy fruit, Stone apple, Bael

Botanical name: *Aegle marmelos* L. Corr. Ex. Schultz. Syn. *Feronia pellucida* Roth, *Crataeva marmelos* L.

Family: Rutaceae

Vernacular names: Bael, Vael (Assamese); Bela, Bilva (Bengali); Bill, Bilum, Bilvaphal (Gujarati); Bel, Bela, Sriphal, Bilva (Hindi); Bilva (Kannada); Bel (Kashmiri); Vilwa maram, Bilwa, Bel, Kuvalam, Koovalam (Malayalam); Bel, Bael (Marathi); Bela (Oriya); Beel (Punjabi); Shree Phal, Bilva, Bilwa, (Sanskrit); Vilva Maram, Vilva Pazham, Vilvam, Sobi, Koovilam, Koovilai, Sivathurumam, Ninmali, Maaluram (Tamil); Madtoum, Beli fruit, Maredu (Telugu); Bel (Urdu).

Origin and distribution: It is indigenous to India. It is widely distributed in Java, Myanmar, Sri Lanka, Pakistan, Thailand and Bangladesh (Bailey, 1963). The trees worshipped in India by the Hindus since time immemorial has been planted frequently in the premises of temples of the Lord Shiva.

Habitat: Bael occurs in dry forests on hills and plains. It copes with a wide range of soil conditions (pH range 5-10), is tolerant of water logging and has an unusually wide temperature tolerance (from -7°C to 48°C).

Habit: The bael fruit tree is deciduous, slow growing, medium size, upto 12-15m tall with short trunk, thick, soft, flaking bark, light brown to green, spreading, sometimes spiny branches, the lower ones drooping. Young suckers bear many stiff, straight spines. A clear, gummy sap, resembling gum arabic, exudes from wounded branches and hangs down in long strands, becoming gradually solid (Fig. 1).

Leaf: Alternate leaves, pale green, trifoliate, borne singly or in 2's or 3's, are composed of three to five oval, pointed, shallowly toothed leaflets, 4-10cm long, 2-5cm wide, the terminal one with a long petiole. New foliage is glossy and pinkish-maroon. Mature leaves emit a disagreeable odor when bruised.

Flower: Fragrant flowers, in clusters of four to seven along the young branchlets, have four recurved, bisexual, actinomorphic, ebracteate. hypogynous, stalked; calyx, gamosepalous, five lobed, pubescent; corolla polypetalous, with five fleshy petals, imbricate, green outside, yellowish inside; androecium, protandrous, polyandrous, 50 or more greenish yellow stamens, dehiscing longitudinally; gynoecium, light green, having capitate stigma and terminal style (Fig. 2).

Flowering, pollination and fruiting

It flowers during mid April to August with optimum floral density in the months of May and June soon after the new leaves appear on the fruit ripens in 10-11 months from bloom *i.e.*, March to June of the following year. Production of large amount of pollen, high pollen ovule ratio and insect visitors especially the honey bees (*Apis indica* and *A. dorsata*) are indicative of insect pollination and out crossing nature of this tree species (Fig. 5A, 5B, 5C, 5D, 5E and 5F). There was no fruit set in bagged flowers eliminating the chances of self pollination. The seed set was 29-35 per cent (Sharma *et al.*, 2011).

Singhal *et al.* (2011) reported that a variety of insects visited the flowers during peak blooming. Insect species visiting the flowers included honey bees (*Apis dorsata* and *Apis spp.*), hover fly moth (*Amata cyssea*), butterflies (*Anaphaeis aurota* and *Pieris brassicae*), yellow wasp (*Polestis herbreus*), black ant (*Solenopsis geminata*), carpenter bee (*Xylocopa pubescens*) and an unidentified weevil (Fig. 6A, 6B, 6C, 6D, 6E, 6F and 6G). Insects visited the flowers during day time between 06:00 and 19:00h when pollen availability was the maximum. Honey bee (*Apis dorsata*) was the most active and frequent visitor during peak blooming. The bees carried 5–8 visits/flower/hour. It had been estimated that during the functional life of a flower, 240–384 bees visited each flower. The bees carried 20-38 pollen grains on their legs, thorax, mouth parts and wings. Compared to other pollinators, *Apis dorsata* visited the flower more frequently, for more duration and carried the highest pollen load. Hover fly moth (*Amata cyssea*), butterflies (*Anaphaeis aurota*, *Pieris brassicae*) and carpenter bee (*Xylocopa pubescens*) carried less pollen load. Yellow wasp (*Polestis herbreus*) and black ant (*Solenopsis geminata*) were merely the visitors as no pollen grains were recorded from their body parts. *Apis* bee, carpenter bee, moth and butterflies landed directly on anthers. Hover fly and weevil insect first sit on the petals, moved slowly toward the anthers and then, made contact with stigma. Honey bees were most active and carried a complete revolving circle on the anthers. For collection of pollen grains, bees landed over the flower and grasped the anther filaments with their mandibles and in the process some pollen grain attached with the body parts.

It produced large number of flowers, but un-pollinated flowers fell down within 3-5 days of anthesis. Those get pollinated showed series of developmental changes leading to fruit set initiation. Freshly opened flowers possessed greenish white corolla and pistil, and pale yellow turgid stamens. After 12h of pollination, the anthers turned light brown and shriveled up. During the next 12h, petals and stamens turned outwards. After 36h, stigma withered slowly, losing its receptivity after 48h. During this time, petal lobes and stamens also turned reddish brown. After 72h, petals started withering, hanging on ovary for sometime before falling off.

After 84h, stigma and style also fell down and fruit formation started. It had been noticed that sweet fragrance waned off as the flowers senescence. All the flowers in a panicle opened within 5-6 days. First sign of flower opening was indicated by the separation of a posterior petal at an acute angle, followed by the opening of other petals one by one exposing the anthers and stigma. Anthesis process completed in approximately 5-8min and flowers after anthesis lasted for 30-40h. Anthesis starts at 06:00am and continued upto14:00h noon but the maximum number of flowers anthesize between 08:00 and 12:00h. Peak of anthesis was slightly earlier (09:00h) in the cultivated trees compared to the wild trees (10:00h). Anthesis time also varied with climatic conditions. In the individuals growing in hot and dry regions anthesis started earlier (05:00h) compared to those growing in the moist and hilly regions (06:00-07:00h). But the duration of anthesis remained the same. Anther dehiscence was synchronous with anthesis and all the anthers in a flower dehisced synchronously presenting the pollen in mass. Anthers dehisced through longitudinal slits which appeared near the basal end of the anther lobe and proceeded toward the top. Stigma which was non-receptive in bud stage acquired receptivity as soon as the flower opened. Stigmatic receptivity was synchronous with anthesis and anther dehiscence. Receptivity of stigma lasted for nearly 48h (Singhal *et al.*, 2011).

Fruit: The fruit, round, pyriform, oval, or oblong, 5-20cm in diameter, may have a thin, hard, woody shell or a more or less soft rind, gray-green until the fruit is fully ripe, when it turns yellowish. The shell is so hard and must be cracked with a hammer or machete. It is dotted with aromatic, minute oil glands. Inside, there is a hard central core and 8 to 20 faintly defined triangular segments, with thin, dark-orange walls, filled with aromatic, pale-orange, pasty, sweet, resinous, more or less astringent, pulp. Embedded in the pulp are 10 to 15 seeds, flattened-oblong, about 1cm long, bearing woolly hairs and each enclosed in a sac of adhesive, transparent mucilage that solidifies on drying (Fig. 3a and 3b).

Seed: Numerous, embedded in the pulp, hairy oblong, compressed, white and flat. Seeds are encapsulated in slimy mucilage (Fig. 4).

Seed collection and extraction: The seeds are perishable and recalcitrant and cannot withstand storage for long and have to be sown immediately after collection. During March - May, collect the ripe fruits from the branches on the tree and open the rind. Fallen fruits are not to be collected. Separate the seeds from the fruit pulp and wash them thoroughly until the mucilage is completely removed; then dry in the sun for a few days and they may be coated with red lead if necessary to keep off ants. One hundred gram seed contains 550 seeds. The seed viability is 1-6months. The seed germination is 56 per cent.

Propagation: The seed does not retain viability for long and regeneration by rootsuckers appears to be the chief mode of propagation naturally. It also multiplied through ring grafting, shield budding and layering.

Nursery: Soak the seeds in water for about 12h and then, dibble into polybags or raised beds. Fresh seeds may also be sown in nursery beds during June-July and covered with earth, at a spacing of less than 30cm in a row and 10cm between the rows. The beds should be regularly watered. 56-80 per cent of the seeds germinate in a period of 21 days. After maintaining in the nursery for 3-12 months, plant the seedlings in the main field. As the growth is slow in the initial stages, it may be desirable to retain the seedlings with undergrowth for another 1-2 years.

Fig. 1. Plant

Fig. 2. Flower

Fig. 3a. Immature fruit

Fig. 3b. Mature fruit

Fig. 4. Seeds

Vegetative propagation: Ring grafting, where a ring of bark with a healthy bud is removed from the suitable scion. This is placed on the stock plant where a similar ring of bark is freshly cut and removed from very near the collar. Rootstocks should be 12 to 18 months old with stem of 0.75cm thickness. In the case of scion, auxillary bud is selected after leaf fall and bud initiation. The ring with the bud scion is placed around the similar ring made on the stock. In 30 to 45 days after grafting the bud develops into a shoot indicating a fusion with the rootstock. Fifteen days after grafting, the top portion of the stock can be cut off to facilitate early union. The seedlings are ready for planting out in next season. As the growth is slow in the seedling stage, it may be desirable to retain seedlings with undergrowth for another one or two years, if necessary.

Micro-propagation: Ajithkumar and Seeni (1998) achieved rapid clonal multiplication by enhanced axillary bud proliferation in single node segment of a twenty five years old tree on MS medium supplemented with BAP(2.5mg/l)+ IAA (1mg/l).

Economic part used: Leaves, flowers, tender fruits, fruits, seeds, roots, root bark, fruit rind, gummy exudates

Alkaloids: Root contains auraptene, marmin, umbelliferone and lupeol. Alcoholic extract of root gave psoralen, xanthotoxin, dimethoxy coumarin, scopoletin, tembamide marmin, skimmianine and a glycoside skimmin.

The presence of decursinol, an alkaloid haplopine, skimmianine fagonine, marmesin, marmin, xanthotaxin, umbelliferone and lupeol is also reported from root bark.

Wood contains vinyl acetate, xanthtoxol-8-O-β-D-glucopyranoside, furoquinoline alkaloid, dictaminine, marmelin, β- sitosterol.

Fruit and stem contains gum.

Fruit rind contains tannin.

Fruit contains xanthotoxol, coumarins scoparone, scopoletin, umbelliferone, marmesin, skimmin, allaimperatorin, marmesolin and β-sitosterol, marmelide and psoralen.

Seeds contain fatty oil.

Bark contains tannin.

Leaves yield an essential oil, four alkaloids besides aegelenine and aegeline; also condensed tannins, leucoanthocyanins, anthocyanins, flavonoid glycosides, skimmianine, β-sitosterol, rutin, marmesinin.

Medicinal uses: The fresh ripe pulp of the higher quality cultivars, and the "sherbet" made from it, are taken for their mild laxative, tonic and digestive effects. It is employed in the treatment of leucoderma. Marmelosin derived from the pulp is given as a laxative and diuretic. In large doses, it lowers the rate of respiration, depresses heart action and causes sleepiness.

Because of the astringency, especially of the wild fruits, the unripe bael is most prized as a means of halting diarrhoea and dysentery, which are prevalent in India in the summer months.

A bitter, light yellow oil extracted from the seeds is given as a purgative. The bitter, pungent leaf juice, mixed with honey, is given to allay catarrh and fever. With black pepper added, it is taken to relieve jaundice and constipation accompanied by edema. The leaf decoction is said to alleviate asthma. A hot poultice of the leaves is considered an effective treatment for ophthahnia and various inflammations, also febrile delirium and acute bronchitis.

A decoction of the flowers is used as eye lotion and given as an antiemetic. The bark decoction is administered in cases of malaria. Decoctions of the root are taken to relieve palpitations of the heart, indigestion and bowel inflammations; also to overcome vomiting.

The fruit, roots and leaves have antibiotic activity. The root, leaves and bark are used in treating snakebite.

Agricultural uses: Sridhar *et al.* (2000) reported that *Aegle marmelos* leaf and root extracts at five per cent spray effectively control caterpillar in red gram. Swaminathan and Gururajan (2005) reported that application of *Gliricidia* leaves @10t.ha^{-1} with combination of five per cent of *Aegle marmelos* spraying after 30 and 45 days in cowpea and green gram increased yield by 2.14t.ha^{-1}. The bael oil may be used as botanical insecticide against different stored grain pests causing infestation in stored wheat and pulses. The bael oil enhanced feeding deterrence of *Callosobruchus chinensis, Rhizophertha dominica* and *Sitophilus oryzae.* This oil significantly reduced oviposition and adult emergence of *Callosobruchus chinensis* in treated cowpea (Kumar *et al.*, 2008). The combined application of *Gliricidia sepium* (Green leaf manure @10t.ha^{-1} incorporated during field preparation and allowed 45 days for decomposition) + *Aegle marmelos* (seed soaked in 5% solution for 30 minutes) recorded the highest percentage (90.52) of seed germination in bhendi cv Arka Anamika (Anbukkarasi *et al.*, 2011).

Other uses

Fruit: The fruit pulp has detergent action and has been used for washing clothes. The gum enveloping the seeds is most abundant in wild fruits and especially when they are unripe. It is commonly used as household glue and is employed as an adhesive by jewelers. Sometimes, it is resorted to as a soap substitute. It is mixed with lime plaster for water proofing wells and is added to cement when building walls. Artists add it to their water colors, and it may be applied as a protective coating on paintings. The limonene rich oil has been distilled from the rind for scenting hair oil. The shell of hard fruits has been fashioned into pill and snuff boxes, sometimes decorated with gold and silver. The rind of the unripe fruit is employed in tanning and also yields a yellow dye for calico and silk fabrics.

Fig. 5A. Inflorescence. Note the presence of a honey bee on one flower

Fig. 5B. Flowers with 4-5 petals

Fig. 5C. Flower with 6 petals

Fig. 5D. Honey bee on flower

Fig. 5E. Butter fly on flower

Fig. 5F. Wasp on the flower

Fig. 5 A-F. Flowers and pollinators of *Aegle marmelos* (Linn.) Correa.

Fig. 6A. A tree showing defoliation; proceeding from the top of tree towards the lower branches

Fig. 6B. Flower opening; separation of a posterior petal at an acute angle (arrowed)

Fig. 6C. A freshly opened flower stretches outward exposing the anthers and stigma to wind as well as insect pollinators

Fig. 6D. Flower after pollination; brownish corolla and pistil, and turgid stamens. (Floral visitors)

Fig. 6E. Apis bee (unidentified)

Fig. 6F. Black ant (*Solenopsis geminata*)

Fig. 6G. Butterfly (*Anaphaeis aurota*)

Fig. 6A-G. Phenology and pollination system of *Aegle marmelos* (Linn.)

Leaves: In the Hindu culture, the leaves are indispensable offerings to the 'Lord Shiva'. The leaves and twigs are lopped for fodder.

Flowers: Cologne is obtained by distillation from the flowers.

Wood: The wood is strongly aromatic when freshly cut. It is best utilized for carving, small scale turnery, tool and knife handles, pestles and combs, taking a fine polish.

REFERENCES

Anbukkarasi, V., A. Sadasakthi and S. Mariappan 2011. Studies on performance of bhendi under "Vrkshayurvedic farming". Lambert Academic Publishing GmbH & Co. KG, Saarbruckeen, Germany. p.148.

Ajithkumar, D. and S. Seeni 1998. Rapid clonal multiplication through *in vitro* axillary shoot proliferation *of Aegle Marmelos* - a medicinal tree. *Plant Cell Reports* 17: 422-426.

Bailey, L. H. 1963. How plants get their names. Dover Publications, Inc., New York.

Kumar, R., A. Kumar, C. S. Prasad, N. K. Dubey and R. Samant 2008. Insecticidal activity *Aegle marmelos* (L). Correa essential oil against four stored grain insect pests. *Internat. J. Food Safety,* 10: 39-49.

Sharma, L., A. Rana and S. Chauhan 2011. Reproductive biology of *Aegle marmelos* (L.) Correa (Rutaceae*). Internat. J. Plant Reproductive Biology* 3(2): 151-154.

Singhal, V. K., A. Salwan, P. Kumar and J. Kaur 2011. Phenology, pollination and breeding system of *Aegle marmelos* (Linn.) correa (Rutaceae) from India. *New Forests* 42:85-100.

Sridhar, S., S. Arumugasamy, H. Saraswathy, K. Vijayalakshmi 2000. Organic vegetable gardening. Centre of Indian Knowledge system. Chennai.

Swaminathan, C. and B. Gururajan 2005. Revisiting and researching an Indian systems of agriculture through vrkshayurveda. 93rd Indian science congress" integrated rural development and science and technology. Acharya N.G. Ranga Agri. Univ. Rajendranagar. Andhra Pradhesh, Jan-3-7 .pp.35-39.

5.2. *Alstonia scholaris*

Common names: Blackboard tree, Indian devil tree, Dita bark, Milkwood pine, White cheesewood, Pulai, birrba, Milk wood, Milky pine.

Botanical name: *Alstonia scholaris* R. Br. Syn. *Echites pala* Buch-Ham. ex Spreng, *Echites scholaris* L. *Pala scholaris* (L.) Roberty, *Tabernaemontana alternifolia* Roxb.

Family : Apocyanaceae

Vernacular names: Chatiyan (Assamese); Chattim (Bengali); Saptaparna, Satvana (Gujarati); Chattiyan, Satawana, Saitan ka jat (Hindi); Maddale, Hale, Eleyalaga (Kannada); Daivaphal, Ezhilam Paala, Pala (Malayalam); Satveen,

Saitan, Satvin, Saptaparna, Shatin (Marathi); Chbatiana, Chatiana (Oriya); Sathi, Satanna (Punjabi); Yakshippala, Daivappala, Saptaparna, Sapthachada, Visalatvak, Chatraparna, Suktiparna, Sarada (Sanskrit); Eezhilai paalai, Eezhilai kalli, Pey maram (Tamil); Edakula Ponna (Telugu).

Origin and distribution

It is native to the following regions:

- China: Guangxi, Yunnan.
- Indian subcontinent: India, Nepal, Sri Lanka, Pakistan
- South East Asia: Cambodia, Myanmar, Thailand, Vietnam, Indonesia, Malaysia, Papua New Guinea, Philippines
- Australia: Queensland

The species is found in the sub-Himalayan tract from Yamuna eastwards, ascending upto 1000m. It occurs in tropical, subtropical and moist deciduous forests in India and is widely cultivated as avenue tree throughout India.

Habitat: It is naturalized in tropical and subtropical climates. Altitude: 0-900m, Mean annual temperature: 12-32°C, Mean annual rainfall: 1200-1400mm, Soil type: Favourable soils include alluvial, basaltic red earth, yellow earth with grey-brown topsoil, stony red earth on basic volcanic soils, sandy grey earth, brown earth from a volcanic mixture of rocks.

Habit: It is a medium sized evergreen tree, usually 12-18m high, sometimes upto 27m high, with close-set canopy (Fig. 1).

Stem: Bark is rough, greyish white, yellowish inside, and exudes bitter latex when injured.

Leaf: Leaves in whorls of 4-8 in the upper axils; leaf stalks 1-1.5cm long, the lamina obovate to elliptical or elliptical-lanceolate, glabrous or sparsely hairy, tapering towards the base, 11.5-23 x 4-7.5cm. Upper surface is dark green, the lower green- white with 25-40 pairs of lateral veins on each side of the midrib and 2-6mm apart. The tip of the leaf is rounded or shortly pointed, tapering towards the base.

Flower: Cymes are dense, fragrant, greenish-white or greyish-yellow, pubescent and umbrella- shaped; peduncle is 4-7cm long. Pedicels are usually as long as or shorter than calyx. The corolla is white and tube like, 6-10mm; lobes are broadly ovate or broadly obovate, 2-4.5mm, overlapping to the left. The ovaries are distinct and pubescent. The follicles are distinct and linear (Fig. 2).

Flowering, pollination and fruiting: The trees are often deciduous at irregular intervals. They do not flower at every leaf change, but only after marked periods of dry weather. The large branches provide favourable nesting sites for wild bees. Flowering and fruiting occur from March to July, extending to August in subtropical climate (Anon., 2008). Pollination is by insects; when flowering, butterflies and bees often surround trees.

Fig. 1. Plant

Fig. 2. Flower

Fig. 3a. Immature fruit

Fig. 3b. Mature fruit

Fig. 4. Seed

Fruit: Follicles (fruits) are narrowly cylindrical, fascicled, with seeds possessing brown hair. Its slender pendulous fruit (pod) which hangs down in clusters from the branches (Venkatesh, 1976) can be as long as 30-60cm (Fig. 3a and 3b).

Seeds: Seeds 6mm long, linear-oblong, flattened, rounded and with a fringe of hairs at both ends, the hairs longer than the seed (Fig. 4).

Propagation: By seeds and grafts. One hundred gram seed contains 3000-6000 seeds.

Seed germination: Based on the seed size, this species may show orthodox seed storage behaviour. Seeds remain viable for more than a year, but the percentage of germination declines from 90 per cent to about 40 per cent after one year (Anon., 2008).

Seed collection: Seeds are difficult to collect, as the fruits open while still on the tree. Fruits may be collected for seeds during summer before splitting of thin and wiry pods.

Seed dispersal: The fruits open on the tree and the seeds, which have a tuft of silky hairs at each end, are dispersed by wind.

Vegetative propagation : Cleft grafting is found to be most appropriate.

Economic part used: Bark, root.

Alkaloids: Ditamine, echitenine and echitamine.

Medicinal uses: In India, the bark is used for medicinal purposes, ranging from malaria and epilepsy to skin conditions and asthma. In Ayurveda, it is used as a bitter and as an astringent herb for treating skin disorders, malarial fever, urticaria, chronic dysentery, diarrhoea, in snake bite and for upper purification process of Panchakarma. The milky juice of the tree is applied to ulcers. At one time, a decoction of the bark was used to treat diarrhoea and malaria as a tonic, febrifuge, emmenagogue, anticholeric and vulnerary. A decoction of the leaves was used for beriberi. Ayurveda recommends for bowel complaints.

Other uses : In Sri Lanka, its light wood is used for coffins. In Borneo, the wood close to the root is very light and of white colour, and is used for net floats, household utensils, trenchers, corks, etc.

REFERENCES

Anonymous. 2008. Agro-techniques of selected medicinal plants, Vol. 1. National Medicinal Plants Board, New Delhi.

Venkatesh, C. 1976. Our tree neighbours, National Council for Educational Research and Training, New Delhi, p. 84.

5.3. *Artocarpus hirsutus*

Common name: Wild Jack fruit

Botanical name: *Artocarpus hirsutus*

Family: Moraceae

Vernacular names: Ranphanas (Hindi); Hebbalasu, Adavi halasu (Kannada); Jeev kadgi (Konkani); Ayani, Anjili, Ayini plavu, Annili (Malayalam); Parphanas (Marathi); Lakucha (Sanskrit); Kaattupalaa, Ayini pala (Tamil); Adavi panasa (Telugu); Deevuda gujji (Tulu).

Origin and distribution: It is endemic to the Western Ghats of the southern states of Karnataka, Kerala, Maharashtra and Tamil Nadu.

Habitat: It is found from the sea level to 1300m above sea level in semi-evergreen and evergreen forests, especially along the rivers.

Habit: This is a lofty evergreen tree, upto 40m tall, which exudes a milky white, sticky latex when injured and the bark is brownish grey (Fig.1).

Leaf: Leaves are arranged in alternate fashion, lamina three lobed and velvetty when young, entire and glabrous on maturity, broadly elliptic in shape, margins entire or wavy and hairy beneath.

Flower: Flowers are monoecious on axillary pedunculate receptacles; the male receptacles narrowly cylindric, upto 15 x 1cm; the female receptacles 15 x 9cm. Perianth lobes two, shortly connate below in male flowers; in female tubular, one stamen (Fig. 2).

Flowering, pollination and fruiting: Flowering and fruiting occurs during January - May. The flowers are cross-pollinated, but pollination is not required for fruit development.

Fruit: Fruits are called as syncarps or sorosis, globose to ovoid, covered with spine like projections and orange yellow when ripe (Fig. 3a, 3b and 3c).

Seed: Seeds are many, smooth, 1.8cm long and ovoid in shape (Fig. 4).

Propagation: It can be propagated both by seeds as well as vegetative means.

Seed propagation: Fresh seeds are collected during May - July. Seeds are reported to have viability for only two weeks. Therefore, fresh seeds are sown immediately. Germination starts after a week and continues for two months. About 65-70 per cent germination can be expected.

Vegetative propagation: In case of vegetative means, base of the stem cuttings are dipped in rootex powder and planted in sandy mother bed. They are then covered with a polythene spread sheet to induce humidity. Planting stem cuttings in a mist chamber gives better result in rooting. Rooting is observed in 70-80 per cent of stem cuttings. Two months old rooted cuttings are transplanted.

Economic part used: Fruits.

Medicinal uses: The ripe fruit is sweet in taste and is used to reduce the body heat. Fruits are also reported to be aphrodisiac. Infusion of the bark is applied to cure pimples, cracks on skin. Powdered bark is used to heal sores. Dry leaves are used to treat swelling of lymph glands.

Fig.1. Plant

Fig.2. Male flower

Fig.3a. Mature fruit

Fig.3b. Ripe fruit

Fig.3c. Fruit opened

Fig. 4. Seeds

Other uses: Its ripe fruit is edible; ripened cooked fruit is used in the kitchen; leaf, bark, ripe fruit and latex are used as medicine; fresh leaves are used as fodder for elephants; fallen leaves (during decaying process release nutrients like nitrogen, phosphorus and potassium) are reported as the best organic mulching for growth of plants; wood is used in house constructions, boat building and in making furniture. The latex from the bark has an insect repellent property and in some places, it is used as wood coating to prevent insect attack. The wood, wood peelings, small branches and stumps of the tree may be used as firewood or fuel because of their good calorific value.

They invite varieties of birds, butterflies and other creatures which facilitate the process of pollination in gardens. This tree will also act as good windbreaks.

5.4. *Boswellia serrata*

Common names: Indian Frankincense, Indian Oblibanum tree

Botanical name: *Boswellia serrata* Roxb. ex Colebr. Syn. *B. thurifera* Roxb., *B. glabra* Roxb., *B. serrata* Roxb. var. *glabra* (Roxb.) Bennet

Family : Burseraceae

Vernacular names: Salai, loban (Hindi, Marathi); Bili dhoopa (Kannada); Kunthrikam (Malayalam); Shallaki (Sanskrit); Ven kungiliam (Tamil); Tella guggilamu (Telugu).

Origin and distribution: This is endemic to India. In India, it is reported from Punjab, Uttar Pradesh, Madhya Pradesh, Rajasthan, Maharashtra, Andhra Pradesh, Karnataka and Tamil Nadu.

Habitat: It is especially abundant in forests of central India (Venkatesh, 1976). The tree commonly grows in deciduous forests from lower elevations to 1150m above sea level, in dry, rocky ferruginous soil and occasionally forms major component of the deciduous forests.

Habit: This is a deciduous tree that reaches upto 15m in height with spreading crown and somewhat drooping branches. All parts of plants emit faint fragrance (Fig. 1).

Stem: The bark is greenish yellow in colour, smooth; exfoliating in thin, papery flakes and exudes fragrant resin when injured. Branchlets are soft and hairy.

Leaf: The leaves are compound, clustered at the ends of branchlets, upto 20cm long and 6cm broad; individual leaflets are 10-15 pairs, 2.5-7cm long and 1-2cm broad, oblong to lanceolate in shape, base of leaflets is unequal, apex is blunt; petioles 1.2cm long; primary stalk of leaves densely soft hairy.

Flower: Inflorescence branched, 15cm long, borne in leaf axils. Flowers are 1cm across; petals pinkish white with pink tinge (Fig. 2).

Flowering, pollination and fruiting: Flowering and fruiting occurs during January to May. This tree is leaf less when flowering, worth seeing the new foliage that begins when the tree is in fruiting stage.

The pollination mechanism is cross pollination and pollinated by honey bees, particularly *Apis dorsata* and *A. indica*.

Fruit: Fruits are drupe, to 1.5 x 1cm, three gonous, splitting along three valves, yellowish brown (Fig. 3).

Seed: Seeds are three, heart shaped and solitary in each locule.

Fig. 1. Plant

Fig. 2. Flower

Fig. 3. Fruit

Propagation: By seeds and vegetative means.

Seed propagation: Fresh seeds are obtained from dried fruits (before dehiscence) during April- May. They weigh 13,400-25,600 seeds/kg. Seeds are stored in dry tins, for not more than 6-9 months. Since seeds have poor viability and can be stored only for six months, fresh seeds should be sown immediately during the months of April - July. Overnight soaking of the seeds in cold or mild hot water before sowing in mother bed will enhance germination percentage. Germination begins after a week and continues for another couple of weeks. 40-70 per cent of germination was observed in this method. One hundred gram seed contains 1400-

1500 seeds. Eight to ten week old seedlings are transplanted in poly bags or containers. Six month old seedlings are transplanted in the field.

Seed storage: Seeds are stored in dry tins, for not more than 6-9 months.

Vegetative propagation

Stem cuttings: Stem cuttings (15-20cm long) with finger thickness can be planted. Tip of the cuttings should be covered with cow dung or red earth or can be coated with white oil paint. Sprouting starts within 15 days while rooting begins after a month. About 76 per cent of the cuttings sprout and 70 to 80 per cent establish.

Root suckers: Suckers are selected from trees of 45 to 60cm girth and roots of 5 to 7.5cm girth. Suckers are planted in June or July. Sprout initiation is observed in about a fortnight to 30 days.

Economic part used: Bark

Alkaloids: Boswellic acid, boswellin.

Medicinal uses: Effective for back pain. It is also used in skin diseases, jaundice, ulcers, piles, diaarrhoea and dysentery. They improve blood flow in the joints. And they also block chemical reactions that set the stage for inflammation to occur in chronic intestinal disorders such as Crohn's disease and ulcerative colitis.

Other uses: This is an important tree in tricky urban landscaping situations. Its uniform branching with round shape open crown best suits for town planning.

REFERENCE

Venkatesh, C. 1976. Our tree neighbours, National Council for Educational Research and Training, New Delhi, p. 84.

5.5. *Butea monosperma*

Common name: Flame of the forest, Bastard teak, Parrot tree

Botanical name: *Butea monosperma* (Lam) Kuntze. Syn. *B. frondosa* Koen. ex Roxb., *Erythrina monosperma* Lam.

Family: Fabaceae

Vernacular names: Polash, Polax, (Assamese); Kinshuk, Polash, Polashi (Bengali); Pauk (Burmese); Khakda (Gujarati); Chichra tesu, Desuka jhad, Dhaak, Palaash, Chalcha, Kankrei (Hindi); Mathuga, Thoras (Kannada); Muriku, Shamata (Malayalam); Palaash (Marathi); Keshu (Punjabi); Palasah, Palasa (Sanskrit); Porasum, Parasu (Tamil); Modugu (Telugu); Palashpapra (Urdu).

Origin and distribution: Native to tropical and sub-tropical parts of the Indian subcontinent and southeast Asia, ranging across India, Bangladesh, Nepal, Pakistan, Sri Lanka, Myanmar, Thailand, Laos, Cambodia, Vietnam, Malaysia and western Indonesia.

Habitat: It is common in mixed dry deciduous forests and in open grass lands and scrub forests. This tree is recommended for afforestation in water logged areas.The plant can be grown on black cotton soils, clay loams and even on saline soils. It survives in areas with as little as 450mm annual rainfall or as much as 4500mm.

Habit: It is a medium sized dry season deciduous tree (Fig. 1).

Stem: Deciduous small to medium sized tree with showy, flame coloured flowers, open crowned, with a twisted trunk. The bark is fibrous and grey or light brown in colour. It is growing to 15m tall. It is a slow growing tree; young trees have a growth rate of a few feet per year.

Leaf: The leaves are trifoliate, rough textured and large with an 8–16cm petiole and three leaflets, each leaflet 10–20cm long. The tree sheds its leaves during November - December and new leaves appear in April.

Flower: The flowers are 2.5cm long, bright orange red and produced in racemes upto 15cm long. These flowers, which are scentless, are massed along the ends of the stalks - dark velvetty green like cup shaped calices and the brilliance of the stiff, bright flowers is shown off to perfection by this deep, contrasting colour. Each flower consists of five petals comprising one standard, two smaller wings and a much curved beak shaped keel. It is this keel which gives it the name of Parrot tree (Fig. 2).

Flowering, pollination and fruiting: The flowering and fruiting occurs during February-April. Phenological studies indicated that although the species shows a regular flowering season, all trees do not flower every year. Flowers are typically papilionaceous; the stigma is wet papillate and the style is hollow. The flowers show characteristics of bird pollination being large and bright orange red in colour with copious amounts of nectar, and exhibiting diurnal anthesis. Although the flowers are frequented by as many as seven species of birds belonging to six families, only one species, the purple sunbird (*Nectarinia asiatica*), is the effective pollinator. The flowers are also pollinated by the three striped squirrel (*Funambulus tristiatus*). Unlike other flower visitors, these two pollinators forage the nectar from the open side of the keel (legitimate path) during which pollen grains are deposited on their body parts. After the first visit of a sunbird or a squirrel, virgin flowers showed pollen load on the stigma and developed into fruits. *B. monosperma* shows a weak form of self incompatibility. Fruit set following manual self-pollination (5.25%) was comparable with open pollination (approx. 5%), but was significantly lower than manual cross pollination (22.51%). This indicates that there is a high degree of geitonogamous pollination in this species, which may lead to a weakening of self-incompatibility as a means of reproductive assurance (Tandon *et al.*, 2003).

Fruit: The fruit is a pod 15-20cm long and 4-5cm broad. The fruit is a flat legume; young pods have a lot of hair a velvetty cover. The mature pods hang down like peculiar legumes. Pods are splitting round the single apical seed, lowest part indehiscent (Fig. 3).

Seed: The seeds are flat, brownish, reniform and curved (Fig. 4).

Propagation: By seeds and root suckers.

Fig. 1. Plant Fig. 2. Flower

Fig. 3. Fruit Fig. 4. Seed

Seed propagation: The seeds mature in May. Ripe pods are collected off the trees by stick in May. Seeding time is April- June. The seeds are extracted from the pods during April-June. Then, the seeds are to be sown immediately after collection, as they lose viability within 5-6 months. The seeds can directly sow in rows 3-5m apart. Pod segments containing seeds are sown in polybags in May. In nursery sowing, 15cm spacing must be given within and between the rows. Normal germination in the nursery takes 15 days; 75-80 per cent is the germination success. One hundred gram seed contains 1000-1500 seeds.

Vegetative propagation: For stump planting, stumps with about 5cm long shoot and about a 30cm long root obtained from one year old seedlings (root suckers) are used. Stumps have to be prepared at the onset of rains. Stump plants show higher percentage (80%) of survival and satisfactory growth. July-August is the favourable time for transplanting.

Economic part used: Stem, leaves, seeds

Alkaloids: The gum contains leucocyanictin. Bark also contains kinnotannic acid and gallic acid, flavonoid, glycosides, butin, butrin, isobutrin, lupeol and β-sitosterol. Flowers contain flavones and flavonoid constituents including butrin, isobutrin, butein, butin, coreopsin, sulphurein and others, chalcones and amino acids. Seeds contain a fixed oil, albuminoid substances, lectins a highly active anthelmintic principle, palasonin, free amino acids, alkaloids and enzymes.

Medicinal uses: The bark is aphrodisiac, laxative and anthelmintic; useful in dysentery, piles and hydrocele; weak decoction of bark is useful in catarrh, cold and cough. Gum of the bark is an excellent astringent to the bowels; useful in diarrhoea, dysentery, stomatitis, cough, pterygium and corneal opacities; cures excessive perspiration. The seeds are laxative and anthelmintic; paste of it is a good remedy for Ascaris and ringworm. The leaves are astringent, carminative, anthelmintic, aphrodisiac and tonic; used externally to disperse boils, pimples, worms and piles. Flowers are astringent to the bowels, depurative and diuretic; cures leprosy, strangury, gout, skin diseases and burning sensation (Rani and Reddy, 2009).

Other uses: It is used for timber, resin, fodder and dye. The wood is dirty white and soft and, being durable under water and is used for well curbs and water scoops. Good charcoal can be made from it.

REFERENCES

Rani, A. S. and K. J. Reddy 2009. Folklore medicinal uses of some indigenous plants among the tribes of Telangana region, Andhra Pradesh, India. *Ad. Plant Sci.*, 22(1):199-204.

Tandon, R., K. R. Shivanna and H. Y. M. Ram 2003. Reproductive biology of *Butea monosperma* (Fabaceae). *Annals Botany,* 92: 715-723.

5.6. *Caesalpinia sappan*

Common names: Sappan wood, False sandal wood, Indian brazil wood, Indian redwood

Botanical name: *Caesalpinia sappan* L. (Syn. *Biancaea sappan* (L.) Tod.)

Family: Caesalpiniaceae

Vernacular names: Bokmo, Bakan, Patungam, Vakum, Vakam, Patunga (Hindi); Patanga-Chekke Sappanga (Kannada); Pathanga, Rakthasara, Sranga,

Ranjana, Pattaranjaka, Pathura, Kuchandana (Sanskrit); Pathimugam, Pathangam, Pathanga, East India chemmaram (Tamil).

Origin and distribution

Native: China, India, Malaysia, Myanmar, Thailand

Exotic : Indonesia, Papua New Guinea, Philippines, Solomon Islands, Sri Lanka, Taiwan, China, United States of America.

In India, it is commonly cultivated in Tamil Nadu, Kerala, Karnataka, Andhra Pradesh and West Bengal.

Habitat: It grows well in all kinds of soil and lush growth is obtained in red soil. It withstands any amount of drought and defoliates only for a short period of 10-15 days. Under natural conditions, it grows mostly in hilly areas with clayey soil and calcareous rocks at low and medium altitudes. In Peninsular Malaysia, it grows best on sandy riverbanks. It does not tolerate too wet soil conditions.

Habit: It is a small to medium sized, thorny tree, growing upto 10m with conspicuous spines on the stem and leaf rachis (Fig. 1).

Stem: Branchlets are dull, lenticellate, usually armed with paired, recurved stipular prickles; trunk upto 14cm in diameter; bark with distinct ridges and many prickles, greyish brown; young twigs and buds hairy, brownish.

Leaf: Leaves stipulate, bipinnate, alternate, 20-45cm long, 10-20cm broad, with 8-16 pairs, upto 20cm long pinnae; pinnae with prickles at the base and with 10-20 pairs of oblong, 10-20mm x 6-10mm long, sub-sessile leaflets, very oblique at base, rounded to emarginated at apex.

Flower: Flowers in terminal panicles, racemes pubescent, primary peduncles 30-40cm long, bracts ovate acuminate, about 6mm long, flowers fragrant, 2-3cm long, pentsa merous; sepals glabrous, petals pubescent, the superior one smaller; calyx tube 3mm long; corolla yellow, uppermost lobes cuneate, other obovate, all clawed and gland punctate; stamens ten, filaments densely tomentose in the lower half; ovary superior, pubescent (Fig. 2).

Flowering, pollination and fruiting: Flowering can occur after one year of growth and usually during the rainy season, fruiting about six months later. It was pollinated by bees and butterflies. The pods will develop 5-15days after fertilization and matures in three months.

Fruit: Fruit is a dehiscent pod, glabrous, thick, flattened, obliquely oblong, prominently beaked, woody, polished-brown, 7-10cm x 3-4cm, 2-3 seeded (Fig. 3).

Seed: Seeds are ellipsoid, flattened, 18-20mm x 10-12mm, dark brown seeds (Fig. 4).

Propagation: Propagated by seed or cuttings which should be 2.5cm diameter and 10cm long.

Fig. 1. Plant

Fig. 2. Flower

Fig. 3a. Mature fruit

Fig. 3b. Dry fruit

Fig. 4. Seeds

Seed dormancy and germination: Usually mature pods burst open in the dry season and scatter the seeds, which remain dormant until the start of the rainy season. Pods are gathered, pounded and put into cold water. After 2-3 hours, the mixture is rubbed and mixed with a solution of iron sulphate. Seed germination occurs readily, but is enhanced by dipping seeds wrapped in cotton cloth in boiling water for five seconds. Germination rate is then 90 per cent.

The seeds should be treated with diluted H_2SO_4 for six minutes. Concentrated H_2SO_4 and soaking for a longer time in acid may damage the radicle and the plumule. This should be followed by washing well in running water. The seeds should be shade dried for 24h and again soaked in 300ppm GA_3 for six hours to promote rapid germination (71%). If seed treatment is not carried out, the seeds have the germination of 51 per cent.

Seed collection: Seeds for raising the nursery are collected in November-December.

Seed storage: Seeds are viable upto two years from harvesting.

Economic part used: Stem core

Alkaloids: The pod shell contains four per cent and bark of the plant contains 1.8 per cent tannins along with aromatic compounds such as brazilin, sappancha-lcone, ceasalpin-J, caesalpin-P, and protosappanin A and B (Anon., 2008).

Medicinal uses: It is emmenagogue, haemostatic and antiinflammatory. It is one of the ingredients of indigenous drug 'lucol', which is administered for the treatment of non-specific leucorrhoea. The heartwood is used in traditional medicine as a treatment for contusion and thrombosis. The paste of the wood is used in curing rheumatism, haemorrhages and wounds. The plant has anti-cancerous and anti-diarrhoeal properties.

It possesses medicinal abilities as an anti-bacterial and anti-coagulant properties. It is widely used to treat leprosy, mania, fits and diarrhoea (Rani and Reddy, 2009).

Other uses: It also produces a valued type of reddish dye called brazilin, used for dyeing fabric as well as making red paints and inks. It can be used as a fuel wood also.

REFERENCES

Anonymous 2008. Agro-techniques of selected medicinal plants, Vol. 1. National Medicinal Plants Board, New Delhi.

Rani, A. S. and K. J. Reddy 2009. Folklore medicinal uses of some indigenous plants among the tribes of Telangana region, Andhra Pradesh, India. *Ad. Plant Sci.*, 22(1):199-204.

5.7. *Commiphora wightii*

Common names: Indian Bedellium tree, Gum Gugal

Botanical name: *Commiphora wightii* (Arn.) Bhandari Syn. *C. mukul* Wt. and Arn., *Balsamodendron mukul* Hook. ex Stocks

Family: Burseraceae

Vernacular names: Guggul (Assamese, Bengali, Gujarati, Hindi, Punjabi); Guggule (Kannada); Guggul Dhoop, Kanth Gan (Kashmiri); Itinnil (Malayalam); Guggul, Mahishaksh (Marathi); Guggulu (Oriya); Koushikaha, Devadhupa (Sanskrit); Maishakshi, Gukkal (Tamil); Kondamamidi, Guggul, Makishakshi guggulu, Guggipannu (Telugu); Muqil (Urdu).

Origin and distribution: It is a small tree indigenous to India, growing wild in the semi-arid states of Rajasthan, Gujarat and Karnataka

Habitat: The plant grows wild in the arid, rocky tracts, also in coastal areas on sandy and sandy loam soils. The climate required is typically semi-arid, characterised by aridity of the atmosphere and extremity of temperature both in the summer and winter with average rainfall 356mm from July to September. The pH of soil may vary from 4 to 8.

Habit: Spiny shrub or small tree with many branches, usually growing two or three metre high (Fig. 1) and a healthy tree yields 250-500g of resin in one season, and typically begin yielding resin after five years. The plant remains leafless during winter season, *i.e.*, from October to March. New leaves sprout during April, are short lived and do not fall until September. A short spell of rainfall initiates leaf formation. It is an oleo-gum resin of pale brown or dull green colour, exudes from bark during winter season (November–February) (Anon., 2008).

Stem: It is much- branched, dioecious, upto 6m tall with brown coloured, spine scented knotty, crooked and spirally ascending branches ending in sharp spines. Bark shiny, ash to yellowish white coming off in rough flakes exposing the greenish underbark, which also peels off in thin papery rolls.

Leaf: Leaves small, sessile, rhomboid-(ob)ovate, 1-3 leaflets, highly aromatic, leathery, shinning green on top and greyish below with irregularly toothed edges, 1 to 5cm length and 0.5 to 2.5cm broad.

Flower: Flowers small, unisexual, sessile, brownish red, occurring singly or in groups of 2-3, 8-10 lobed disc and an oblong-ovoid ovary; stamen 8-10 (Fig. 2).

Flowering, pollination and fruiting: The flowering and fruiting take place throughout the year. However, maximum fruiting is observed from January to April.

In this species, andromonoecious, female and male plants were recorded. Female plants set seed irrespective of the presence or absence of pollen. Hand pollination experiments and embryological studies have confirmed the occurrence of non-pseudogamous apomixis, nucellar polyembryony and autonomous endosperm formation (Gupta, 1996).

Fruit: Fruits are pulpy drupe, ovoid, upto 1cm long, marked with two white longitudinal lines, mucronate, mesocarp yellow, rarely orange four lined and fused at the base; epicarp dehiscing from the base upwards, when ripe it become red and split into two (Fig. 3).

Seed: Seed generally contain an under developed embryo (Fig. 4).

Propagation: By seed, cuttings and layers

Seed propagation: Black coloured seeds may be used as planting material, although germination averages around 40 per cent only. Yellowish white seeds are non-viable. One to four seedlings emerge from a single seed due to polyembryony. However, due to poor germination two seeds are sown in each polybags. The seeds sown in the months of June and July in well worked soil beds, germinate at the onset of monsoon rains within 7–10 days after sowing. One month old seedling are transferred into plastic bags after two years, these seedling attain a height of 30-50cm and suitable for field planting.

Seed maturation and germination: Seeds are the result of apomixis so their formation is very irregular. Seed maturation depends upon the development of cellular endosperm. The endocarp of the fruit is very hard and does not allow entry of water, exchange of gases and expansion of embryo. So the seeds if not taken out from the endocarp or if the endocarp is not made porous, the germination of seed may not take place at all. So, under natural conditions germination percentage is very low.

Seed collection: Seeds can be collected during February-March.

Seed treatment: Better seed germination can be obtained by mechanical scarification (with sand paper) of seed coat and keeping these scarified seeds under running water for about 24h.

Vegetative propagation

Stem cutting: It is successfully propagated through leafless stem cuttings prepared in May. Cuttings are obtained from healthy and disease free branches and usually planted during late summer, when the plant is almost leafless. With the onset of monsoon, when foliation growth/flowering starts, the plant becomes physiologically active and cutting also show signs of sprouting in 25-50 days. Stem cuttings planted without any treatment show approximately 90 per cent rooting. However, cuttings can be pre-treated with guggulu solution @100ppm or IBA @ 10ppm to hasten rooting and to achieve better survival rate (Anon., 2008). The plants after establishing in the nursery beds are transferred into plastic bags and then planted in the field during rainy season.

Air layering : For air layering, 5-8 years old mother plants are selected in July-August. Firm and healthy branch, about 60-90cm long is selected and 5cm ring on the wood is marked. The bark of the marked portion is removed from the stem. The exposed portion is covered by applying a ball of adhesive soil, holding it securely together with coir fibre, bandaging firmly all round the branch. This bandaged portion is supplied with water drop by drop through the drip system. Temperature, moisture and oxygen directly influence the rate of cell division and enlargement, lack of light stimulate the production of root. Growing points are formed and develop rapidly on young wood. In two to three months when young adventitious roots are seen protruding through the layering, it may be severed from the parent plant and planted.

Fig. 1. Plant

Fig. 2. Flower

Fig. 3. Fruit

Fig. 4. Seed

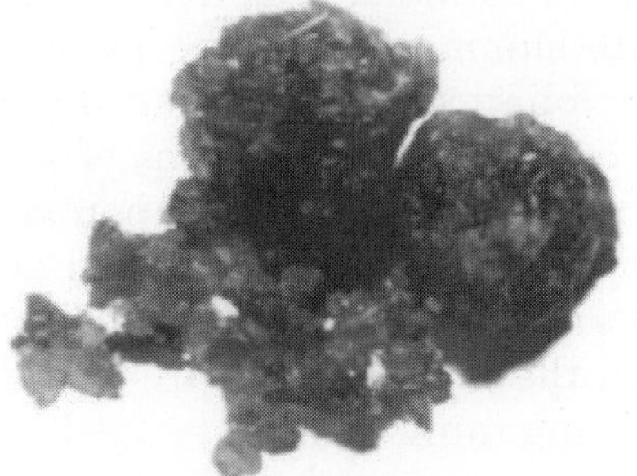

Fig. 5. Resin

Economic part used: Gum exudates from stem (Fig. 5).

Alkaloids: The gum resin contains 3.2 per cent gum and 19.5 per cent mineral matter chiefly consisting of silicon dioxide, magnesium, calcium, iron, and aluminium. It also contains about 1.5 per cent essential oil, which contains 6.5 per cent myrcene and 11 per cent dinyrcene. A number of steroids have also been isolated from guggul like Z-guggul sterone and E-guggulo sterol I, II, and III (Anon., 2008).

Medicinal uses: Used for curing obesity, arthritis, bone fracture, muscle pain, swellings, anemia, aphrodisiac, diabetes, eczema and tooth problems.

REFERENCES

Anonymous 2008. Agro-techniques of selected medicinal plants, Vol. 1. National Medicinal Plants Board, New Delhi.

Gupta, P., K. R. Shivanna and H. Y. M. Ram 1996. Apomixis and polyembryony in the guggul plant, *Commiphora wightii*. *Annals Botany*, 78: 67-72.

5.8. *Emblica officinalis*

Common name: Indian Goose berry, Aonla

Botanical name: *Emblica officinalis* Gaertn. Syn. *Phyllanthus emblica*

Family: Euphorbiaceae

Vernacular names: Amlakhi (Assamese); Amloki (Bengali); Aamla (Gujarati, Hindi, Oriya); Nellikayi (Kannada); Embali, Amli (Kashmiri); Nellikka (Malayalam); Heikru (Manipuri); Aavnlaa (Marathi); Aula (Punjabi); Amalika (Sanskrit); Nelli, Amalagam, Alagam, Ambal, Amarikam, Thaththari, Miruthupala, Meethunthu (Tamil); Usiri, Usrikayi (Telugu); Amla, Amlaj (Urdu).

Origin and distribution: Aonla is indigenous to tropical South-east Asia, particularly central and southern India it is under cultivation since ancient times (Firminger, 1947). Under natural habitat, it is found in dry deciduous forests of India. In India, its cultivation is very common in the eastern districts of Uttar Pradesh particularly Pratapgarh and Varanasi. It is also grown in states like Haryana, Himachal Pradesh, Maharashtra and some parts of Karnataka.

Habitat: The mature plants can tolerate both freezing temperature as well as temperatures as high as 46°C. It has been observed that warm season is conducive for the initiation of floral buds. The young seedlings are frost and drought sensitive. The soils having a pH of 7-9.5 are ideal for its cultivation. It grows in a wide range of soils including mildly alkaline, shallow, gravely soils, but prefers deep, moist and loamy to alluvial soils.

Habit: The tree is small to medium in size, reaching 8 to 18m in height, with a crooked trunk and spreading branches. The branchlets are glabrous or finely pubescent, 10–20cm long, usually deciduous (Fig. 1).

Leaf: The leaves are simple, sub-sessile and closely set along branchlets, light green, resembling pinnate leaves.

Flower: The flowers are minute and axillary. The male flowers in aonla appears in clusters in leaf axils over the entire branchlets, while female flowers are positioned mainly at the distil end on same of these branchlets (Bajpai, 1968). The male flowers light in colour whereas female flowers are usually greenish turning brown later. Aonla consists of perianth, androecium and gynoecium. The androecium comprises of three stamens each opening vertically. The small green perianth is usually six segmented (5-70) constitutes the female flower. Ovary is tricarpellate, disk usually evident and lanceolate, short style, stigma either once or twice bifurcated and distally dilated (Ram, 1971:1974) (Fig. 2).

Flowering, pollination and fruiting: Flowering occurs during spring and fruiting period is autumn. Aonla trees flower twice in South India, once in February to March and again in June to July with very poor crop in later case. Mixed planting of four to five varieties is must for proper pollination. The aonla pollen being light and non-sticky, pollination may occur through wind. Besides wind, cross pollination is accomplished by wasps and honey bees (Reddi and Janaki Bai, 1979) and more so in late evening hours by honey bees (Bajpai, 1957). In Himachal Pradesh, *Apis mellifera* was observed to be more efficient than *A. cerana*, which visited aonla tree during flowering mainly to correct nectar and the crop is a minor source of honey.

When hot winds blow fruitset is often a problem. Bajpai (1968) observed that flower bud differentiation in cv. Banarasi takes place in the first week of March. Male flowers appear in clusters on the axil of leaf all over the branchlet while female flowers on the upper end of a few branchlets only. The maximum number of male flowers open between 6-7pm and dehiscence of anthers occurs soon or about 10-15 minutes after anthesis. The female flowers open gradually and it takes 72h to open completely. The stigma becomes receptive on the third day of anthesis. Self incompatibility is not found, but poor fruit set may be due to higher percentage of staminate flowers. Sometimes only 0.06 per cent female flower is seen in Banarasi variety which can be increased by Tri-iodo-benzoic acid (TIBA) spray before flowering.

Fruit: The fruits are fleshy, nearly spherical, 2cm in diameter, light greenish yellow, quite smooth and hard on appearance with six vertical stripes or furrows which ultimately splits into three portions, each containing usually two seeds (Venkatesh, 1976) (Fig. 3).

Seed : Seeds are in unequal pairs, 3.9-4.5mm long (Fig. 4).

Propagation : This can be propagated by seeds as well as by vegetative means *viz.,* budded or grafted saplings.

Fig. 1. Plant

Fig. 2. Flower

Fig. 3. Fruit

Fig. 4. Seed

Seed propagation: One hundred gram seed contains 6800-8900 seeds. Seedlings raised from seeds are also used as rootstocks.

Seed collection: The berries are harvested by hand after climbing to upper branches bearing the fruits. The ripe fruits are collected during January and dry in the sun.

Seed dormancy: Freshly harvested seeds of aonla do not germinate even if exposed to favourable conditions of germination owing to seed dormancy. Dormancy may be because of internal (physiological) factors affecting embryo or morphological factor such as hard, thick testa, or due to incorrect storage or handling (secondary dormancy). Such seeds may require special treatments like stratification, scarification, soaking in water, growth regulators etc. for overcoming dormancy.

Bhujbal (1975) reported highest germination per cent (92.50) with minimum period when dried stones of aonla were treated with 500ppm GA_3. Stratify the fresh seeds in sand moist to 60 per cent with KNO_3 5g lit^{-1} kept at 5°C for ten days to remove the morphological dormancy. Dry storage of fresh seeds for ten months can also remove this dormancy. Dhankar and Singh (1996) observed that treatments with GA_3 at 250ppm were found effective for better germination.

Seed germination: After removing the fleshy portion of the fruit, soak the seeds in water for 24h and wrap in a wet gunny cloth till roots emerge. Soaking the seeds in hot water for 5min. hastens the germination. Once the fleshy part of the fruit is removed, the seeds will be viable for a month. The fresh seeds are sown immediately due to its low viability on storage. Sowing at the time of commencement of rainy season will give best results.

After one month, the seedlings are transplanted in polybags. Seeds germinate in about three to four days and there is about 70 to 80 per cent germination. Seedlings raised from seeds are used as rootstocks.

Seed development and maturation: Seeds attain physiological maturity 23 weeks after germination of fruit dormancy of five months when the fruit colour turns to yellowish green and seed colour chestnut brown.

Seed extraction: Amla seeds can be extracted by soaking the fruits in 30 per cent brine solution during night followed by drying during the day and repeated for three days. The mesocarp following soaking in brine solution remained fresh and can be used for pickle/ by products preparation.

Seed grading: Using 8/64" round perforated metal sieve, maximum seed recovery can be used to process seeds. The size graded seeds can be further upgraded by density grading using water to remove the light weighted empty seeds.

Seed storage: Seed treated with carbendazim 50% WP @ 2g/kg seed and halogen formulation (Bleaching powder + $CaCO_3$ + Arappu leaf powder @ 5:4:1) @ 3g/kg seed can be stored upto 18 months in cloth bag and at 24 months in moisture proof container.

Vegetative propagation: The advantage of the vegetative method of propagation lies in the fact that vegetatively propagated plants start fruiting in 5-6 years, while seedling trees take at least 10 years to fruit. Grafting, layering and budding are practiced. Shield/patch budding was tried successfully in aonla. It is important that proper mother plants are selected and buds are selected from a branch that has a good number of female flowers. Though it gives fairly higher percentage of success, it is cumbersome, time and labour consuming process and can only be done on seedling rootstocks of age, more than one year. Softwood grafting is reported to be easy, convenient to practice, involves simple skills and is a quick method of grafting. Amin (1978) tried softwood grafting in aonla and obtained success of 73.3 per cent in the month of August at Anand.

Economic part used : Fruits, seeds, flowers, roots and bark.

Alkaloids: Emblicanin A & B, puniglucanin, pedunculagin, 2-keto-gluconolact-one (Vitamin-C equivalents), ellageic acid, hexahydroxy-diphenic acid and conjugates.

Medicinal uses: The gallic acid present in fruit has anti-oxidant properties. The fruit is useful in haemorrhage, leucoderma and dysentery. Cures vomit, cough, bronchitis, tooth ache, head ache, gas problems, eye disorders, hair falling.

The seeds are used for alleviating asthma and stomach disorders.

Past work done on propagation of Aonla

Srimathi (1997) carried out the investigations in amla (*Emblica officinalis* Gaertn) to standardize the seed technological practices for elite seeding production and the findings are given below.

(a) The flowering season was January - February and the fruiting season was November - February. The type of inflorescence is racemose and the fruit set was 13.0 per cent.

(b) Seeds reached physiological maturity 22-24 weeks after the termination of fruit dormancy period of five months when the colour of the fruit turned to greenish yellow. Endocarp at this stage was translucent with chestnut colour of seed which had maximum dry weight, germination and vigour.

(c) Significant variation in seed germination and vigour was observed among individual trees and years of collection indicating the importance of plus tree selection for seed collection. The fruit bearing was regular.

(d) Fruits collected from top of the crown are best for seed purpose and also collected from base of branches are found good. The quality of seed was superior in the fruits collected from north direction. Middle aged trees of about 25 years outperformed the young trees by registering the highest germination and seedling vigour. Seeds collected from Bhavanisagar edged over other sources in respect to seed germination and vigour.

(e) Size grading of fruits was found beneficial where large size fruits gave higher recovery of good quality seeds. Collection of greenish yellow fruits either from the crown or from ground soon after fall was ideal for getting large quantity of quality seeds.

(f) Extraction of seeds by three repeated alternate cycles of soaking in 30 per cent sodium chloride solution gave more recovery of filled seeds (sinkers) with high germination and vigour.

(g) Grading of seeds could be done using 8/64" round perforated metal sieve for obtaining good quality seeds. By water floatation method, empty seeds could be eliminated. Further, upgradation of sinkers could be done by seed size grading.

(h) Graded seeds (8/64" retained) treated with thiram @ 4g + carbaryl 400mg kg^{-1}, packed in 700 gauge polyethylene bag and stored at 5°C maintained viability and vigour for 24 months with minimum deterioration.

(i) Seed dormancy (morpho-physiological nature) can be broken by stratifying in 60 per cent moist sand by adding 0.5 per cent KNO_3 solution, at 5°C for 10 days.

(j) For elite seedling production, graded seeds should be sown in pot mixture containing silt: sand: Farm Yard Manure in 2:1:1 ratio taken in large container (22.5 x 15cm). Addition of vesicular arbuscular mycorrhizae improved the seedling growth.

Studies on seed germination, softwood grafting and micro propagation were carried at Department of Horticulture, University of Agricultural Sciences, Dharwad during 2005-2006 by Nayaka (2006). In seed germination studies, the seeds scarified with concentrated sulphuric acid for 30 seconds were found to be the best one with very high per cent germination (80.39), taking least number of days for 50 per cent germination (12.27) and highest vigour index (1223) (Table 1 and 2). Propagation studies revealed that, season of grafting had profound influence on graft success. Maximum success per cent of 87.50 was recorded in November grafting followed by 66.69 per cent in case of December. Least success was recorded during June grafting (16.52%) (Table 3). Influences of rootstock age on success of grafts were significant. The higher per cent successful grafts (62.88) were observed in eight months old rootstocks followed by four months (56.25%) old rootstocks (Table 4).

Table 1. Effect of different seed treatments on days taken for 50 per cent germination and per cent germination in *Phyllanthus emblica* L.

Tr. No.	Treatments	Per cent germination	Days taken for 50 per cent germination
T_1	Water soaking for 24h	55.20 (48.01)	15.22
T_2	Soaking seeds in cow's urine for 24h	60.34 (50.99)	14.76
T_3	Soaking seeds in thiourea (2%) for 24h	67.47 (55.26)	14.35
T_4	Gibberellic acid 500ppm for 12h	68.00 (55.58)	13.36
T_5	Gibberellic acid 500ppm for 24h	72.46 (58.38)	12.69
T_6	Stratification at 5°C for 10 days	34.01 (35.69)	16.70
T_7	Acid scarification for 30 seconds	80.39 (63.75)	12.27
T_8	Control	32.33 (34.67)	16.68
	S.Em±	0.19	0.37
	CD at 1%	1.54	0.78

Values in the parenthesis indicate arcsine transformed value

Table 2. Effect of different seed treatments on root length, seedling length and vigour index in aonla

Tr. No.	Treatments	Root length (cm)	Seedling length (cm)	Vigour index
T_1	Water soaking for 24h	6.89	13.77	760
T_2	Soaking seeds in cow urine for 24h	6.83	11.27	679
T_3	Soaking seeds in thiourea (2%) for 24h	6.77	11.01	743
T_4	Gibberellic acid 500 ppm for 12h	7.13	13.56	982
T_5	Gibberellic acid 500 ppm for 24h	7.68	15.23	1035
T_6	Stratification at 5°C for 10 days	5.23	9.60	326

T_7	Acid scarification for 30 seconds	7.07	15.22	1223
T_8	Control	5.23	12.94	418
	S.Em±	0.20	0.32	19.11
	CD at 1%	0.81	1.33	78.95

Table 3. Effect of season on success of grafting, average number of leaves and shoot length in aonla

Tr. No.	Month of grafting	Per cent graft-take	Mean number of leaves		Average shoot length (cm)	
			30 DAG	45 DAG	30 DAG	45 DAG
T_1	November 2005	87.50 (69.33)	8.7	13.00	2.43	2.62
T_2	December 2005	66.69 (54.78)	7.7	11.53	1.93	2.17
T_3	January 2006	50.04 (45.05)	7.2	10.87	1.83	2.07
T_4	February 2006	31.25 (34.00)	6.3	9.47	1.93	2.13
T_5	March 2006	43.43 (41.25)	5.4	6.20	2.78	2.83
T_6	April 2006	31.27 (34.01)	5.2	5.97	2.99	3.13
T_7	May 2006	25.10 (30.08)	3.8	5.73	2.71	2.84
T_8	June 2006	16.52 (23.99)	3.2	5.13	1.68	1.83
	S.Em±	0.18	0.09	0.14	0.09	0.21
	CD at 1%	0.55	0.27	0.41	0.26	0.62

Values in the parenthesis indicate arcsine transformed value; DAG – Days after grafting

Table 4. Effect of age of rootstock on success of grafting, average number of leaves and shoot length in aonla

Tr. No.	Month of grafting	Per cent graft-take	Mean number of leaves		Average shoot length (cm)	
			30 DAG	45 DAG	30 DAG	45 DAG
T_1	3 months	34.62 (36.08)	7.07	8.87	1.74	2.47
T_2	4 months	56.25 (48.62)	9.73	10.02	2.92	3.31
T_3	5 months	45.00 (42.15)	6.60	9.60	2.31	2.64
T_4	6 months	44.62 (41.93)	6.67	8.60	2.52	2.88
T_5	7 months	38.92 (38.62)	6.47	9.37	2.25	2.43
T_6	8 months	62.88 (52.49)	6.60	8.47	2.41	2.57
T_7	9 months	26.77 (31.18)	7.27	7.87	1.63	1.88
T_8	10 months	42.88 (40.93)	7.33	9.00	1.52	1.72
	S.Em±	0.19	0.12	0.30	0.07	0.06
	CD at 1%	0.56	0.35	0.89	0.20	0.18

Values in the parenthesis indicate arcsine transformed value; DAG – Days after grafting

REFERENCES

Amin, R. S. 1978. Softwood grafting – a new technique for hardwood plants. *Curr. Sci.*, 47: 468-469.

Bajpai, P. N. 1957. Blossom bilogy and fruit set in *Phyllanthus emblica* L. *Indian J. Hort.*, 14:99-102.

Bajpai, P. N. 1968. Studies on flowering and fruit development in Aonla (*Emblica officinalis* Gaertn.). *Hort. Adv.*, 7: 38-67.

Bhujbal, B. G. 1975. Improvement of seed propagation of aonla. *J. Mahatma Phule Agricultural University*, 6: 73-75.

Dhankar, D. S. and Singh 1996, Seed germination and seedling growth in amla (*Phyllanthus emblica*) as influenced by gibberellic acid and thiourea. *Crop Res.*, 12(3) : 363-366.

Firminger, T. A. 1947. Firminger's manual of gardening for India, Thacker Spink Co. Ltd., Calcutta.

Nayaka, G. D. 2006. Propagation studies in anola (*Phyllanthus emblica* L.), M.Sc (Hort.) Thesis, University of Agricultural Sciences, Dharwad.

Ram, S. 1971. Studies on physiology of fruit growth in aonla (*Emblica officinalis* Gaertn.). Ph.D Thesis, Kanpur University, Kanpur.

Ram, S. 1974. Aonla tree you must grow. *Intl. Agric.*, 12:6-8.

Reddi, C. S. and Janaki Bai, 1979. A study on the pollination in *Emblica officinalis* Gaertn. *New Bot.*, 4: 155-160.

Srimathi, P. 1997. Research focus on seed collection, processing and storage of Amla (*Emblica officinalis* Gaertn.) Jamun (*Syzygium cuminii* Skeels) and Ber (*Zizyphus mauritiana* Lamk). Ph.D Thesis, TNAU, Coimbatore.

Venkatesh, C. 1976. Our tree Neighbours, National Council for Educational Research and Training, New Delhi, p.84.

5.9. *Ephedra gerardiana*

Common name: Ma Huang

Botanical name: *Ephedra gerardiana* Wall. Ex J.A. May.; Syn. *E. vulgaris E. distachya* L., *E. vulgaris* sensu Hook. f., *E. wallichii* Stapf., *E. gerardiana* var. *congesta* C.Y. Cheng

Family: Ephedraceae

Vernacular names: Ain, Khanta, Somlata (Hindi); Chhepat, Somlata (Ladakhi); Buchchur (Malayalam); Soma, Somlata (Sanskrit).

Origin and distribution: It is native to China and the Himalayas. It is widely distributed in Afghanistan, China, India, Nepal, Pakistan and Tajikstan. In China, found at 3700-5300m elevation on dry and rocky slopes. In India, it is found in drier regions of temperate and alpine Himalaya from Kashmir to Sikkim, Chamba, Lahul, Spiti and Ladakh (Chauhan,1999; Polunin and Stainton, 1987).

Habitat: It is found on sandy seashores and in temperate climates of both hemispheres.

Habit: An evergreen shrub, it grows to a height of 1m with a spread of upto 3m; woody stems buried in soil, parts above ground scabrous, rarely almost smooth.

Stem: The stem is slender, erect, grooved, green and branching; branchlets directed upward, dark green, later brown, short, stout, usually with 1-3 nodes, internodes 1-1.5cm long, 1.5-2mm wide, longitudinally furrowed, often scabridulous.

Leaf: Leaves are opposite, 2-3mm, connate for 2/3 their length. The leaves are small, scale like, articulate and sheathing at each joint. Leaves reduced to minute two toothed sheaths.

Fig. 1a. Male flower

Fig. 1b. Female flower

Fig. 2. Seeds

Flower: Dioecious, the male flowers (Fig. 1a) have prominent, yellow pollen sacks, male flowers few together in ovate spikes. Female flowers (Fig. 1b) 1-2, in short spike pollen cones solitary or two at nodes, small, 2-5mm long, 2mm wide, sessile; bracts in three or four pairs; staminal column exserted for 1/2 its length, with eight sessile anthers. Seed cones solitary, sessile or shortly pedunculate, subglobose at maturity, 5-7mm in diameter; bracts in two or three pairs, connate for 1/4-1/3 their length, the apical pair for 2/3 their length, red and fleshy at maturity; integument tube short, 1mm. (Fig. 1a and 1b).

Flowering and pollination : Flowering occurs in summer. Pollination is in July.

Fruit: Fruits 7-10mm long, berry, ovoid, red, surrounded by succulent bracts with a single seed in each.

Seed: Seeds one or two, oblong or obovoid- oblong, 4-6mm long, 1.5-3mm wide, exerted, apex obtuse or sub-acuminate (Fig. 2).

Seed maturity: Seed maturity occurs during August-September.

Propagation: By seeds

Seed germination: Seeds sown in sand medium recorded maximum germination with shortest period than those on soil and filter paper (Bhagat *et al.,* 1992).

Economic part used: Dried stems and roots

Alkaloids: Its dried twigs yield alkaloids viz., ephedrine and pseudo-ephedrine

Medicinal uses: Used for treatment of asthma, heart stimulant and fever.

REFERENCES

Bhagat, S., J. S. Lalhal and Ombir Singh. 1992. Germination and longevity of *Ephedra geradiana* seeds. *Van Vigyan*, 30(1): 12-18.

Chauhan, N. S. 1999. Medicinal and aromatic plants of Himachal Pradesh, Indus Publishing Company, New Delhi, pp. 223-224.

Polunin, O. and A. Stainton. 1987. Concise flowers of the *Himalaya*, Oxford University Press, Delhi, p. 228.

5.10. *Garcinia indica*

Common names: Indian Tallow tree, Kokum, Butter tree

Botanical name: *Garcinia indica* (Thou.) Chois. Syn. *Garcinia purpurea*

Family: Clusiaceae

Vernacular names: Kokam (Gujarati); Bhirand Kokum, Kokam, Ratambi (Hindi); Punarpuli (Kannada and Tulu); Kokam(Konkani); Murugul (Malayalam); Bhirand, Kokam, Kokambi (Marathi); Tintali (Oriya); Vrikshamala, Amlabija, Amlapura, Amlashaka (Sanskrit); Murgal, Murgalmara (Tamil).

Origin and distribution: This species is indigenous to the Western Ghats of India, along the West coast. Tropical evergreen tree mainly found along coastal belt of Konkan region of Ratnagiri district of Maharashtra, Goa, Uttara Kannada, Udupi and Dakshina Kannada districts of Karnataka and Kasaragod areas of Kerala.

Habitat: This is found in forest lands, riversides and wasteland and sometimes cultivated. In its natural habitat, it comes up well in hot humid climate under partial shade. The optimum temperature ranging from 10-38ºC with an annual rainfall of 250cm is ideal for its cultivation. The soil should be well drained, deep with good water holding capacity. The loamy soils rich in organic matter are ideal for its cultivation with soil pH ranging 5-6.

Habit: This is a slender tree with sloping branches, reaching heights of 15m (Fig. 1).

Stem: The thin bark is lined. Bark light brown, smooth, fissured when old; blaze reddish. Branches horizontal; branchlets opposite, sub-angular, glabrous.

Leaf: Leaves simple, opposite, decussate; petiole 0.7-1.5cm long, planoconvex, glabrous, slightly sheathing at base; lamina 6-12 x 1.5-4.3cm, narrow-elliptic to narrow-obovate, apex obtuse or acute to slightly acuminate, base acute to attenuate, margin slightly repand, chartaceous to sub-coriaceous, glabrous; secondary nerves 4-7 pairs, with many inter-secondary nerves; tertiary nerves obscure.

Flower: The plant is dioecious nature where the sex of the plant is confirmed only after completion of juvenile period of seven to ten years. Fifty per cent of the plants will be male and hence, unproductive. Because of seedling progenies, there is large variation in quality and quantity of fruits and due to late bearing and harvesting period extends in rainy season leading to loss of fruits. Flowers polygamodioecious, male flowers in axillary and terminal clusters; female flowers solitary, terminal or axillary (Fig. 2).

Flowering, pollination and fruiting: Female plants of seedling origin flowers in 8-10 years which needs sufficient pollinizer for good fruit set and yield. The tree flowers from October to November and continues upto February. The fruits are ready for harvest during the months of March to June. The fruit colour changes from greenish to reddish at maturity. Karale (2002) reported that seedling trees require 7-8 years to flower and fruit while grafted plants start bearing at the age of 4-5 years. Pollination is through wind.

Fruit: Dark purple fruit is round, about 4cm in diameter with 5-8 seeds. The fruits are picked when ripe; the rind is then, removed and soaked in the juice of the pulp and then sun dried (Fig. 3).

Seed: Seeds are 5-8, flat, smooth, shiny and brown (Fig. 4).

Propagation: It can be propagated through seeds, grafts (inarching, side grafting and softwood grafting) and cuttings.

Seed propagation: The seeds can either be dried in coal ash or they can be tied in a cloth bag and watered for eight days. For early germination, seeds may be soaked in water for about two days. The pre-treated seeds should be sown in seed pans and kept in a polyhouse, as they fail to germinate outside. Watering should be done daily. Germination is initiated after 30 days and continues till 60 days. The seedlings can be transferred to polythene bags containing soil, sand and farm yard manure in the ratio of 2:1:1 after three months. The seeds do not germinate readily in open fields and are susceptible to insect or fungal attack. They require hot and moist climate for their germination.

Fig. 1. Plant

Fig. 2. Flower

Fig. 3. Fruit

Fig. 4. Seeds

Seed collection: For raising seedlings, fruits are collected from early maturing, heavy yielding, elite trees having bold sized fruits. The fruits turn dark purple when ripe. These are hand plucked from the trees during May-June. Sometimes shaking of branches is done to make the fruits fall to the ground. The fruits are collected for seeds and depulped to separate the seeds. After extraction, seeds are washed thoroughly in running water. Then, the seeds are mixed with ash and dried in the shade for 3-4 days.

Seed extraction: Seeds extracted from fresh fruits during first week of June.

Seed weight: 2000-3000 seeds/kg

Viability period: Short lived

Germination: 80-90%. Temperature of 30 to 35°C is ideal for germination.

Germination period: 30-60 days

Pre treatments: Wet packing or drying in coal ash for 8 days prior to sowing is advised.

Season of planting: June - July

Age at planting: 12-14 months

Vegetative propagation

Cuttings: Semi-hardwood cuttings with two nodes are used for planting. Planting should be done in the poly house.

Grafts: The soft wood grafting has been standardized. It is observed that the mature scion stick of six months old can be preferred for soft wood grafting. Prior defoliation is not required for this technique in this crop and the age of rootstock should be more than 22 weeks. October is the best month for softwood grafting and the graft could be successfully maintained in poly house or under open sun after grafting until planted in the main field. Haldankar *et al.* (1987) reported that in Kokam once the rootstocks attained a graftable size, which takes about 22 weeks, further increase in age did not influence the percentage of success significantly. The percentage of success ranged between 70 and 78 per cent, and the highest success (78%) was noted with 26 weeks old rootstock in softwood grafting.

Economic part used: Fruits and seeds.

Alkaloids: Anthocyanins (Cyanidin-3-sambubioside, Cyanidin-3-glucoside), Fatty Acids (palmitic, stearic, oleic, linoleic acid), Hydroxy citric acid (HCA), Garcinol.

Medicinal uses: Kokum fruits are useful in loss of appetite, indigestion, constipation, calculi, thirst, asthma, liver enlargement, spleenitis, heart weakness, cold, cough, hiccough, flatulence, general debility etc. Kokum butter is useful as an infusion, or by direct application in skin ailments such as rashes caused by allergies.

Other uses: Fruits make a refreshing sweet drink. Concentrated juice also prepared and used for making curry, and ready to serve juices with and without sugar. Kokum seeds are used for oil extraction. That oil is called Kokum butter and used in curries, cosmetics, medicines and costly confectionery preparations in foreign countries.

REFERENCES

Haldankar, P. M., M. J. Sahi and G. D. Joshi 1987, Softwood grafting in Kokam. *Indian Hort.*, 32(3): 10-11.

Karale, A. R. 2002. Kokam. In: Fruits of India-Tropical and subtropical. Volume-2 (Eds.) T.K. Bose *et al.*, Naya Udyog, Kolkata. p. 743-750.

5.11. *Helicteres isora*

Common name: East Indian screw tree, Nut leaved screw tree, Red isora, Spiral bush

Botanical name: *Helicteres isora* L. Syn. *Helicteres grewiaefolia* DC., *Helicteres roxburghii* G.Don, *Helicteres chrysocalyx* Miq. ex Mast.

Family: Sterculiaceae

Vernacular names: Antamora (Bengali); Marosi, Marophalli (Hindi); Kavargi (Kannada); Valambiri (Malayalam); Muradasinge (Marathi); Avarttani, Mriga-shringa (Sanskrit); Yedampuri (left helix),Valampuri, Kaiwa (right helix) (Tamil); Adasamti (Telugu).

Origin and distribution: It is distributed from India, Pakistan, Nepal and Sri Lanka through Myanmar, Thailand, Indo-China, Malaysia and Indonesia to southern China and northern Australia. It is occasionally grown in Java.

Habitat: It is a gregarious species common in evergreen forests and secondary jungles along roads and forest edges. In Java, it is found in relatively dry areas upto 300m altitude, the habitats including teak forests, brush woods and roadsides. In Thailand, it is found in deciduous forests and scrub areas. Fertile soils rich in humus are preferred; in sandy and lateritic soils, the plants are stunted, much branched, and with a very thin bark.

Habit: It is a sub-deciduous shrub or small tree that can grow upto 2-4(-8)m tall. Its bark is finely wrinkled, lenticellate and pale. All young parts are often densely clothed with long, yellowish and stellate hairs (Fig. 1).

Stem: Shrubs consisting of branchlets which are brown stellate-tomentose.

Leaf: The leaves are simple and arranged in two vertical rows. The stipules are slender, measure 3-10mm long and caducous. The petiole is upto 4cm long. The blade is rounded to obovate, measuring 5-21cm x 3-18cm, and with rounded to sub-cordate base but sometimes oblique. The margins are irregularly serrate, with rounded to acuminate apex, palmately 3-5 veined and sometimes slightly lobed near the apex. They are variably covered with simple and stellate hairs, often densely on the underside.

Flower: The inflorescence is an axillary fascicle of usually 2 (-5) flowered cymes with small, sessile and basal glands. The peduncle is 2-5mm long, and with linear bracts 3-5mm long. The flowers are bisexual, irregular, about 2.5cm in diameter and fragrant. The pedicel is upto 1cm long. The bracteoles are linear. The sepal is 1-2cm long, tubular, yellow, and with five triangular and unequal lobes. The two inferior lobes are connate almost to the apex while the three superior lobes are separated. There are five petals which are unequal, clawed, 3-4cm long and bluish at anthesis but turn bright red. The two broader lower petals measuring about 3 x 1.5cm with a short claw is winged or widened towards the apex while the three upper petals are narrower, measuring about 1 x 0.5cm, and with a claw upto 2.5cm long which is auricled near the apex. The gynandrophore is 3-6cm long with red apical glands. The staminal tube is 2.5mm long, making an angle of 90° with gynandrophore, terminated by ten stamens (filaments connate in lower 2/3) and, more inwardly inserted and often hidden by the stamens. There are five staminodes altogether. The filaments are short, and with two celled anther which are parallel to each other. The pistil is with a five celled, red glandular ovary, and five sub-connate short styles.

Flowering, pollination and fruiting: Flowering peaks from September-December (Apate *et al.*, 2009). Fruiting occurs from December - March. The red flowers are pollinated mainly birds of the sunbird family. They are also visited by many butterflies and hymenoptera.

Fruit: It is popularly known as 'Screw fruit' due to its left or right twisting. Fruits are greenish brown in colour, beaked, cylindrical, and spirally twisted on ripening. The fruit is 4-8cm long consists of five twisted follicles. Each follicle is about 20-25 seeded (Fig. 2).

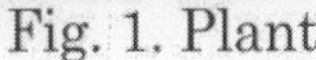

Fig. 1. Plant

Fig. 2. Fruit

Seed: The seed is angular-sub rhomboid, measuring about 2.5 x 1.5 x 2.7mm, finely warty, smooth and dark brown.

Propagation: It can be propagated both by seeds and cuttings.

Seed propagation: Generally seeds are sown in sand beds during July-September. Germination starts after a month and 6-8cm tall seedlings are transplanted in polybags or containers. One year old seedlings can be planted in the field. Gangare *et al.* (1998) recommended top of paper method for better and quick germination.

Vegetative propagation: Semi-hardwood cuttings of pencil thickness measuring 35-40cm are selected for propagation. Stem cuttings are planted in raised pure sand beds (without mixing any fertilizer) in mist chambers during February - March. Rooting is initiated within 25-30 days and can be transplanted after three months.

Economic part used: Fruits, roots and bark

Alkaloids: From the roots betulic acid, daucosterol, sito-sterol, isorin were isolated. Cucurbitacin B and iso-cucurbitacin B were isolated (Qu *et al.,* 1991).

Medicinal uses: The fruits are cytotoxic, astringent, demulcent, diuretic and antigalactagogue, root decoction stimulates the uterus, is demulcent and astringent in action. The fruits are reported to be useful in haemorrhages, epistaxis, ophthalmitis, colic, flatulence, verminosis, wounds, ulcers and for intestinal complaints. Fruits are also used to improve appetite. Root juice is useful in the treatment of diabetes and is used in anti-diarrhoeal and anti-dysentric formulations. Root bark is reported to be used in scabies, gastropathy, diarrhoea and dysentery. The fruits are also useful in acute and chronic bronchitis. Stem bark is used in diarrhoea, dysentery, scabies and biliousness and is useful in gripping of the bowels. Fried pods are given to children to kill intestinal worms.

Other uses: It is a handy shrub with delightful flowers and dense foliage. It boasts plenty of beautiful, scarlet coloured, erect flowers that imitate tiny birds resting in slender branches. It flowers profusely throughout the year and keeps the garden graceful.

REFERENCES

Apate, S. A., G. S. Kulkarni and M. M. Patwardhan. 2009. Study of vegetation of some plateaus near by Ratnagiri (Maharashtra). *Ad. Plant Sci.*, 22 (1): 209-212.

Gangare, K. C., Patil and Y. B. Suryawanshi. 1998. Seed dormancy and germination studies in some medicinal plant species. *Seed Tech. News*, 28: 59.

Qu, W. H., J. G. Li and M. S. Wang. 1991. Chemical studies on the *Helicteres isora. Zhongguo Yaoke Daxue Xuebao,* 22(4): 203-206.

5.12. *Lawsonia inermis*

Common name: Henna, Egyptian privet.

Botanical name: *Lawsonia inermis* L.; Syn. *L. alba*

Family: Lythraceae

Vernacular names: Mehndi (Hindi); Mayilanji (Malayalam); Goranti (Kannada); Madayantika (Sanskrit); Maruthani (Tamil); Goranta, Gorata, Gorinta, Gorintaku, Krommi (Telugu).

Origin and distribution: The henna plant is native to tropical and sub-tropical regions of Africa, southern Asia, and northern Australia in semi-arid zones. Henna is commercially cultivated in UAE, Morocco, Algeria, Yemen, Tunisia, Libya, Saudi Arabia, Egypt, India, Iraq, Iran, Pakistan, Bangladesh, Afghanistan, Turkey, Somalia and Sudan. Presently, the Pali district of Rajasthan is the most heavily cultivated.

Habitat: This plant is hardy by nature and can tolerate a wide range of climatic conditions. However, hot and dry climate with a temperature range of 20-30°C is considered to be ideal. It can even grow with low rainfall ranging from 200-450mm. It can grow on soils ranging from light loam to clay loam, but does best on heavy soils with good amount of organic matter and retentive soil moisture. It is found growing on soils with a pH ranging from 4.3 to 8.0. It can also grow on rocky, gravelly, light or heavy soils or any other soil that is unfit for cultivation. However, saline and acidic soils and soils having water logging problem are not suitable for its cultivation.

Habit: Henna is a tall shrub or small tree, 2.6m high. It is glabrous, multi-branched with spine tipped branchlets (Fig. 1).

Leaf: Leaves are opposite, entire, glabrous, sub-sessile, elliptical, and broadly lanceolate (1.5-5.0cm x 0.5-2cm), acuminate, having depressed veins on the dorsal surface.

Flower: Henna flowers are fragrant and have four sepals and a 2mm calyx tube with 3mm spread lobes. Petals are obovate, white or red stamens inserted in pairs on the rim of the calyx tube. Ovary is four celled, style upto 5mm long and erect (Fig.2).

Flowering, pollination and fruiting: Flowering and fruiting occurs during March to June. It is pollinated by honeybees (Abrol, 2012).

Fruit: Fruits are small, globose, brownish capsules, 4-8mm in diameter, with 32-49 seeds per fruit, and open irregularly into four splits (Fig. 3a and 3b).

Seed: Seeds trigono-pyramidal, about 2.5mm long, externally sub-tuberculate.

Seed dispersal: The plant is frequently spread through the digestive processes of birds.

Propagation: Seeds and cuttings can be used for propagation. However, seeds are preferred.

Seed propagation: The seeds are kept in moist cloth for 6-8 days till they sprout. The seeds are spread randomly in well prepared nursery beds during March-April. The raised nursery beds are prepared in shaded areas. The seed beds are kept moist by light irrigation. The seeds are germinated (20%) after 10-20 days. The neem leaf solution has to be sprayed at an interval of 15-20 days to maintain health of the seedlings. When the seedlings attain a height of 30-45cm, they may be transplanted.

Economic part used: Leaves

Alkaloids: Lawsone.

Medicinal uses: Henna has been used since the Bronze Age to dye skin (including body art), hair, fingernails, leather, silk and wool. Henna flowers have been used to create perfume since ancient times, and henna perfume is experiencing a resurgence. Henna repels some insect pests and mildew.

Fig. 1. Plant

Fig. 2. Flower

Fig. 3a. Immature fruit

Fig. 3b. Mature fruit

Agricultural uses: The plant extract of *Lawsonia inermis* showed 53.6 per cent reduction over control for the incidence of chillies fruit rot caused by *Colletotrichum capsici* (Rajavel *et al.*, 2001).

REFERENCES

Abrol, D. P. 2012. Pollination biology, biodiversity conservation and agricultural production. Springer London, p.792.

Rajavel, R., V. Valluvaparidasan and C. Ushamalini. 2001. Management of chillies fruit rot by using botanicals. *Allelopathy J.*, 8 (2):277.

5.13. *Mimusops elengi* L.

Common name: Bullet wood tree, Red coondoo

Botanical name: *Mimusops elengi* L.

Family: Sapotaceae

Vernacular names: Bakul (Sanskrit and Hindi); Bakula, Ranjala, Ranjaka (Kannada); Elanji (Malayalam); Wowli (Marathi); Mahizham (Tamil); Bakulamu, Pogada (Telugu).

Origin and distribution: It is native to India, Sri Lanka, Andaman Islands, Myanmar and Indo-China, but is commonly planted as an ornamental tree throughout the tropics, also in Africa, where it has been recorded from e.g. Ghana, Tanzania, Mozambique, Réunion and Mauritius. It has become naturalized locally in Réunion. Its presence has been noticed in north India, western Ghats of southern India and also in Andaman Islands. Sporadic population has been reported from Meghalaya.

Habitat: It grows in semi evergreen to evergreen forests of India. It can also be seen among the rivarine or riparian vegetation of moist deciduous forests. Adaptable to almost all kinds of soils, drought, heat and pollution, successfully planted in any kinds of localities other than water logged areas. It can be planted in sunny localities or in partial shade between buildings.

Habit: It is a medium sized, lactiferous, evergreen tree upto 15m tall (Fig. 1).

Stem: Bole upto 100cm in diameter, often short and divided into several large main branches but sometimes branchless for upto 15(–20)m, buttresses absent or upto 2m high; bark surface becoming deeply fissured and sometimes peeling off in thin scales, grey, brown or dark red to blackish, inner bark fibrous, pink or reddish, with scanty watery or white sticky latex; crown dense, rounded and spreading, glossy dark green.

Leaf: Leaves arranged spirally, more or less in tufts at the ends of branches, simple; stipules minute and caducous; petiole 1-3.5cm long, grooved above; blade ovate to elliptical or oblong-elliptical, 4.5-17 × 2-7cm, rounded at base, acuminate at apex, margins often wavy and upcurled, glabrous, with 10-20 pairs of lateral veins.

Flower: Flowers in fascicles of upto six in the leaf axils, bisexual or functionally unisexual, regular, fragrant; pedicel 1–1.5cm long; sepals in two whorls of four; corolla white, with a short tube and eight lobes each deeply divided into three, 1cm long; stamens eight, alternating with eight staminodes; ovary superior, (6-)8 celled (Fig. 2).

Flowering, pollination and fruiting: January- April is the flowering period. Stray fruiting starts in February, but will be in peak from the month of May and ripens during June - August. The stigma is receptive before the pollen is released, stimulating cross pollination. It is pollinated by insects and wind.

Fruit: Fruits are fleshy, edible, ovoid, single seeded berries, 3cm long, turn to bright orange red when mature (Fig. 3). It encloses one (rarely two) seeds.

Seed: Seeds are albuminous, fleshly cut milky latex exudes out from the seed. The seeds are 1.7 -1.9cm in length and 1.2-1.5cm in breadth. It is compressed, obliquely ovoid, hilum elongated and lateral. They are shining, light brown to blackish in colour (Fig. 4).

Seed dispersal: The seed is known to be dispersed by bats, but monkeys, squirrels and wild pigs probably also eat the fruits.

Propagation: It is propagated through seeds and cuttings.

Seed propagation: Seeds are collected during June - August months. Fleshy pulps of fruits are removed and sun dried for a day or two. Seed can be stored for about nine months and needs after ripening during the first month of storage. There are about 2000 dry seeds/kg. Sowing should be done immediately as the seeds are reported to have less viability. Sun dried seeds are soaked in cow dung slurry for four days to soften the hard outer coat of the seeds. Seeds can also be given a mild boiled water treatment to soften the hard testa. Sowing can be done in sandy mother beds or directly in poly bags/ containers. Seed germinates in 17–82 days and the germination rate is 70-90 per cent. Growth of seedlings is slow. In case of mother beds, three month old seedlings are transplanted in containers. One year old seedlings can be transplanted in the main field.

Propagation through cuttings: The rooting success of 10-15cm long cuttings with a diameter of 0.5- 1cm is 70- 90 per cent.

Economic part used: Bark, flower and fruits.

Alkaloids: Several triterpenes (e.g. β-amyrin, lupeol, α-taraxerol and ursolic acid) have been isolated from this plant as well as steroidal glycosides.

Medicinal uses: Bark, flowers and fruits are reported to be acrid, astringent, cooling and anthelmintic. Bark is said to be useful against urethrorrhoea, diarrhoea and dysentery and is also used as a gargle. Lotion prepared out of flowers is used to cure wounds and ulcers. Dried flowers are made into powder and given internally as brain tonic; the same powder is used as a snuff to relieve headache and teeth problems (Malik *et al.*, 2009). Unripe fruits are reported to be used as mastigatory to help fix loose teeth. The seed is used traditionally for curing piles, headache, constipation, spermicidal etc. The seed oil is used in cooking and as a luminant.

Fig. 1. Plant

Fig. 2. Flower

Fig. 3. Fruit

Fig. 4. Seeds

Other uses: In most parts of India, it is planted in gardens, temples and along roadsides for its fragrant flowers and edible fruits. Its tall, spreading canopy and short leaders (main trunks) make this successful avenue tree. It attracts variety of birds, thus creating natural elements in the urban settings.

REFERENCE

Malik, V., I. Mohammad, A. Kashyap and U. Kumar. 2009. Traditional medicinal plants used by the inhabitants of foot hills of Siwalik Himalaya of district Saharanpur. *Ad. Plant Sci.*, 22(1): 191-193.

5.14. *Nothapodytes nimmoniana*

Common name: Stinking tree

Botanical name: *Nothapodytes nimmoniana* (Grah.) Mabb. Syn. *Mappia foetida, Nothapodytes foetida, Stemonurus foetidus, Premna nimmoniana*

Family: Icacinaceae

Vernacular names: Durvasane mara (Kannada); Kalgur, Kalagaura (Marathi); Arali, Perum pulagi, Kal kurinji (Tamil)

Origin and distribution: The species is native to warmer regions of South India. It is found in China, Sri Lanka, Myanmar, Thailand, Malaysia and China. It is widely distributed in many parts of the Western Ghats from Konkan southwards to Nilgiris, Anamalais and is common in Uttara Kannada. It is also found in some parts of Assam, the Himalayan foothills, Sri Lanka, Myanmar and Thailand. They are vulnerable to indiscriminate harvest for trade and an estimated 20 per cent of the population of this species is believed to have declined over the last decade. The species has a threat status of endangered / vulnerable.

Habitat: It is a small sized tree found as an under storey in evergreen to moist deciduous forest types. It prefers forest edges as habitat.

Habit: It grows upto 15m height in undisturbed evergreen forests, but appears shrubby in secondary or disturbed forests. It sheds its leaves, become leafless and remains dormant in summer months, whereas new leaves arise with the advent of monsoon (Fig. 1).

Stem: Bark is rough, about 5mm thick, grey coloured with peculiar lenticels. Branchlets are slightly angled, corky, with prominent leaf scars.

Leaf: Leaves simple, alternate, spiral, slightly leathery, broadly egg shaped to elliptic oblong, 1-25cm long and 4-12cm wide. Leaf base is often unequal; tip is pointed to long point. Leaves are crowded at the ends of branchlets. Leaf stalks are 3-6cm long.

Flower: Inflorescences are terminal; many flowered, shorter than leaves. Flowers are creamy yellow, foul smelling, about 5mm across, in flat topped clusters at the end of branches (Fig. 2). Petals are hairy inside. The species is polygamous in nature. Hombegowda *et al.* (2002) reported that the population shows a wide array of sexuality with male, female, hermaphrodite, monoecious, andromonoecious and gynomonoecious individuals, indicating that the species is polygamodioecious. Both male and female flowers bear ovary with two ovules. Ovary is monocarpellary, with two ovules inside.

Male flowers, in all breeding types are large, 9.2mm in length, 9.4mm in diameter, dull yellowish in colour and have rudimentary pistil. Stamens are large, 6.8mm in length and five in number. All anthers are functional. Male flowers dehisce during early morning (06.00-07.00h) and maximum pollen production is observed during the noon (12.00-14.00h). Longevity of male flowers is only two days.

Female flowers are small, 6.2mm in length, 6.9mm in diameter, dull yellowish in colour and with rudimentary stamens that produce no pollen grains. The pistils are 5.7mm in length. Anthesis of female flowers occurs during early morning (06.00–08.00h). Longevity of female flowers varies from two to eight days depending upon their pollination.

Bisexual flowers are intermediate to male and female flowers in size. They are 7.4mm in length, 7.2mm in diameter and are dull yellow in colour. Both stamens and pistils are medium in size and are 3.0mm and 5.0mm in length, respectively. In bisexual flowers, only 1-2 anthers produce pollen grains and stigma is active.

Flowering begins during July to August and most of the early flowering trees are dioecious, whereas late flowering trees are monoecious, hermaphrodite and a mixture of other breeding types. The flowers open mostly on sunny days. The number of flowers per inflorescence in male, female and bisexual plants varies greatly. Male inflorescence carries relatively higher number of flowers (350-530) than those bearing female flowers alone (70-185) and bisexual flowers (90-210). Further, the production of lesser number of flowers per inflorescence in bisexual trees and female trees might be a device to save energy for fruit production, whereas higher number of flowers per inflorescence among male trees may help in pollinator attraction and pollen movement between female and male trees.

Flowering, pollination and fruiting: The plant usually flowers during August and the fruits ripen till November, however, flowering and fruiting periods vary in different regions. Hand pollination of female and bisexual flowers with pollen of male and bisexual flowers of the same tree as well as male flowers of other trees showed good pollen germination and set fruits, suggesting that the female and bisexual flowers produced on polygamous trees are self-compatible. Eight of the 30 bisexual and female buds (on mixed inflorescences of polygamous trees) netted to keep out visiting insects set fruit, indicating 26.7 per cent natural self pollination (Sharma *et al.,* 2010a). In the disturbed open forests of Northern Western Ghats near Pune and Raigarh district, the fruits ripe in November-December, however in the tall forests of Kolhapur and Sindhudurg district of Maharashtra, they can be harvested in January – February; indicates that rainfall might be playing a critical role.

Sharma *et al.* (2010b) also studied the floral biology and her findings are given below.

1. A single male flower remained open for one day, while a female flower for 1-4 days depending on pollination success: if pollinated on the first or second day of opening, the flower begins to senesce and if pollen has not landed on the stigma, the flower remained open longer.
2. Male flowers also offer more reward than females in terms of pollen in addition to nectar. Pollen viability lasted for >10 days from the day of opening. There were nectar hunters and pollen foragers on the flowers. Seventeen species were encountered across the two sites, with flies being the most frequent visitors. Two species of calliphoridae, one species of syrphidae, the short tongued halictid bee, the stingless bee *Trigona iridipennis, Apis dorsata, Apis florea,* Sepsid flies, Lycaenid butterflies and five species of ants. The empididae, scathophagidae, mirid bug and pompilid wasp appeared to be predators foraging on flower parts. Of these, only *T. iridipennis, A. dorsata* and two calliphoridae members showed pollen load on their body and were treated as the four possible pollinators.
3. Female inflorescences attracted significantly fewer number of pollinators compared to male and mixed inflorescences. Pollinators also showed a weakly significant preference for foraging during the morning. Male flowers and morning hours showed a highly significant interactive effect.

Fruit: Fruits are drupe, smooth, oblong to ellipsoid, about 2 x 1cm, smooth, purplish black when ripe, with a single seeds (Fig. 3).

Seed : The seeds are albuminous.

Seed dispersal: Birds such as black bulbul, red whiskered bulbul, yellow browed bulbul, white chicked barbet and mammals like giant squirrel etc. are known to eat and disperse the fruits.

Propagation : By seeds and stem cuttings.

Seed propagation: A prerequisite for successful germination is that the seeds should be necessarily handpicked from the plant and not to be collected from the ground. Fallen seeds immediately get infected by a fungus thereby drastically affecting germination success.

Fruit collection and seed storage

1. Ripened seeds are black in colour.
2. Fruits should be collected in the period of November - January.
3. Fruits should be depulped and dried in shade.
4. Seeds should be stored in airtight containers under ordinary room temperature, if not used immediately.
5. For nearly 100 per cent viability, storage should not exceed 45 days.

Fig. 1. Plant

Fig. 2. Flower

Fig. 3. Fruit

Seed treatment

1. Treatment with cow dung slurry – Soak the seeds with seed coat intact in fresh cow dung slurry for 24h. Then, remove the seeds from slurry and then, sow them.
2. Cold water treatment – Remove seed coat with pointed forceps without harming the seeds inside. Soak the seeds in cold water for minimum 12h.
3. Sharma *et al.* (2000) studied seed germination behaviour by providing physical treatments and growth hormonal treatments. Among the physical treatments, piercing the micropyle end and mechanical scarification showed improvement in germination upto 24 per cent and with 5 per cent H_2SO_4, the improvement was upto 30 per cent. Among the growth hormonal treatments, 100ppm GA_3 resulted the highest germination of 70 per cent against five to six per cent in control. GA_3 100ppm was found to be the best in terms of germination enhancement and seedling growth survival percentage when compared with all other treatments.

Seed bed preparation

1. Seeds should be sown immediately after pretreatment in a raised nursery bed size of 1.2 x 7.0 x $0.3m^3$
2. Seed bed should be drenched with 0.1 per cent carbendazim to protect against fungal attack.
3. Seeds should be sown at the depth of 2cm in flat position.
4. Moist conditions should be maintained.

Seed bed protection: Spray the seedlings and drench the soil with 0.2 per cent solution of bavistin once in 10 days to control fungal attack.

Seedling transplantation: Seedling should be of height 8-10cm before they are transferred to polybag (size 6" x 9"), filled with soil: sand: farm yard manure (FYM) in 2:1:1 proportion along with 10g of finely powdered swine manure.

Fertilizer application: After two months of growth, 1g DAP is supplied to seedlings as top dress.

Vegetative propagation

1. Season for cutting is February to March.
2. Branches of pencil thickness girth are given a slant cut at the basal end.
3. Application of IBA at 2000ppm in dust form.
4. The rooting media should be sand: coir @ 1:1 ratio.
5. The propagation chamber should be provided with temperature range 28 - 33°C and 90 per cent relative humidity.
6. Cuttings should be hardened for about 30 days before they are out planted.

Economic part used : Wood chips.

Alkaloids: The active principle of the wood-camptothecin (CPT), is known as a potent drug that breaks single strand DNA in the mammalian systems and is found to be useful in the treatment of tumours. CPT is also known to inhibit retroviruses such as human immuno deficiency virus (HIV) and equine infectious anaemia virus. It is believed that CPT is the third most important alkaloid sought after by the pharmaceutical companies around the world.

Medicinal uses : Camptothecins are used against variety of tumours, HIV-I, parasitic trypanosomes, malaria and also known to exhibit antibacterial activity.

REFERENCES

Hombegowda, H. C., R. Vasudeva, G. P. Mathachen, R. U. Shaanker and K. N. Ganeshaiah 2002. Breeding types in *Nothapodytes nimmoniana* Graham.: An important medicinal tree. *Curr. Sci.*, 83:1077–1078.

Sharma, S. N., S. C. Puri, G. H. Srivastava and B. L. Kaul 2000. Enhancement of seed germination in *Nothapodytes foetida*. *J. Med. Arom. Pl. Sci.*, 22: 206-210.

Sharma, M. V., R. U. Shaanker, R. Vasudeva and K. R. Shivanna 2010a. Functional dioecy in *Nothapodytes nimmoniana*, a distylous species in the Western Ghat. *Curr. Sci.*, 99(10): 1444-1449.

Sharma, M. V., R. U. Shaanker, S. R. Leather R. Vasudeva and K. R. Shivanna 2010b. Floral resources, pollinators and fruiting in a threatened tropical deciduous tree. *J. Plant Ecology* Advance Access published November 25, 2010 p. 1-9.

5.15. *Pterocarpus santalinus*

Common names: Red Sanders, Red Sandalwood

Botanical name: *Pterocarpus santalinus* (L.) f.

Family : Fabaceae

Vernacular names: Lal chandan, Rakta chandan (Hindi); Agaru, Honne (Kannada); Patrangam, Tilaparnni (Malayalam); Tambada chandana (Marathi); Rakta chandana (Sanskrit); Ratha sandanam, Chenkunkumam, Sivappu chandanam, Sandana vengai (Tamil); Agaru gandhamu, Yerra chandanamu, Rakta chandanam (Telugu).

Origin and distribution: It is native to India and distributed in peninsular India and Sri Lanka. It occurs gregariously in patches in tropical dry deciduous forests, towards south-eastern Ghats, Endemic to the hills of Cudappah, Nellore, Chittor and Prakasam districts of Eastern Ghats of Andhra Pradesh and also to the adjoining regions of Karnataka and Tamil Nadu. In Karnataka it has been reported from Devarayandurga, Sandoor and Karpakapalli MPCAs. In Tamil Nadu it was reported from the drier zones of Kanchipuram, Salem and Dharmapuri districts.

Habitat: Requires around 100mm of rain in each of the two annual monsoons, relative humidity ranging from 32 to 79 per cent. Often rocky terrain at altitudes ranging from 150-900m often on precipitous hill slopes. At lower altitudes, it occurs on all aspects; but in higher hills, it shows a tendency to confine itself to cooler Northern and Eastern aspects. It grows on rocky soils consisting of quartzite shales, lime stone and laterite. The tree grows well particularly in laterite loam, requires perfect drainage and hence it is found mainly on stony or gravelly soil. Under water logged conditions the trees fail to stand stiff.

Habit: A deciduous medium sized tree.Trees are about 12m in height. Leaf fall occurs during January to March; New foliage occurs during February to March (Fig.1).

Stem: Stem erect and dense, rounded crown. The prominent, crocodile back like bark exuding blood red thick gum when cut freshly. Bark blackish brown, 1-1.5cm thick, deeply cleft into rectangular plates by deep vertical and horizontal cracks, blaze pale yellow with numerous pink streaks exuding copious red sticky thick gum. Branch lets are drooping, glabrous.

Leaf: Three glossy, dark green leaflets and compressed, circular, winged fruits with an apical beak are few distinguishing field characters. Leaves tri- foliate, 10 – 18cm long; rachis swollen at base; leaflets usually three, rarely four or five, broadly ovate or orbicular, base round or sub – cordate, apex obtuse or emarginated, margin entire, coriaceous, shiny, glabrous, distinctly stalked.

Flower: Flowers are bisexual, stalked in axillary simple or sparingly branched racemes, yellow, about 2cm long, fragrant (Fig. 2).

Flowering, pollination and fruiting: Flowering occurs from February to April, after new leaves sprout. Pods are formed rapidly but get ripened in next February- March. The flowers are papilionaceous, nectariferous, self-compatible and pollinated by *Apis dorsata* in the natural habitat and also by other bees in other habitats.

Fruit: Pods obliquely orbicular, 3-4cm diameter, including the wing, flat, gradually narrowed into a short tip which is about 1cm long and single seeded (Fig.3).

Seed : Seeds one rarely two, more or less kidney shaped, 1 - 1.5cm long, smooth, reddish brown (Fig.4).

Propagation : It is propagated through seeds and stem cuttings.

Seed propagation : The dry, mature and healthy pods are collected from trees during February- June and dry them in the sun for three days. One hundred gram fruit contains 100 fruits. The raised or flat nursery beds are prepared with 12m long and more than a metre broad with well drained soil. The entire pod with wing are sown and covered with a thin layer of soil or hay and water profusely. July-September is the ideal time for sowing. The seeds start germinating in 15-20 days and the germination success varies from 10-80 per cent. When the seedlings are 5-10cm tall, they may be transplanted into polythene bags kept under shade, duly watered for a month. Later the seedlings are shifted from the shade. They are maintained for another 3-4 months. At the end of this period roots are developed. During the rainy season the seedlings are transplanted into the main field.

Seed dispersal: The fruit is like a shuttle cock, the sepals enlarging in the fruit to form papery wings helpful in dispersal (Venkatesh, 1976).

Seed treatment: Before sowing, the pods are soaked in cold water for three days to increase the germination success. Alternately the pods are immersed in cow dung slurry for 48h before sowing. Seed treatment with GA_3 + Benzyl adenine @ 250ppm also helps in early and optimum germination.

Seed storage: Seeds are stored in bamboo baskets or gunny bags till used for planting. There is no loss of viability for at least eight months. Seeds lose viability, if kept for more than a year.

Fig. 1. Plant Fig. 2. Flower

Fig. 3. Fruit Fig. 4. Seed

Vegetative propagation: The stem cuttings are planted in polythene bags. Then, these saplings are planted in field during rainy season, at a spacing of 3-4m in $30m^3$ pits.

Economic part used : Whole plant

Alkaloids: Bark, heartwood, and sapwood contain santalin A and B, isopterocarpolone, pterrocarptriol, isopterocarpene, pterocarpdiolone, pterocarpol, acyl olealonic aldehyde and acetyl oleanolic acid.

Medicinal uses : It is one of the ingredients of 'Triphaladya ghrta', used in chronic fever. The heartwood is rubbed with water, honey, ghee and oil and applied as collyrium alleviates defects of vision. It is also used in treating skin diseases, bone fracture, leprosy, spider poisoning, scorpion sting, hiccough, ulcers, general debility and mental aberrations. Wood paste applied on boils and other skin eruptions, infections, inflammations and on forehead for headache. Decoction of fruits used in chronic dysentery and also used by the 'Kani' tribe in dermatological conditions including psoriasis. Wood and bark brew taken orally relieves chronic dysentery, worms, blood vomiting, weak vision and hallucination. Wood powder is used in haemorrhage, bleeding piles and inflammations.

Other uses: It is commonly planted as an ornamental tree.

REFERENCE

Venkatesh, C. 1976. Our tree Neighbours, National Council for Educational Research and Training, New Delhi, p.84.

5.16. *Salvadora persica*

Common names: Mustard tree, Salt bush, Tooth brush tree

Botanical name: *Salvadora persica* (L.) Garcin

Family: Salvadoraceae

Vernacular names: Jhal, Mah-Khakhin Kickni, Miraj, Pelu, Pilva (Bengali); Kharijal, Piludi (Gujarati); Jhak, Kharjal, Araak tree, Peelu tree, Meswak (Hindi); Goni-mara (Kannada); Pivla, Khakan (Marathi); Kotungo, Pilu (Oriya); Kalawa, Karkol, Perungoli, Ughaiputtai, Vivay (Tamil); Ghunia, Varagogu (Telugu).

Origin and distribution: It is widely distributed in Algeria, Angola, Cameroon, Chad, Egypt, Eritrea, Ethiopia, India, Iran, Israel, Jordan, Kenya, Libyan Arab Jamahiriya, Malawi, Mali, Mauritania, Mozambique, Niger, Nigeria, Oman, Pakistan, Saudi Arabia, Senegal, Somalia, South Africa, Sri Lanka, Sudan, Syrian Arab Republic, Tanzania, Uganda, Yemen, Republic of Zambia and Zimbabwe.

Habitat: This is widespread, notably in thorn shrubs, desert flood plains, river and stream bank vegetation, and grassy savannahs. Prefers areas where groundwater is readily available, by riverbanks, on perimeters of waterholes, in seasonally wet sites, and along drainage lines in arid zones. Also found in valleys, on dunes and on termite mounds. The tree is able to tolerate a very dry environment with mean annual rainfall of less than 200mm. It is highly salt tolerant, it can grow on coastal regions and inland saline soils. Prefers clays but is found on loam, black soils and sand. It is adapted to alkaline or very saline soils, usually clay rich, and soils without salt. It is a halophyte and accumulates salt in the leaves (Rao *et al.*, 1999).

Habit: *Salvadora persica* is an evergreen shrub or small tree to 6-7m; its bark is scabrous and cracked, whitish with pendulous extremities. It is grown in plantations or hedges. It is generally a slow growing tree. From seed, the juvenile period is about four years (Fig. 1).

Root: The root produced numerous suckers creating a thicket. The root bark of the tree is similar to sand, and the inner surfaces are even lighter shade of brown. It has a pleasant fragrance, as well as warm and pungent taste.

Stem: Main trunk erect or trailing with profusely branched, wide crown of crooked, straggling and drooping branches; young branches green in colour; bark slightly rough, greyish brown on main stem, paler elsewhere. Coppicing is advantageous for the tree's used as a fuel, and the branches are repeatedly cut to produce short stems that are harvested for tooth brushes.

Leaf: Leaves oblong elliptic to almost circular, 3 x 7cm, light to dark green, rather fleshy, sometimes with wart like glandular dots and dense, rather loose hairs; apex broadly tapering to rounded, sharp tipped; base broadly tapering; margin entire; petiole upto 10mm long; leaves in opposite pairs, main nerves are in 5-6 pairs, and the petioles 1.3- 2.2cm long.

The leaf is unilateral. Both epidermises compose of small polygonal cells and the upper epidermis is covered with a raised thick striated cuticle. Both epidermises bear many unicellular conical covering trichomes of different sizes and they are easily detachable. The stomata are oval and they are very small in size and of the parasitic type. They are distributed on the lower epidermis. The upper epidermis is underlain by a hypodermis which is composed of large broad cells having thick cell walls. The hypodermis is discontinuous and it is interrupted by one or two layers of palisade tissues of different sizes having wavy cell walls. All cells are fairly rich in chlorophyll and granular structures of various shapes including greenish cluster crystals. The spongy tissues compose of large cells of irregular shapes. They also contain granular structures and they embed a group of adjacent sclereids with angular shapes and thick pitted cell walls. The spongy tissues are also traversed by vascular tissues that contain reticulately and spirally thickened vessels and pitted tracheids.

Flower: The flowers are greenish yellow in color, in axillary and terminal compound lax panicles 5–12.5cm long, numerous in the upper axils, pedicels 1.5-3mm long, bracts beneath the pedicels, ovate and very caducous. Calyx is 1.25mm long, glabrous, cleft half-way down, lobes rounded. Corolla is very thin, 3mm long, deeply cleft, persistent, lobes are 2.5mm long, oblong, obtuse and much reflexed. Stamens are shorter than corolla, but exerted, owing to the corolla lobes being reflexed (Fig. 2).

Flowering, pollination and fruiting: In India flowering occurs from December and January-June. The first fruit ripen three months from flowering. It is highly cross pollinated crop (Balasubramanian and Bole, 1993).

Fruit: Fruit is drupe, spherical, fleshy, 5-10mm in diameter, pink to scarlet when mature and single seeded. During ripening the fruit turns from white to pink or purple red with a persistent yellow calyx (Fig. 3).

Seed: The single brown seed is 1-4mm in diameter, sub-globose and smooth (Fig. 4).

Seed dispersal: Seeds dispersed by animals and man after they eat the fruit.

Propagation: By seeds

Fig. 1. Plant

Fig. 2. Flower

Fig. 3. Fruit

Fig. 4. Seed

Seed propagation: Readily germinates from seed. Seeds exhibit no dormancy but the fruit pulp contains germination inhibitors that should be removed before sowing. Inhibitors in the fruit pulp will prevent germination if the seeds are not extracted. If the seeds have not been properly dried and stored under optimal conditions, storage life and viability will be reduced. The process of seed germination starts with imbibitions in water at 30-35°C for 24-72h, but under saline conditions the absorption of water is dependent upon osmotic pressure of the media and cell sap. Soaked, depulped seeds of *S. persica* will germinate in 24h. Seeds have been raised in the nursery for upto three years prior to transplanting in the field. There are about 3400 seeds/kg.

Seed storage: Seed storage is orthodox; seeds can be stored with low moisture content. Dry the seeds to 15-20 per cent relative humidity and store at -20ºC, or as cool as possible. This species has oily seeds which will absorb relatively less water than similar sized non-oily seeds. Be aware that when these seeds are dried to the recommended 15 per cent relative humidity, the equilibrium moisture content will be lower than for non-oily seeds.

Economic part used: All parts

Chemical compounds: It contains ten different natural chemical compounds, considered essential for good oral and dental hygiene. They are: fluoride, silica, tannic acid, resins, alkaloids (salvadorine), volatile oils (sinigrin), sulfur, vitamin C, sodium bicarbonate, chlorides, calcium, benzyl iso thiocyanate (BIT) and others including salicylic acids, sterols, trimethylamine, saponins, flavenoids.

Medicinal uses: Tooth brushes made from roots and small branches of about 3-5mm diameter have been used for over 1000 years, especially by Islamic populations in India, Arabia and Africa. Several agents occurring in the bark and wood have been suggested as aids in prevention of dental caries, such as antimicrobial agents that suppress bacterial growth and the formation of plaque. The tooth stick is also said to relieve toothache and gum disease. Roots also are used for cleaning teeth and for relieving toothache. Decoctions of leaves are used as a mouthwash, and masticated leaves for tooth and gum problems. A decoction of the root is used to treat gonorrhoea, spleen trouble and general stomach ache. Roots are also used for chest diseases or pounded and used as a poultice to heal boils. The bark is scratched and the latex used for treating sores. Seeds are used as a tonic, and seed oil is used on the skin for rheumatism.

Other uses

Fruits have a sweet, agreeable, aromatic, slightly pungent and peppery taste. They can be eaten raw, cooked or dried and stored. Fruit with or without seeds is said to contain 1.7-1.86 per cent sugars when ripe. Fermented drinks are also made from the fruit. The leaf is somewhat bitter and aromatic, with a taste likened to mustard. The leaves are also cooked as a sauce and eaten with couscous or as a green vegetable. Tender shoots, seeds and seed oil are also edible. Edible salts are obtained from ashes.

Leaves and young shoots are browsed by all stock, but normally cattle do not occur in the driest part of the *S. persica* distribution range and hence, it tends to be valued more as a camel, sheep and goat forage. Leaves make good fodder as their water content is high (15-36%). The high salt content of the leaves is said to affect the taste of milk, but the leaves are said to increase lactation in cows.

S. persica is reported as a good source of nectar.

The wood is sometimes used for firewood and charcoal. The wood is soft, white, easy to work and is not liable to termite attack. Used for coffins and clubs. Resin that drips from the tree is supposed to useful for making varnish. Seeds of *S. persica* contain 30-40 per cent of greenish yellow, non-edible oil that has over 50 per cent lauric and myristic acids. It has a high melting point and a disagreeable odour that disappears on purification. The most important aspect of the oil is the presence of a low percentage of C8 and C10 fatty acids that are of great economic significance. The oil is an alternative source of oil for soap and detergent industries.

Planted as shelterbelts and windbreaks to protect the farm habitation, gardens and orchards. Planted in sand dune reclamation and also useful for reclaiming saline soils.

REFERENCES

Balasubramanian, P. and P. V. Bole. 1993. Fruiting phenology and seasonality in tropical dry evergreen forest in Pt. calimere wildlife sanctuary. *J Bombay Natural History Society*, 90: 163-177.

Rao, G. G., V. N. Polra and V. R. Babu. 1999. Salt tolerant studies in *Salvadora persica*; a facultative halophyte. *Indian J. Soil Conservation*. 27: 55-63.

5.17. *Saraca asoca*

Common name: Asok Tree

Botanical name: *Saraca asoca* Roxb. Syn. *Saraca indica* Linn.

Family: Caesalpinaceae

Vernacular names: Ashoka (Assamese, Bengali, Gujarati, Hindi, Oriya); Ashoka mara, Seetha mara, Ashuker, Ahusunkar, Ashoka, Kankalimara (Kannada); Ashok (Kashmiri, Punjabi); Asokam, Asjogam (Malayalam); Ashok, Jaswant (Marathi); Asokah, Gatasokah, Kakeli (Sanskrit); Asoka, Asogu, Asogam (Tamil); Ashokamu, Ashokapatta, Vanjalamu (Telugu).

Origin and distribution: The species is distributed globally in Indo-Malayan region.

Habitat: Commonly found wild along streams or in the shade of evergreen forests.

Habit: It is a medium sized handsome evergreen tree upto 9m in height with numerous spreading and drooping glabrous branches. Trees with dense crown; branchlets stout, short with crowded leaves (Fig. 1).

Stem: It has blackish bark and reddish brown wood. Bark is almost black and warty. Wood is fort, and light reddish brown.

Leaf: Leaves alternate, abruptly paripinnate, upto 30cm long with rachis often ending in a subulate appendage; leaflets petiolulate, uppermost often the largest, 3-22 x 1-8cm oblong of lanceolate, coriaceous, glabrous, cuneate, rounded or rarely, sub-cordate at base, acute or shortly acuminate at apex; petiolules upto 0.5cm long, glandular at base.

Flower: Flowers are orange red in colour, very fragrant, dense, terminal, upto 15cm long corymbs, bract and bracteoled; Calyx petaloid, tubular, four lobed, upto 1.5cm long with ovate of obovate, rounded lobes; corolla absent; stamens 6-8, with filaments upto 0.2cm long and anthers yellowish upto 0.2cm long; pistil upto 2.5cm long, with stripe and ovary upto 0.4cm long, style filiform, curved into a ring and minute, capitate stigma (Fig. 2).

Fig. 1. Plant

Fig. 2. Flower

Fig. 3. Fruit

Fig. 4. Seed

Flowering, pollination and fruiting: Flowering and fruiting occur from July to October. Stigma with large lobes clearly indicated on increase in the receptive surface for effective cross pollination by insects (Mukerji *et al.*, 2000).

Fruit: Pods upto 18 x 4cm, oval or oblong, flat, leathery, thick, rounded or cuneate at base, shortly beaked at apex, thickened along the margins. Ripe fruits are dark purple and black in colour, pendulous, long and persist on the tree for quite some time (Fig. 3).

Seed: Seeds 4-8, 3.5cm long, ellipsoid-oblong, slightly compressed (Fig. 4).

Propagation: Propagated through seeds.

Seed collection: Ripened seeds can be collected during December–January from more than five to six year old plants. Ripe pods are collected from the tree by shaking off the branches. Mature fruits falling on ground can also be collected. The pods are dried under shade to release the seed. Insect attacked seeds should be discarded.

Seed storage: The seeds will remain viable only for about two months. Seeds retain viability upto six months in sealed tins when store under cold conditions.

Seed treatment: The seeds may be soaked in water for 12h before sowing, which improves the germination percentage.

Seed germination: Fresh seeds are dibbled into raised nursery beds. Germination takes about 15 to 20 days and varies from 50 to 100 per cent. About 80 per cent of the seedlings established. The seedlings are most often damaged by rodents and hence need protection. The seedlings are planted into polythene bag of size 22.5 x 17.5cm filled with potting mixture.

The seeds can also be sown in seed pans containing sand, soil and farm yard manure in the ratio of 1:1:1. The seed pans should be kept in polyhouses and watered daily. Germination starts after 30 days and continues till 60 days. These seedlings can be transplanted after 3-4 months into polythene bags. These polythene bags should be kept in the shade and irrigated every day.

Economic part used: Flower, bark

Alkaloids: Major constituents in the stem bark of Asoka are tannins (0.57%–7.85%).

Medicinal uses: Crushed flowers and leaves are rubbed on the skin to get relief from skin diseases. The seeds are strengthening and the ash of plant is good for external application in rheum arthritis. It is considered as best female tonic. Leaves juice is mixed with cumin seeds consumed to treat stomach ache. Bark is used in gynaecological problems such as leucorrhoea and dysfunctional bleeding.

Other uses: It is grown as an avenue tree due to its foliage and fragrant flowers.

REFERENCE

Mukerji. K. G., B. P. Chamola, A. K. Sharma. 2000. Glimpses in botany, APH Publishing Corporation, New Delhi, p. 430

5.18. *Symplocos cochinchinensis*

Common name: Pachotti, Symplocos bark

Botanical name: *Symplocos cochinchinensis* S. Moore Syn. *S. spicata* Roxb.

Family: Symplocaceae

Vernacular names: Mugam (Assamese); Lodha, Lodhra (Bengali); Lodhaz (Gujarati); Lodha (Hindi); Lodhra, Boodaganni, Manjatte (Kannada); Kath (Kashmiri); Pachotti (Malayalam); Lodhrah (Marathi); Lodhar (Punjabi); Rodhra, Lodhra (Sanskrit); Vellilathi, Vellilothram, Kambli vetti, Vaspa maram (Tamil); Lodhuga (Telugu); Lodh, Lodhpathani (Urdu).

Origin and distribution

Global: India to Malaysia, China and Japan

National: Widely distributed in evergreen forests from Himalayan to southern India and Shola forests of Western Ghats.

Regional: Very common in the moist to evergreen forests of Karnataka, Kerala and Tamil Nadu.

Habitat: Shola borders, slopes and along riversides from wet deciduous to shola forests.

Habit: Medium sized trees, 8-18m tall, about 1 m girth (Fig. 1).

Stem: Stems and branch lets always covered with greenish yellow lichen patches. Bark is smooth, light, grey, thin, about 4 mm thick. Blaze is creamy white. Wood is white, soft and even grained. Branch lets are hairless.

Leaf: Mature leaves turn yellow and form a yellow carpet under the tree. Fresh leaves taste sour. The new flush of leaves in purplish pink. Leaves are alternate, egg shaped to elliptic lanceolate, 5-16 x 3-6cm, base acute, apex rounded to acute acuminate, margin toothed with rounded to saw like teeth, leathery, glossy, distinctly stalked, hairless, lateral nerves 6-8 pairs.

Flower: Inflorescence a spike, raceme, or panicle, sometimes condensed to clusters, usually in the upper leaf axils; flowers subtended by a bract and two bracteoles, actinomorphic, bisexual (rarely by reduction unisexual), often fragrant; calyx with short tube, the limb 3-5 lobed; corolla sympetalous, often divided nearly to the base, whitish, bluish or purplish; stamens many, connate in a long monadelphous tube or only at the very base and then monadelphous or pentadelphous; anthers globose, two celled, lengthwise dehiscent, introrse; ovary inferior, two to five celled, style one, stigma punctiform or peltate; ovules two to four in each cell (Nooteboom, 1991) (Fig. 2).

Flowering, pollination and fruiting: Flowering occurs during July to October and fruiting occurs between February to May. Stray flowers and fruits seen almost throughout the year. Pollination is probably carried out by insects such as bees and bumble bees, but self pollination has also been suggested.

Fruit: Drupes are globose or conical, about 8 x 6mm, faintly ribbed, smooth, crowned with an apical ring, ripening purplish blue (Fig. 3).

Seed : Seeds are 1-3 in number per fruit, oblong in shape, straight or curved in each developed cell, with copious endosperm (Fig. 4).

Propagation : By seeds and stem cuttings.

Economic part used : Bark

Medicinal uses : Stem bark is used in the treatment of uterine complaints, vaginal diseases and menstrual disorders. Useful in treating diarrhoea, dysentery, eye diseases, fever, cough, ulcers, swellings, menorrhagia and leucorrhoea. A decoction of the bark is used as gargle for giving firmness to spongy and bleeding gums.

Fig. 1. Plant

Fig. 2. Flower

Fig. 3. Fruit

Fig. 4. Seed

REFERENCES

Nooteboom, H. P. 1991. *Symplocos* Jacq. [Internet] Record from proseabase. (Eds.) Lemmens, R.H.M.J. and Wulijarni-Soetjipto, N. PROSEA (Plant Resources of South East Asia) Foundation, Bogor, Indonesia. http://www.proseanet.org. Accessed from Internet: 18-Jan-2013.